Yorick Hardy
Willi-Hans Steeb

Classical and Quantum Computing

with C++ and Java Simulations

Birkhäuser Verlag
Basel · Boston · Berlin

Authors:

Yorick Hardy and Willi-Hans Steeb
International School for Scientific Computing
Rand Afrikaans University
P.O. Box 524
Auckland Park 2006
South Africa

2000 Mathematical Subject Classification 68Q01; 81P68

A CIP catalogue record for this book is available from the
Library of Congress, Washington D.C., USA

Deutsche Bibliothek Cataloging-in-Publication Data
Hardy, Yorick:
Classical and quantum computing with C++ and Java simulations /
Yorick Hardy ; Willi-Hans Steeb. - Basel ; Boston ; Berlin : Birkhäuser,
2001
   ISBN 978-3-7643-6610-0

ISBN 978-3-7643-6610-0   Birkhäuser Verlag, Basel – Boston – Berlin

© 2001 Birkhäuser Verlag, P.O. Box 133, CH-4010 Basel, Switzerland
Member of the BertelsmannSpringer Publishing Group
Cover design: Micha Lotrovsky, 4106 Therwil, Switzerland
Printed on acid-free paper produced from chlorine-free pulp. TCF ∞

ISBN 978-3-7643-6610-0

www.birkhauser-science.com

9 8 7 6 5 4 3 2 1

Contents

6   Latches and Registers

7   Synchronous Circuits

8   Recursion

9   Abstract Data Types

II Quantum Computing

List of Tables

List of Figures

List of Symbols

$\emptyset$	empty set
$\mathbf{N}$	natural numbers
$\mathbf{N}_0$	$\mathbf{N} \cup \{0\}$
$\mathbf{Z}$	integers
$\mathbf{Q}$	rational numbers
$\mathbf{R}$	real numbers
$\mathbf{R}^+$	nonnegative real numbers
$\mathbf{C}$	complex numbers
$\mathbf{R}^n$	n-dimensional Euclidean space
$\mathbf{C}^n$	n-dimensional complex linear space
$\mathcal{H}$	Hilbert space
i	$:= \sqrt{-1}$
$\Re z$	real part of the complex number z
$\Im z$	imaginary part of the complex number z
$A \subset B$	subset A of set B
$A \cap B$	the intersection of the sets A and B
$A \cup B$	the union of the sets A and B
$f \circ g$	composition of two mappings $(f \circ g)(x) = f(g(x))$
$\psi, |\psi\rangle$	wave function
t	independent time variable
x	independent space variable
$\mathbf{x} \in \mathbf{R}^n$	element $\mathbf{x}$ of $\mathbf{R}^n$
$\|\cdot\|$	norm
$\mathbf{x} \times \mathbf{y}$	vector product
$\otimes$	Kronecker product, tensor product
$\wedge$	exterior product (Grassmann product, wedge product)
$\langle , \rangle, \quad \langle | \rangle$	scalar product (inner product)
det	determinant of a square matrix
tr	trace of a square matrix
$\{ , \}$	Poisson product
$[,]$	commutator
$[,]_+$	anticommutator
δ_{jk}	Kronecker delta
δ	delta function
$\mathrm{sgn}(x)$	the sign of x, 1 if $x > 0$, -1 if $x < 0$, 0 if $x = 0$

λ	eigenvalue
ϵ	real parameter
I	unit operator, unit matrix
U	unitary operator, unitary matrix
Π	projection operator, projection matrix
H	Hamilton function
$\hat{H}$	Hamilton operator
V	potential
$b_j, b_j^\dagger$	Bose operators
$c_j, c_j^\dagger$	Fermi operators
$\mathbf{p}$	momentum
$\hat{\mathbf{p}}$	momentum operator
$\mathbf{L}$	angular momentum
$\hat{\mathbf{L}}$	angular momentum operator
$|\beta\rangle$	Bose coherent state
D	differential operator $\partial/\partial x$
Ω_+	Møller operator
$Y_{lm}(\theta, \phi)$	spherical harmonics
$\cdot$	AND operation in Boolean algebra
$+$	OR operation in Boolean algebra
$\oplus$	XOR operation in Boolean algebra
$\overline{A}$	negation of A in Boolean algebra
$\lfloor x \rfloor$	the greatest integer which is not greater than x

Preface

Scientific computing is not numerical analysis, the analysis of algorithms, high performance computing or computer graphics. It consists instead of the combination of all these fields and others to craft solution strategies for applied problems. It is the original application area of computers and remains the most important. From meteorology to plasma physics, environmental protection, nuclear energy, genetic engineering, symbolic computation, network optimization, financial applications and many other fields, scientific applications are larger, more ambitious, more complex and more necessary. More and more universities introduce a Department of Scientific Computing or a Department of Computational Science. The components of this new department include Applied Mathematics, Theoretical Physics, Computer Science and Electronic Engineering. This book can serve as a text book in Scientific Computing. It contains all the techniques (including quantum computing). Most of the chapters include C++ and Java simulations.

Chapter 1 covers the description of algorithms and informal verification techniques. Some basic concepts for computing are also introduced, such as alphabets and words, and total and partial functions.

Chapter 2 discusses Boolean algebra. The definition of a Boolean algebra is given, and various properties of the algebra are introduced. The chapter focuses on how Boolean algebra can be used to implement a computation. Methods are discussed to obtain efficient implementations.

Chapter 3 deals with number representation for computing devices. This includes the different implementations of integers, and the representation of real numbers. Conversion between different representations of numbers is also described.

Chapter 4 gives an overview of logic gates, which serve as the building blocks for implementing functions in digital electronics. All of the commonly used gates such as AND, OR, XOR and their negations are discussed.

Chapter 5 shows how to use the gates introduced in Chapter 4 to build circuits for specific purposes. The circuits described are important components in computing devices. The arithmetic operations such as addition and multiplication are described, as well as methods to increase the efficiency of the implementations. Various techniques for programming circuits are also considered, such as programmable logic devices and programmable gate arrays.

Chapter 6 is about latches. Latches serve as memory for a computing device. We consider three different types of latches. Using the latches, registers can be constructed which provide memory capability in a more useful form.

Chapter 7 considers synchronous circuits. To perform a computation, certain operations must be applied at specific times. This chapter describes how the timing of operations can be achieved.

Chapter 8 illustrates the technique of recursion and its usefulness through a number of problems with recursive solutions. The implementation of recursion is also discussed. The description of recursive functions is also given, which is important for discussions on computability.

Chapter 9 serves as an introduction to the concept of abstract data types. The data types support a specific organization of information. Three examples are provided.

Chapter 10 is devoted to classical error detection and correction. Specifically the chapter deals with Hamming codes to correct single bit errors, and the technique of weighted checksums. Besides these techniques, the noiseless coding theorem (which gives bounds on how reliably information can be transmitted with limited resources) is also discussed in detail.

Chapter 11 deals with cryptography. Methods to encrypt information using a private key are considered. The public key cryptography technique, which uses a computationally difficult system to provide security and only public keys are exchanged, is described.

Chapter 12 introduces computing models in the form of state machines. A finite automaton is described, and improvements are considered leading to the Turing machine. An example program provided aids in the understanding of the operation of a Turing machine.

Chapter 13 discusses the concepts of computability in terms of Turing machines, functions, and complexity. Complexity is described in terms of the repetitiveness of bit strings.

Chapter 14 provides an extensive discussion of neural networking techniques. The different models are illustrated through example programs. The networks discussed are the single layer perceptron and multilayer perceptron models.

Chapter 15 is concerned with the technique of random searches through a solution space for an optimum in the form of genetic algorithms. Several problems illustrate the strengths of variations to the genetic algorithm.

Chapter 16 considers the theoretical background needed for quantum computing. Hilbert spaces are defined and linear operators and important properties are discussed. The postulates of quantum mechanics give the formal description of quantum systems, their evolution and the operations that can be performed on them.

Chapter 17 describes the fundamentals of quantum computation. The basic unit of storage, how a quantum computation evolves, what operations can be performed and some important operations are discussed.

Chapter 18 deals with the different approaches proposed to explain the process of measurement in quantum mechanics.

Chapter 19 is about quantum state machines. Extensions to classical state machines are considered which can model quantum computations.

Chapter 20 is devoted to the description of the teleportation algorithm. The algorithm is an important illustration of what can be achieved using the properties of quantum mechanics.

Chapter 21 covers six quantum algorithms which display a significant advantage over current classical algorithms. The problems include Deutsch's, problem which cannot be solved classically, secure key distribution for cryptography, factoring and database searching.

Chapter 22 discusses quantum information theory. The Von Neumann entropy is introduced, and measurement of entanglement is considered. Finally bounds on communication of quantum states with limited resources are considered in the form of the quantum noiseless coding theorem and the Holevo bound.

Chapter 23 shows how to avoid errors in quantum states due to interaction with the environment. Some of the techniques from classical error correction apply to quantum error correction, but a new theory is developed to tolerate new errors possible in the quantum computation model.

Chapter 24 explains some approaches to the physical implementation of a quantum computing device. The device must have some properties such as the ability to store quantum information and support the operations on quantum systems.

Chapter 25 lists sites on the internet where more information on quantum computation can be found.

Ends of proofs are indicated by ♠. Ends of examples are indicated by ♣. Any useful suggestions and comments are welcome. The e-mail addresses of the authors are:

    whs@na.rau.ac.za
    steeb_wh@yahoo.com
    yorickhardy@yahoo.com

The web pages of the authors are

    http://zeus.rau.ac.za
    http://issc.rau.ac.za

Part I

Classical Computing

Chapter 1
Algorithms

1.1 Algorithms

An *algorithm* [48, 63, 77, 97, 115] is a precise description of how to solve a problem. For example algorithms can be used to describe how to add and subtract numbers or to prove theorems. Usually algorithms are constructed with some basic accepted knowledge and inference rules or instructions. Thus programs in programming languages such as C++ and Java are algorithms. Thus an algorithm is a map $f : E \rightarrow A$ of the input data E to the set of output data A.

Knuth [104] describes an algorithm as a finite set of rules which gives a sequence of operations for solving a specific type of problem similar to a recipe or procedure. According to Knuth [104] an algorithm has the following properties

1. **Finiteness.** An algorithm must always terminate after a finite number of steps.

2. **Definiteness.** Each step of an algorithm must be precisely defined; the actions to be carried out must be rigorously and unambiguously specified for each case.

3. **Input.** An algorithm has zero or more inputs, i.e., quantities which are given to it initially before the algorithm begins.

4. **Output.** An algorithm has one or more outputs, i.e., quantities which have a specified relation to the inputs.

5. **Effectiveness.** This means that all of the operations to be performed in the algorithm must be sufficiently basic that they can, in principle, be done exactly and in a finite length of time by a man using pencil and paper.

Not every function can be realized by an algorithm. For example, the task of adding two arbitrary real numbers does not satisfy finiteness.

Example. The *Euclidean algorithm* is a method to find the *greatest common divisor* (GCD) of two integers. The GCD d of two integers a and b is the integer that divides a and b, and if $c < d$ is a divisor of a and b then c divides d.

1. Let $a' := a$ and $b' := b$

2. Let r and q be integers such that $a' = qb' + r$ and $0 \leq r < b'$

3. If r is not zero

    (a) Let $a' := b'$ and $b' := r$

    (b) Goto 2

4. The GCD is b'.

To illustrate this we find the GCD of 21 and 18:

- 1) $a' = 21$ and $b' = 18$

- 2) $q = 1$ and $r = 3$

- 3) $a' = 18$ and $b' = 3$

- 2) $q = 6$ and $r = 0$

- 4) The GCD is 3.

Now we find the GCD of 113 and 49:

- 1) $a' = 113$ and $b' = 49$

- 2) $q = 2$ and $r = 15$

- 3) $a' = 49$ and $b' = 15$

- 2) $q = 3$ and $r = 4$

- 3) $a' = 15$ and $b' = 4$

- 2) $q = 3$ and $r = 3$

- 3) $a' = 4$ and $b' = 3$

- 2) $q = 1$ and $r = 1$

- 3) $a' = 3$ and $b' = 1$

- 2) $q = 3$ and $r = 0$

- 4) The GCD is 1.

♣

Definition. *Execution* of an algorithm refers to the following (or execution) of the steps given in the algorithm.

Definition. *Termination* of an algorithm is when an algorithm finishes, there is nothing more to be done.

An algorithm executes uniquely if, for a given input, the termination of the algorithm is always the same, i.e. the variables, memory, state, output and position of termination in the algorithm are always the same.

Definition. An algorithm is said to be *deterministic* if the algorithm execution is uniquely determined by the input.

Example. The Euclidean algorithm is deterministic, in other words for any given a and b the algorithm will always give the same result (the GCD of the given values).
♣

Definition. An algorithm which is not deterministic is said to be *non-deterministic*.

Example. An algorithm which follows a certain path with probability is non-deterministic. For example a learning algorithm can use probabilities as follows:

- Suppose there are n options (paths of execution) available.

- The algorithm assigns probabilities $p_1, p_2, \ldots, p_n$ according to merit for each of the options, where

$$p_i \geq 0, \qquad \sum_{i=1}^{n} p_i = 1.$$

- The algorithm calculates the *Shannon entropy*

$$S := -\sum_{i=1}^{n} p_i \log_2(p_i)$$

which is a measure of the information available about the options.

- The algorithm then chooses an option (according to the given probabilities).

- The outcome of the event is used for learning where the learning is weighted using S.

♣

1.2   Algorithm Verification

To illustrate the method of program verification we first introduce *mathematical induction.*

If P is a property with respect to natural numbers **N** and

- P holds for 1,

- if P holds for k then P holds for $k+1$,

then P holds for all natural numbers.

So if P holds for 1 it also holds for 2 and therefore also for 3 and so on.

Example.

- For $n = 1$

$$\sum_{i=1}^{n} i = 1 = \frac{n(n+1)}{2}$$

- Suppose

$$\sum_{i=1}^{k} i = \frac{k(k+1)}{2}$$

then

$$\sum_{i=1}^{k+1} i = \sum_{i=1}^{k} i + k + 1 = \frac{k(k+1)}{2} + k + 1 = \frac{(k+1)(k+2)}{2}$$

Thus

$$\sum_{i=1}^{n} i = \frac{n(n+1)}{2}$$

for n a natural number.

♣

Example. Let $x \geq 1$ and $n \in \mathbf{N}$.

- For $n = 1$
$$(1+x)^n \geq 1 + nx$$

- Suppose
$$(1+x)^k \geq 1 + kx$$

  then
$$(1+x)^{k+1} = (1+x)^k(1+x) \geq (1+kx)(1+x)$$

  now
$$(1+kx)(1+x) = 1 + (k+1)x + x^2 \geq 1 + (k+1)x$$

Thus
$$(1+x)^n \geq 1 + nx$$

for n a natural number.

♣

This method allows us to verify that a property is true for all natural numbers by building on initial truths. The same method can be extended for algorithm verification. A program starts with some conditions known to be true and verification is the process of determining whether certain desired properties always hold during execution to give the desired result.

Definition. An *assertion* is a statement about a particular condition at a certain point in an algorithm.

Definition. A *precondition* is an assertion at the beginning of a sub-algorithm.

Definition. A *postcondition* is an assertion at the end of a sub-algorithm.

Definition. An *invariant* is a condition which is always true at a particular point in an algorithm.

To prove that an algorithm is correct it is necessary to prove that the postcondition for each sub-algorithm holds if the precondition for the same sub-algorithm holds. It is also necessary to prove that invariants hold at their positions in the algorithm. The preconditions, postconditions and invariants must of course reflect the purpose of the algorithm.

Example. We consider the algorithm to add n numbers $x_1, x_2, \ldots, x_n$:

1. Let sum be a variable which takes numerical values initially set to 0.

2. Precondition: $x_1, \ldots, x_n$ are the n numbers to be added

3. For each of the n numbers x_i $(i = 1, 2, \ldots, n)$ do $sum := sum + x_i$
   Invariant:

$$sum = \sum_{j=1}^{i} x_j$$

4. Postcondition:

$$sum = \sum_{j=1}^{n} x_j$$

5. The desired result is given by sum.

Example. The correctness of the *Euclidean algorithm* is based on a simple invariant.

1. Let $a' := a$ and $b' := b$

2. Invariant: $\text{GCD}(a, b) = \text{GCD}(a', b')$
   Let r and q be integers such that $a' = qb' + r$ and $r < b'$

3. If r is not zero

   (a) Let $a' := b'$ and $b' := r$

   (b) Goto 2

4. The GCD is b'.

To prove the invariant holds we need to prove that after step 2

$$\text{GCD}(a', b') = \text{GCD}(b', r).$$

Obviously $\text{GCD}(b', r)$ divides $\text{GCD}(a', b')$. The reverse argument is also easy. $\text{GCD}(a', b')$ divides both a' and b' and therefore divides r. When $r = 0$ the GCD is b'.

In C and C++ the function assert in the header file assert.h is provided to help with debugging. The function takes one argument which must be an expression with numerical value. The function assert aborts the program and prints an error message if the expression evaluates to 0.

Example. We can use assert to make sure that, whenever a program calls the function sum, the function adds at least one number.

```cpp
// sum.cpp
#include <iostream>
#include <assert.h>

using namespace std;

double sum(int n,double x[])
{
  // precondition: x is an array of n doubles

  assert(n > 0);

  int i;
  double sum = 0.0;

  for(i=0;i < n;i++) sum += x[i]; // invariant sum = x[0]+...+x[i]

  // postcondition: sum = x[0]+...+x[n-1]
  return sum;
}

void main(void)
{
  double x[5] = { 0.5,0.3,7.0,-0.3,0.5 };

  cout << "sum=" << sum(5,x) << endl;
  cout << "sum=" << sum(-1,x) << endl;
}
```

The output is:

```
sum=8
Assertion failed: n>0, file sum.cpp, line 9

abnormal program termination
```

1.3 Random Algorithms

Random or *stochastic* algorithms use random numbers to try solve a problem. Generally the technique is used where approximations are acceptable and a completely accurate answer is difficult to obtain.

Examples include Monte Carlo methods, genetic algorithms, simulated annealing and neural networks. These algorithms are usually non-deterministic.

Random algorithms exist for numerical integration but other numerical methods are generally better for not too large dimensions.

In C++ the functions `rand` and

```
void srand(unsigned)
```

in `stdlib.h` and

```
unsigned time(time_t *)
```

in `time.h` can be used to generate uniformly distributed random numbers. The function call

```
srand(time(NULL))
```

initializes the random number generator. The function `rand()` generates a random number between 0 and `RAND_MAX`. Note that the random number sequences generated in this way by the computer are not truly random and are eventually periodic. The number sequences have properties which make them appropriate approximations for random number sequences for use in algorithms. The statement `double(rand())/RAND_MAX` takes the integer returned by `rand()` and casts it to type `double` so that the division by `RAND_MAX` gives a random number of type `double` in the unit interval $[0, 1]$.

Example. To calculate the value of π we use the fact that the area of a quadrant of the unit circle

$$S := \{ (x,y) \mid x^2 + y^2 \le 1 \}$$

is $\frac{\pi}{4}$. By generating random coordinates in the first quadrant of the unit circle the proportion of coordinates in the unit circle to the total number of coordinates generated approximates the area.

A few examples of the output are given below

```
pi=3.13994
pi=3.13806
pi=3.14156
pi=3.13744
```

```cpp
// calcpi.cpp

#include <iostream>
#include <time.h>
#include <stdlib.h>

using namespace std;

void main(void)
{
 const int n = 500000;
 double x,y,pi;
 int i;

 // initialize the counter of the number of points
 // found in the unit circle to zero
 int in_count=0;

 // initialize the random number with a
 // seed value given by the current time
 srand(time(NULL));

 for(i=0;i<n;i++)
 {
  x = double(rand())/RAND_MAX;
  y = double(rand())/RAND_MAX;

  if(x*x+y*y<=1)
   in_count++;
 }

 pi = 4.0*double(in_count)/n;
 cout << "pi=" << pi << endl;
}
```

Example. Annealing [164] is the process of cooling a molten substance with the objective of condensing matter into a crystaline solid. Annealing can be regarded as an optimization process. The configuration of the system during annealing is defined by the set of atomic positions r_i. A configuration of the system is weighted by its Boltzmann probability factor,

$$e^{-E(r_i)/kT}$$

where $E(r_i)$ is the energy of the configuration, k is the Boltzmann constant, and T is the temperature. When a substance is subjected to annealing, it is maintained at each temperature for a time long enough to reach thermal equilibrium.

The iterative improvement technique for combinatorial optimization has been compared to rapid quenching of molten metals. During rapid quenching of a molten substance, energy is rapidly extracted from the system by contact with a massive cold substrate. Rapid cooling results in metastable system states; in metallurgy, a glassy substance rather than a crystalline solid is obtained as a result of rapid cooling. The analogy between iterative improvement and rapid cooling of metals stems from the fact that iterative improvement and rapid cooling of metals accepts only those system configurations which decrease the cost function. In an annealing(slow cooling) process, a new system configuration that does not improve the cost function is accepted based on the Boltzmann probability factor of the configuration. This criterion for accepting a new system state is called the *Metropolis criterion*. The process of allowing a fluid to attain thermal equilibrium at a temperature is also known as the Metropolis process.

The simulated annealing procedure is presented below. Simulated annealing essentially consists of repeating the Metropolis procedure for different temperatures. The temperature is gradually decreased at each iteration of the simulated annealing algorithm.

If the initial temperature is too low, the process gets quenched very soon and only a local optimum is found. If the initial temperature is too high, the process is very slow. Only a single solution is used for the search and this increases the chance of the solution becoming stuck at a local optimum. The changing of the temperature is based on an external procedure which is unrelated to the current quality of the solution, that is, the rate of change of temperature is independent of the solution quality. These problems can be rectified by using a population instead of a single solution. The annealing mechanism can also be coupled with the quality of the current solution by making the rate of change of temperature sensitive to the solution quality.

In the following program we apply simulated annealing to find the minimum of the function

$$f(x) = x^2 \exp(-x/15) \sin x.$$

```cpp
// anneal.cpp
// simulated annealing
// x range: [0 : 100]

#include <iostream>
#include <math.h>
#include <stdlib.h>
#include <time.h>

using namespace std;

inline double f(double &x)
{
 return sin(x)*x*x*exp(-x/15.0);
}


inline int accept(double &Ecurrent,double &Enew,double &T,double &s)
{
 double dE = Enew - Ecurrent;
 double k = 1380662e-23;

 if(dE < 0.0)
  return 1;
 if(s < exp(-dE/(k*T)))
  return 1;
 else return 0;
}

int main()
{
 cout << "Finding the minimum via simulated annealing:" << endl;
 double xlow = 0.0;    double xhigh = 100.0;
 double Tmax = 500.0; double Tmin = 1.0;
 double Tstep = 0.1;
 double T;

 srand(time(NULL));
 double s = rand()/double(RAND_MAX);

 double xcurrent = s*(xhigh - xlow);
 double Ecurrent = f(xcurrent);

 for(T=Tmax; T>Tmin; T-=Tstep)
 {
  s = rand()/double(RAND_MAX);
  double xnew = s*(xhigh - xlow);
  double Enew  = f(xnew);
  if(accept(Ecurrent,Enew,T,s))
```

```
  {
    xcurrent = xnew;
    Ecurrent = Enew;
  }
}

  cout << "The minimum found is " << Ecurrent << " at x = "
       << xcurrent << endl;

  return 0;
}
```

Typical outputs are given below.

```
Finding the minimum via simulated annealing:
The minimum found is -121.796 at x = 29.8397

Finding the minimum via simulated annealing:
The minimum found is -121.796 at x = 29.8397

Finding the minimum via simulated annealing:
The minimum found is -121.749 at x = 29.874
```

The global minimum of f is found as one of the solutions to the transcendental equation

$$\tan(x^*) = \frac{15x^*}{x^* - 30}$$

in the interval $[0, 100]$ with $x \approx \frac{19\pi}{2}$.

1.4 Total and Partial Functions

Functions and algorithms are closely related since algorithms are used to implement functions in computing and algorithms can be described in terms of functions.

Definition. A function $f : A \to B$ is a *total function* if f associates every $a \in A$ with exactly one image $f(a)$ in B.

Definition. A function $f : A \to B$ is a *partial function* if it is total on some subset A' of A. The set A' is then called the domain of f and is denoted $\mathrm{dom}(f)$. If $a \in A \setminus A'$, $f(a)$ is said to be *undefined* otherwise it is said to be *defined*.

Definition. Suppose $f : A \to B$ and $g : A \to B$. By definition, f and g are *equal* if and only if for each $a \in A$, either

1. both $f(a)$ and $g(a)$ are defined, and $f(a) = g(a)$; or

2. both $f(a)$ and $g(a)$ are undefined.

Definition. A function

$$f : A_1 \times A_2 \times \ldots \times A_n \to B$$

is said to be *n-ary*. Unary, binary and ternary are synonyms for 1-ary, 2-ary and 3-ary respectively. In the expression

$$f(a_1, a_2, \ldots, a_n),$$

we say that $a_1, a_2, \ldots, a_n$ are the *arguments* of f.

Definition. The range of a function $f : A \to B$ is the set $\{\, f(a) \mid a \in \mathrm{dom}(f) \,\}$ and is denoted $\mathrm{rng}(f)$.

Definition. The function $f : A \to B$ is onto if $\mathrm{rng}(f) = B$. f is one to one if $f(a) = f(a')$ implies $a = a'$ for all $a, a' \in A$.

Example. Let $A = \mathbf{N_0} \times \mathbf{N_0}$ and $B = \mathbf{N_0}$. The function $f : A \to B$ defined by

$$f(x, y) = x + y$$

is a total binary function.   ♣

Example. The function $f : A \to B$ defined by

$$g(x, y) = x - y$$

is a partial function with domain $\{\, (x, y) \mid x \geq y \,\}$.   ♣

Definition. A partial or total n-ary function f is said to be *effectively computable* if there is an effective process which, when given any n argument values $x_1, \ldots, x_n$, will either

1. eventually halt, yielding $f(x_1, \ldots, x_n)$ if it is defined, or

2. never halt if $f(x_1, \ldots, x_n)$ is undefined.

Definition. The *characteristic function* of a set A is defined as

$$\chi(x) := \begin{cases} 1 & x \in A \\ 0 & x \notin A \end{cases}.$$

Definition. Let $f : A \to B$ and $g : B \to C$ where f is partial and g is total. The composition $g \circ f : A \to C$ is defined as

$$(g \circ f)(a) := g(f(a)).$$

If f is total then $g \circ f$ is total.

Example. In the theory of Lie transformation groups the following function

$$\exp(\alpha D) f$$

plays a central role, where $f : \mathbf{R}^n \to \mathbf{R}$ is an analytic function, D is the differential operator

$$D := D_1(\mathbf{x}) \frac{\partial}{\partial x_1} + \ldots + D_n(\mathbf{x}) \frac{\partial}{\partial x_n}$$

where $D_i : \mathbf{R}^n \to \mathbf{R}$ are analytic functions and α is a parameter ($\alpha \in \mathbf{R}$). If $n = 1$ and $D := d/dx$ we have

$$\exp(\alpha D) f(x) = f(x + \alpha).$$

Thus the argument x of the function f maps to $x + \alpha$ (translation). The following C++ program shows an implementation of this function.

```cpp
// trans.cpp

#include <iostream>
#include <math.h>

using namespace std;

template <class T> T translation(T (*f)(T),T x,T alpha)
{
 return f(x + alpha);
}

double f1(double x)
{
 return sin(x);
}

int f2(int x)
{
 return x*x;
}

void main()
{
 double x1 = 1.0;
 double alpha1 = 0.5;

 cout << "f1(x1=" << x1 << ") = " << f1(x1) << endl;
 cout << "f1(" << x1 << " + " << alpha1 << ") = "
      << translation(f1,x1,alpha1) << endl;

 int x2 = 5;
 int alpha2 = 3;

 cout << "f2(x2=" << x2 << ") = " << f2(x2) << endl;
 cout << "f2(" << x2 << " + " << alpha2 << ") = "
      << translation(f2,x2,alpha2) << endl;
}
```

1.5   Alphabets and Words

Alphabets and words are used as the inputs and outputs for computing. They are used in complexity and computability analysis. Words can also be considered as sequences of characters, i.e. strings.

Definition. An *alphabet* is any finite set of symbols.

Definition. A *word* over an alphabet Σ is any finite string of symbols from Σ. Σ^* denotes the set of all words over Σ.

Definition. The *length* of a word x is the number of symbols contained in x and is denoted by $|x|$.

Definition. The word of length 0 is called the *null* or *empty* word and is denoted by ϵ or λ.

Definition. Let $x, y \in \Sigma^*$ where $x = a_1 a_2 \ldots a_n$ and $y = b_1 b_2 \ldots b_m$. The *concatenation* of x and y is $xy = a_1 a_2 \ldots a_n b_1 b_2 \ldots b_m$.

Definition. $x \in \Sigma^*$ is a *prefix* of $y \in \Sigma^*$ if there exists $z \in \Sigma^*$ such that $y = xz$.

For any symbol $a \in \Sigma^*$, a^m denotes the word of length m consisting of m a's.

Definition. Let $X, Y \subseteq \Sigma^*$.

- $XY = \{\, xy \mid x \in X,\ y \in Y \,\}$

- 1. $X^0 = \{\epsilon\}$
  2. $X^{n+1} = X^n X$, for $n \geq 0$

- $X^* = \bigcup\limits_{n=0}^{\infty} X^n$

- $X^+ = \bigcup\limits_{n=1}^{\infty} X^n$

The set Σ^n is the set of all words of length n over Σ.

Example. *Lindenmayer systems* or *L-systems* consist of a set of rules for modifying a word to produce a new word. Lindenmayer systems play a role in modelling biological systems. The rules specify for each symbol in the alphabet, a word with which to replace it. This system is called a 0L-system. The *L-language* corresponding to a ruleset is the set of all words derived by successive application of the ruleset to all symbols in the alphabet. An example ruleset for the alphabet $\{0, 1\}$ is

$$0 \to 1, \qquad 1 \to 01.$$

Thus beginning with 0, this produces a series of derivations as follows

$$\{0, 1, 01, 101, 01101, 10101101, \ldots\}.$$

This is the L-language for this ruleset. Each word in the derivation is simply the concatenation of the previous two words in the derivation. We can prove this fact by induction. Let $L(w_j)$ denote the mapping from the bit string w_j to the next derivation using the ruleset and starting from 0, and w_j be the j-th bit string in the derivation. We have

$$
\begin{aligned}
w_0 &= 0 \\
w_1 &= L(w_0) = 1 \\
w_2 &= L(w_1) = L(1) = 01 = w_0 w_1 \\
w_3 &= L(w_2) = L(01) = 101 = w_1 w_2
\end{aligned}
$$

$$\vdots$$

$$w_{j+1} = L(w_j)$$

By induction

$$
\begin{aligned}
w_{j+1} &= L(w_j) \\
&= L(w_{j-2} w_{j-1}) \\
&= L(w_{j-2}) L(w_{j-1}) \\
&= w_{j-1} w_j
\end{aligned}
$$

The following Java program shows how to implement the derivation. We use the `StringBuffer` class which is built into Java. The `StringBuffer` class implements a mutable sequence of characters. The method

`StringBuffer append(String str)`

in class `StringBuffer` appends the `String str` to the `StringBuffer`.

```java
// LSystem.java

class LSystem
{
 public static void map(StringBuffer sold,StringBuffer snew)
 {
  int i;
  for(i=0; i < sold.length(); i++)
  {
   if(sold.charAt(i) == '0') snew.append("1");
   if(sold.charAt(i) == '1') snew.append("01");
  }
 } // end method map

 public static void main(String[] args)
 {
  StringBuffer sold = new StringBuffer("01101");
  StringBuffer snew = new StringBuffer("");

  map(sold,snew);
  System.out.println("snew = " + snew); // 10101101

  StringBuffer s0 = new StringBuffer("0");
  StringBuffer s1 = new StringBuffer("");

  int j;
  for(j=0; j < 6; j++)
  {
   map(s0,s1);
   s0 = s1;
   System.out.println("s = " + s0);
   s1 = new StringBuffer("");
  }
 }
}
```

Example. UTF-8 encoding is an efficient method of coding characters and words from many languages as integers. The encoding uses variable length codes to obtain the efficiency by noting that the most common characters used are from the ASCII character set. The Java language uses the methods `writeUTF()` in class `DataOutputStream` and `readUTF()` in class `DataInputStream` to implement the encoding and decoding to and from UTF-8. ASCII codes (the codes numbered from 1 to 127) are stored in 8 bits with the highest order bit set to zero.

The encoding for a `String` begins with two bytes for the length of the string. The first byte is the high order byte and the second byte is the low order byte. The character encoding follows this. A zero value is encoded as two bytes

$$11000000, 10000000.$$

The bytes are written in left to right order. All ASCII codes from 1 to 127 are written using a single byte with a leading 0 bit,

$$0 \ (0\text{--}6)$$

where (0–6) indicates that the bits indexed by $0, 1, \ldots 6$ are written in the remaining bit positions. All codes in the range 128 to 2047 are encoded as two bytes

$$110(6\text{--}10), \ 10(0\text{--}5).$$

Finally all codes in the range 2048 to 65535 are encoded as three bytes

$$1110(12\text{--}15), \ 10(6\text{--}11), \ 10(0\text{--}5).$$

Thus the string "UTF example" would be encoded as the bytes (in hexadecimal)

$$00, \ 0B, \ 55, \ 54, \ 46, \ 20, \ 65, \ 78, \ 61, \ 6D, \ 70, \ 6C, \ 65.$$

The following Java program uses the above methods to illustrate the encoding.

```java
// UTFexample.java

import java.io.*;

public class UTFexample
{
 public static void main(String[] args) throws IOException
 {
  DataOutputStream output =
  new DataOutputStream(new FileOutputStream("myout.dat"));

  String s = new String("UTF example");
  System.out.println("s = " + s);

  output.writeUTF(s);
  output.flush();
  output.close();

  DataInputStream input =
  new DataInputStream(new FileInputStream("myout.dat"));

  String t = input.readUTF();
  input.close();

  System.out.println("t = " + t);
 }
}
```

Chapter 2
Boolean Algebra

2.1 Introduction

Boolean algebra forms the theoretical basis for classical computing. It can be used to describe the circuits which are used as building blocks for classical computing.

In this chapter we introduce the definitions of Boolean algebra and the rules for manipulation. We introduce the standard forms for manipulation and describe how Boolean algebra can be used to describe functions. Efficiency is an important issue in computing and we describe the methods of Karnaugh maps and Quine-McKluskey to simplify expressions.

At the end of the chapter two programs are given to illustrate the concepts. The first example program uses the properties of Boolean algebra to efficiently implement sets in C++. This implementation reduces the memory requirements for a set since only one bit of information is needed for each element of the set. The second example is an implementation of the Quine-McKluskey method in C++. The Quine-McKluskey method is easier to implement on computer whereas the Karnaugh map method is easier to do by hand.

The smallest Boolean algebra consists of two elements usually labelled 0 and 1 or *false* and *true* but larger Boolean algebras exist.

2.2 Definitions

Definition. A *Boolean algebra* is a closed algebraic system containing a set B of two or more elements and two operations

$$\cdot : B \times B \to B, \qquad + : B \times B \to B$$

with the following properties:

- *Identity Elements.* There exist unique elements $0, 1 \in B$ such that for every $A \in B$

  1. $A + 0 = A$
  2. $A \cdot 1 = A$

- *Commutativity.* For every $A_0, A_1 \in B$

  1. $A_0 + A_1 = A_1 + A_0$
  2. $A_0 \cdot A_1 = A_1 \cdot A_0$

- *Associativity.* For every $A_0, A_1, A_2 \in B$

  1. $A_0 + (A_1 + A_2) = (A_0 + A_1) + A_2$
  2. $A_0 \cdot (A_1 \cdot A_2) = (A_0 \cdot A_1) \cdot A_2$

- *Distributivity.* For every $A_0, A_1, A_2 \in B$

  1. $A_0 + (A_1 \cdot A_2) = (A_0 + A_1) \cdot (A_0 + A_2)$
  2. $A_0 \cdot (A_1 + A_2) = (A_0 \cdot A_1) + (A_0 \cdot A_2)$

- *Complement.* For every $A \in B$ there exists $\overline{A} \in B$ such that

  1. $A + \overline{A} = 1$
  2. $A \cdot \overline{A} = 0$

The operations $\cdot$ and $+$ are referred to as the AND and OR operations respectively. 0 is called the identity element for the OR operation and 1 is called the identity element for the AND operation. The complement will also be referred to as the NOT or negation operation. The AND operation is sometimes referred to as *conjunction*. The OR operation is sometimes referred to as *disjunction*.

From the properties of identity elements and complements we find

$$\overline{0} = 1 \quad \text{and} \quad \overline{1} = 0.$$

Example. The smallest Boolean algebra consists of the identity elements $\{0, 1\}$. The Boolean algebra can be summarised in a table.

A_0	A_1	$A_0 + A_1$	$A_0 \cdot A_1$	$\overline{A_0}$
0	0	0	0	1
0	1	1	0	1
1	0	1	0	0
1	1	1	1	0

♣

Example. The set $P(X)$ (set of all subsets of the finite set X) of a non-empty set X with $\cdot$ the intersection of sets, $+$ the union of sets and the complement with respect to X as negation forms a Boolean algebra with identity elements $0 = \emptyset$ and $1 = X$. This Boolean algebra has $2^{|X|}$ members, where $|X|$ denotes the cardinality (number of elements) of X.

♣

Example. The set A of all functions from the set $\{p_1, p_2, \ldots, p_n\}$ into $\{0, 1\}$ (i.e. a function in the set assigns 0 or 1 to each of $p_1, p_2, \ldots, p_n$) and $\cdot, +$ and negation described pointwise by the definitions in the first example forms a Boolean algebra. For example, if $f_1, f_2 \in A$ then

$$(f_1 + f_2)(p_i) = f_1(p_i) + f_2(p_i)$$

and

$$(f_1 \cdot f_2)(p_i) = f_1(p_i) \cdot f_2(p_i).$$

The Boolean algebra has 2^{2^n} members and is called a *free Boolean algebra* on the generators $p_1, p_2, \ldots, p_n$.

♣

Example. Let $\{p_1, p_2, \ldots\}$ be a countably infinite set. Then we can again form a free Boolean algebra on this generating set by considering finite Boolean expressions in the p_i.

♣

2.3   Rules and Laws of Boolean Algebra

The following are consequences of the definitions.

- *Double negation.* $\overline{\overline{A}} = A$

- *Idempotence.*

    1. $A \cdot A = A$
    2. $A + A = A$

- *Absorption.*

    1. $A + 1 = 1$
    2. $0 \cdot A = 0$
    3. $A_0 + A_0 \cdot A_1 = A_0$
    4. $A_0 \cdot (A_0 + A_1) = A_0$
    5. $A_0 \cdot A_1 + \overline{A_1} = A_0 + \overline{A_1}$
    6. $(A_0 + A_1) \cdot \overline{A_1} = A_0 \cdot \overline{A_1}$

The double negation property is obvious. The idempotence property follows from

1. $A \cdot A = A \cdot A + 0 = (A \cdot A) + (A \cdot \overline{A}) = A \cdot (A + \overline{A}) = A \cdot 1 = A$
2. $A + A = (A + A) \cdot 1 = (A + A) \cdot (A + \overline{A}) = A + (A \cdot \overline{A}) = A + 0 = A$

The absorption properties are derived as follows

1. $A = A + A = A + (A \cdot 1) = (A + A) \cdot (A + 1) = A \cdot (A + 1)$
2. $A = A \cdot A = A \cdot (A + 0) = (A \cdot A) + (A \cdot 0) = A + (A \cdot 0)$
3. $A_0 + A_0 \cdot A_1 = A_0 \cdot 1 + A_0 \cdot A_1 = A_0 \cdot (1 + A_1) = A_0$
4. $A_0 \cdot (A_0 + A_1) = (A_0 \cdot A_0) + (A_0 \cdot A_1) = A_0 + A_0 \cdot A_1$
5. $A_0 \cdot A_1 + \overline{A_1} = (A_0 + \overline{A_1}) \cdot (A_1 + \overline{A_1}) = (A_0 + \overline{A_1}) \cdot 1 = A_0 + \overline{A_1}$
6. $(A_0 + A_1) \cdot \overline{A_1} = (A_0 \cdot \overline{A_1}) + (A_1 \cdot \overline{A_1}) = (A_0 \cdot \overline{A_1}) + 0 = A_0 \cdot \overline{A_1}$

2.4 DeMorgan's Theorem

Another property of Boolean algebra is given by *DeMorgan's theorem*

$$\overline{A_0 \cdot A_1} \equiv \overline{A_0} + \overline{A_1}$$

$$\overline{A_0 + A_1} \equiv \overline{A_0} \cdot \overline{A_1}$$

Thus the left-hand side of the two identities involves two operations and the right-hand side three operations. DeMorgan's theorem can be proved using the properties given above. It describes the relationships between the operations $+, \cdot$ and negation.

$$
\begin{aligned}
(\overline{A_0} + \overline{A_1}) \cdot (A_0 \cdot A_1) &= (\overline{A_0} \cdot A_0 \cdot A_1) + (A_0 \cdot A_1 \cdot \overline{A_1}) \\
&= 0 \cdot A_1 + A_0 \cdot 0 \\
&= 0 + 0 = 0
\end{aligned}
$$

$$
\begin{aligned}
(\overline{A_0} + \overline{A_1}) + (A_0 \cdot A_1) &= (A_0 + \overline{A_0} + \overline{A_1}) \cdot (A_1 + \overline{A_0} + \overline{A_1}) \\
&= (1 + \overline{A_1}) \cdot (1 + \overline{A_0}) \\
&= 1 \cdot 1 = 1
\end{aligned}
$$

This theorem is very important for building combinational circuits consisting of only one type of operation.

2.5 Further Definitions

We will use the set $B = \{0, 1\}$ and the operations AND, OR and complement defined by Table 2.1.

A_0	A_1	$A_0 \cdot A_1$	$A_0 + A_1$	$\overline{A_0}$
0	0	0	0	1
0	1	0	1	1
1	0	0	1	0
1	1	1	1	0

Table 2.1: AND, OR and Complement.

Definition. A *Boolean function* is a map $f : \{0,1\}^n \rightarrow \{0,1\}$ where $\{0,1\}^n$ is the set of all n-tuples consisting of zeros and ones.

Definition. *Boolean variables* are variables which may only take on the values of 0 or 1.

Definition. *Bit* is short for *binary digit* which refers to a 0 or 1.

Definition. A *literal* is a variable or the complement of a variable.

We will use the notation $B^n := B \times B \times \ldots \times B$ (n times). Thus $B^n = \{0,1\}^n$.

Definition. Any function $f : B^n \rightarrow B$ can be represented with a *truth table*.

A_0	A_1	$\ldots$	A_{n-1}	$f(A_0, A_1, \ldots, A_{n-1})$
0	0	$\ldots$	0	$f(A_0 = 0, A_1 = 0, \ldots, A_{n-1} = 0)$
0	0	$\ldots$	1	$f(A_0 = 0, A_1 = 0, \ldots, A_{n-1} = 1)$
$\vdots$	$\vdots$	$\vdots$	$\vdots$	$\vdots$
1	1	$\ldots$	1	$f(A_0 = 1, A_1 = 1, \ldots, A_{n-1} = 1)$

The rows of the table are over all combinations of $A_0, A_1, \ldots, A_{n-1}$.

There are 2^n such combinations. Thus the truth table has 2^n rows.

Definition. Two functions $f : B^n \rightarrow B$, $g : B^n \rightarrow B$ are *equivalent* if

$$f(A_0 = a_0, \ldots, A_{n-1} = a_{n-1}) = g(A_0 = a_0, \ldots, A_{n-1} = a_{n-1})$$

for all $A_0, \ldots, A_{n-1} \in \{0,1\}$.

Definition. A *product form* is an AND of a number of literals $l_1 \cdot l_2 \ldots \cdot l_m$.

Definition. A *sum of products (SOP)* form is an OR of product forms

$$P_1 + P_2 + \ldots + P_m.$$

Definition. *Disjunctive normal form* is a disjunction of conjunctions of literals. This is equivalent to SOP form.

Definition. *Conjunctive normal form* is a conjunction of disjunctions of literals.

Theorem. Any function $f : B^n \to B$ can be represented in SOP form.

To see this we construct product forms $P_j = l_{j,1} \cdot l_{j,2} \cdot \ldots \cdot l_{j,n}$ for each row in the truth table of f where $f = 1$ with $l_{j,i} = A_i$ if the entry for A_i is 1 and $l_{j,i} = \overline{A_i}$ if the entry for A_i is 0. If $f = 1$ in m of the rows of the truth table then

$$f(A_0, A_1, \ldots, A_{n-1}) = P_1 + P_2 + \ldots + P_m.$$

Example. Consider the *parity* function for two bits with truth table Table 2.2.

A_0	A_1	$P(A_0, A_1)$
0	0	1
0	1	0
1	0	0
1	1	1

Table 2.2: Parity Function

Then $P(A_0, A_1) = \overline{A_1} \cdot \overline{A_2} + A_0 \cdot A_1$. ♣

Definition. A *canonical SOP form* is a SOP form over n variables, where each variable or its negation is present in every product form, in other words a Boolean expression E is in canonical SOP form if it can be written as

$$E = l_{1,1} \cdot l_{1,2} \cdot \ldots \cdot l_{1,n} + l_{2,1} \cdot l_{2,2} \cdot \ldots \cdot l_{2,n} + \ldots + l_{m,1} \cdot l_{m,2} \cdot \ldots \cdot l_{m,n}$$

where $l_{i,j} = A_j$ or $l_{i,j} = \overline{A_j}$.

Definition. The *exclusive OR* function is defined by the following table.

A_0	A_1	$A_0 \oplus A_1$
0	0	0
0	1	1
1	0	1
1	1	0

Table 2.3: XOR Truth Table

Some more properties of the XOR operation are given below:

- $A \oplus A = 0$

- $A \oplus \overline{A} = 1$

- $A_0 \oplus A_1 = A_1 \oplus A_0$

- $A_0 \oplus A_1 = \overline{A_0} \oplus \overline{A_1}$

- $(A_0 \oplus A_1) \oplus A_2 = A_0 \oplus (A_1 \oplus A_2)$

- $A_0 \oplus A_1 = A_0 \cdot \overline{A_1} + \overline{A_0} \cdot A_1$

- $(A_0 \cdot \overline{A_1}) \oplus A_0 = (A_0 \oplus \overline{A_1}) \cdot A_0 = A_0 \cdot A_1$

The XOR operation can be used to swap two values a and b (for example integers in C++ and Java):

1. $a := a \oplus b$

2. $b := a \oplus b$

3. $a := a \oplus b$

By analysing the variables at each step in terms of the original a and b the swapping action becomes clear. In the second step we have $(a \oplus b) \oplus b = a \oplus 0 = a$. In the third step we have $(a \oplus b) \oplus a = b \oplus 0 = b$.

In C, C++ and Java the XOR operation is denoted by $\hat{}$. The following C++ program illustrates the swapping.

```
// xor.cpp

#include <iostream>

using namespace std;

void main(void)
{
 int a=23;
 int b=-565;

 cout << "a = " << a << " , b = " << b << endl;
 a ^= b; b ^= a; a ^= b;
 cout << "a = " << a << " , b = " << b << endl;
}
```

The results are

```
a = 23 , b = -565
a = -565 , b = 23
```

Definition. The operation $\overline{A_0 + A_1}$ is called the *NOR* function.

Example. Let $A_0 = 0$ and $A_1 = 0$. Then

$$\overline{A_0 + A_1} = 1$$

♣

Definition. The operation $\overline{A_0 \cdot A_1}$ is called the *NAND* function.

Example. Let $A_0 = 1$ and $A_1 = 0$. Then

$$\overline{A_0 \cdot A_1} = 1$$

♣

Definition. The operation $\overline{A_0 \oplus A_1}$ is called the *XNOR* function.

Example. Let $A_0 = 1$ and $A_1 = 1$. Then

$$\overline{A_0 \oplus A_1} = 1$$

♣

Definition. A *universal set of operations* is a set of operations which can be used to build any Boolean function.

For Boolean functions there exist universal sets of operations with only one element.

For simplicity of implementation, it is useful to know the minimum number of parameters a Boolean function must take in order to be able to build any other Boolean function. Obviously functions taking only a single parameter cannot fulfill this requirement. The minimum number of parameters is thus at least two.

The NAND and NOR operations can be used to build any other function which we will show in the next section.

2.6 Boolean Function Implementation

The physical implementation of a Boolean function is achieved by interconnection of *gates*. A gate is an electronic circuit that produces an output signal (representing 0 or 1) according to the signal states (again representing 0 or 1) of its inputs. The task is then to build an implementation of a Boolean function with gates of prescribed types. This is not always possible, for example the NOT operation cannot implement two bit operations and the OR operation cannot be used to implement the AND and NOT operations. In the previous section it was shown informally that any Boolean function can be implemented with AND, OR and NOT operations (in SOP form). Therefore to show that any Boolean function can be implemented with a set of gates it is sufficient to show that AND, OR and NOT can be implemented with the set of gates.

The NAND gate is sufficient to build an implementation of any Boolean function:

- $\overline{A} = \overline{A \cdot A}$

- $A_0 \cdot A_1 = \overline{\overline{A_0 \cdot A_1}} = \overline{\overline{A_0 \cdot A_1} \cdot \overline{A_0 \cdot A_1}}$

- $A_0 + A_1 = \overline{\overline{A_0} \cdot \overline{A_1}} = \overline{\overline{A_0 \cdot A_0} \cdot \overline{A_1 \cdot A_1}}$

♠

Example. We show now how to implement the NOR operation using only NAND operations. As mentioned earlier De Morgan's laws are important to achieve this.

$$\overline{A_0 + A_1} = \overline{\overline{\overline{A_0 \cdot A_0} \cdot \overline{A_1 \cdot A_1}}} = \overline{\overline{A_0 \cdot A_0} \cdot \overline{A_1 \cdot A_1}} \cdot \overline{\overline{A_0 \cdot A_0} \cdot \overline{A_1 \cdot A_1}}$$

♣

It can also be shown that the NOR gate is sufficient to build an implementation of any Boolean function.

Data are represented by bit strings $a_{n-1}a_{n-2} \ldots a_0$, $a_i \in \{0, 1\}$. Bit strings of length n can represent up to 2^n different data elements. Functions on bit strings are then calculated by

$$f(a_{n-1}a_{n-2}\ldots a_0) = f_{m-1}(a_{n-1}a_{n-2}\ldots a_0)f_{m-2}(a_{n-1}a_{n-2}\ldots a_0)\ldots f_0(a_{n-1}a_{n-2}\ldots a_0)$$

with $f : B^n \to B^m$ and $f_i : B^n \to B$. In other words a function of a bit string of length n gives a bit string of length say m, each bit in the output string is therefore a function of the input bits. It is sufficient then to consider functions with an output of only one bit.

Example. The set

$$\{\, z \mid z \in \mathbf{N}_0,\ 0 \le z < 2^n \,\}$$

can be represented by

$$a_{n-1}a_{n-2}\ldots a_0 \to \sum_{i=0}^{n-1} a_i 2^i.$$

If $n = 32$ the largest integer number we can represent is

$$\sum_{i=1}^{n-1} 2^i = 2^n - 1 = 4294967295.$$

This relates to the data type unsigned long in C and C++. Java has only signed data types. ♣

Example. The set

$$\{\, x \mid x \in \mathbf{R},\ x = b + j\frac{c-b}{2^n-1},\quad j = 0,1,\ldots,2^n-1 \,\}$$

where $b, c \in \mathbf{R}$ and $c > b$ can be represented by

$$a_{n-1}a_{n-2}\ldots a_0 \to b + \frac{c-b}{2^n-1}\sum_{i=0}^{n-1} a_i 2^i.$$

So we find

$$a_{n-1}a_{n-2}\ldots a_0 = 00\ldots 0 \to b$$

and

$$a_{n-1}a_{n-2}\ldots a_0 = 11\ldots 1 \to c.$$

♣

Minimizing the number of gates in an implementation decreases cost and the number of things that can go wrong.

One way to reduce the number of gates is to use the properties of the Boolean algebra to eliminate literals.

Example. The *full adder* (Table 2.4) consists of two outputs (one for the sum and one for the carry) and three inputs (the carry from another adder and the two bits to be added).

C_{in}	A_0	A_1	S	C_{out}
0	0	0	0	0
0	0	1	1	0
0	1	0	1	0
0	1	1	0	1
1	0	0	1	0
1	0	1	0	1
1	1	0	0	1
1	1	1	1	1

Table 2.4: Full Adder

Thus

$$S = \overline{A_0} \cdot A_1 \cdot \overline{C_{in}} + A_0 \cdot \overline{A_1} \cdot \overline{C_{in}} + \overline{A_0} \cdot \overline{A_1} \cdot C_{in} + A_0 \cdot A_1 \cdot C_{in}$$

and

$$C_{out} = A_0 \cdot A_1 \cdot \overline{C_{in}} + \overline{A_0} \cdot A_1 \cdot C_{in} + A_0 \cdot \overline{A_1} \cdot C_{in} + A_0 \cdot A_1 \cdot C_{in}$$

Simplification for C_{out} yields

$$
\begin{aligned}
C_{out} &= A_0 \cdot A_1 \cdot \overline{C_{in}} + \overline{A_0} \cdot A_1 \cdot C_{in} + A_0 \cdot \overline{A_1} \cdot C_{in} + A_0 \cdot A_1 \cdot C_{in} + A_0 \cdot A_1 \cdot C_{in} \\
&\quad + A_0 \cdot A_1 \cdot C_{in} \\
&= A_0 \cdot A_1 \cdot (\overline{C_{in}} + C_{in}) + (\overline{A_0} + A_0) \cdot A_1 \cdot C_{in} + A_0 \cdot (\overline{A_1} + A_1) \cdot C_{in} \\
&= A_0 \cdot A_1 + A_1 \cdot C_{in} + A_0 \cdot C_{in}
\end{aligned}
$$

The *half adder* is a full adder where $C_{in} = 0$ is fixed.   ♣

Karnaugh maps and the Quine-McKluskey method [153] for simplification of Boolean expressions are discussed next.

2.6.1 Karnaugh Maps

Karnaugh maps can be used to simplify Boolean expressions of up to six variables. When an expression has more than four variables the Karnaugh map has dimension greater than 2. For 2, 3 and 4 variables the Karnaugh map is represented by 2×2, 2×4 and 4×4 grids, respectively. The rows and columns of the grid represent the values of variables. Each square contains the expression value for the variables with values given by the row and column.

Example. The Karnaugh map for the carry flag

$$C_{out} = A_0 \cdot A_1 \cdot \overline{C_{in}} + \overline{A_0} \cdot A_1 \cdot C_{in} + A_0 \cdot \overline{A_1} \cdot C_{in} + A_0 \cdot A_1 \cdot C_{in}$$

of the full adder is as follows

$$A_0 A_1$$

C_{in}	00	01	11	10
0	0	0	1	0
1	0	1	1	1

♣

Note that adjacent columns and rows only differ in the assignment of one variable (only one bit differs). This is important for the simplification algorithm to work correctly. Suppose two adjacent squares have the value 1 and only differ in the variable A. Writing the corresponding product forms as $P \cdot A$ and $P \cdot \overline{A}$ the canonical SOP form can be simplified using

$$P \cdot A + P \cdot \overline{A} = P.$$

In fact this can be extended to any 2^n adjacent squares in a row or column. The first column is adjacent to the last column ("wrap around"), and the same applies to rows. The simplification is indicated by circling the adjacent squares involved in the simplification. Overlapping circles are allowed due to the idempotence property. If two circles are "adjacent" in the sense that they cover the same columns(rows) in adjacent rows(columns) they may be joined to form one circle encircling all the appropriate squares. The only restriction is that the number of rows and columns encircled are a power of 2, i.e. 1,2,4,8,.... This is due to the algebraic simplification used. Each set of encircled squares is called a *group* and the squares are said to be *covered*. There are two algorithms for this method.

Algorithm 1.

1. Count the number of adjacencies (adjacent 1-squares) for each 1-square on the Karnaugh map.

2. Select an uncovered 1-square with the fewest number of adjacencies. (An arbitrary choice may be required.)

3. Circle the 1-square so that the circle covers the most uncovered 1-squares.

4. If all the 1-squares are not yet covered goto 2

Algorithm 2.

1. Circle all 1-squares so that the circle covers the most 1-squares.

2. Eliminate all circles that do not contain at least one 1-square that is not covered by another circle.

3. Introduce a minimum number of circles to complete the cover.

The SOP form is the OR of product forms representing the groups of the Karnaugh map. The variable A_i is in the product form if $A_i = 1$ is constant in the group, $\overline{A_i}$ is in the product form if $A_i = 0$ is constant in the group.

Example. The Karnaugh map for the carry flag

$$C_{out} = A_0 \cdot A_1 \cdot \overline{C_{in}} + \overline{A_0} \cdot A_1 \cdot C_{in} + A_0 \cdot \overline{A_1} \cdot C_{in} + A_0 \cdot A_1 \cdot C_{in}$$

is the same after application of either algorithm

$$
\begin{array}{c|c|c|c|c|}
 & \multicolumn{4}{c}{A_0A_1} \\
 & 00 & 01 & 11 & 10 \\
\hline
C_{in} \quad 0 & 0 & 0 & 1 & 0 \\
\hline
1 & 0 & 1 & 1 & 1 \\
\hline
\end{array}
$$

So $E = A_0 \cdot A_1 + A_1 \cdot C_{in} + A_0 \cdot C_{in}$.

An advantage of Karnaugh maps is that simpler expressions can be found when certain inputs cannot occur. A "don't care" symbol is placed in the squares on the Karnaugh map which represent the inputs which cannot occur. These d-squares can be interpreted as 1-squares or 0-squares for optimal grouping.

Example. The truth table for a decimal incrementer (4-bit) with 4 inputs and 4 outputs is given by Table 2.5.

Number in	Number out	I_3	I_2	I_1	I_0	O_3	O_2	O_1	O_0
0	1	0	0	0	0	0	0	0	1
1	2	0	0	0	1	0	0	1	0
2	3	0	0	1	0	0	0	1	1
3	4	0	0	1	1	0	1	0	0
4	5	0	1	0	0	0	1	0	1
5	6	0	1	0	1	0	1	1	0
6	7	0	1	1	0	0	1	1	1
7	8	0	1	1	1	1	0	0	0
8	9	1	0	0	0	1	0	0	1
9	0	1	0	0	1	0	0	0	0
		1	0	1	0	d	d	d	d
		1	0	1	1	d	d	d	d
		1	1	0	0	d	d	d	d
		1	1	0	1	d	d	d	d
		1	1	1	0	d	d	d	d
		1	1	1	1	d	d	d	d

Table 2.5: 4-bit Decimal Incrementer

The Karnaugh map for O_0 and O_3 is:

Therefore $O_0 = \overline{I_0}$ and $O_3 = \overline{I_0} \cdot I_3 + I_0 \cdot I_1 \cdot I_2$.

2.6.2   Quine-McKluskey Method

This method provides an algorithmic description for simplification which lends itself to implementation in programming languages such as C++ and Java. It is also more general than the method of Karnaugh maps and can handle an arbitrary number of Boolean variables. Suppose the Boolean expression to be simplified has the representation in canonical SOP form

$$
\begin{aligned}
E = \quad & P_{0,1}(A_0, A_1, \ldots, A_n) + P_{0,2}(A_0, A_1, \ldots, A_n) + \ldots \\
+ \quad & P_{1,1}(A_0, A_1, \ldots, A_n) + P_{1,2}(A_0, A_1, \ldots, A_n) + \ldots \\
& \vdots \\
+ \quad & P_{n,1}(A_0, A_1, \ldots, A_n) + P_{n,2}(A_0, A_1, \ldots, A_n) + \ldots
\end{aligned}
$$

where $P_{i,j}$ denotes the jth product form with exactly i negated Boolean variables. The method is as follows

1. Let $QM(n) := \{P_{i,j} \mid i = 0, 1, \ldots, n \quad j = 1, 2, \ldots\}$

2. Set $m := n$.

3. Set
   $QM(m-1) := QM(m)$
   and
   $QM_{m,i} := \{P \in QM(m) \mid P \text{ has } m \text{ Boolean variables of which } i \text{ are negated}\}$

4. For each pair of elements
   $e_1 = l_{1,1} \cdot l_{1,2} \cdot \ldots \cdot l_{1,m} \in QM_{m,i}$
   and
   $e_2 = l_{2,1} \cdot l_{2,2} \cdot \ldots \cdot l_{2,m} \in QM_{m,i+1}$ where $i = 0, 1, \ldots, m-1$
   which differ in only one literal $l_{1,j} \neq l_{2,j}$
   set
   $QM(m-1) := (QM(m-1) - \{e_1, e_2\}) \cup \{l_{1,1} \cdot \ldots l_{1,j-1} \cdot l_{1,j+1} \cdot \ldots \cdot l_{1,m}\}$

5. Set $m := m - 1$.

6. If $(m > 0)$ goto step 3

7. E is the OR of all the elements of $QM(0)$.

Example. We consider the two's complement operation on two bits.

I_0	I_1	O_0	O_1
0	0	0	0
0	1	1	1
1	0	1	0
1	1	0	1

Table 2.6: Two's Complement Operation on 2 Bits

Thus $O_1 = \overline{I_0} \cdot I_1 + I_0 \cdot I_1$.

- $m = 2$.
  $QM(2) = \{\overline{I_0} \cdot I_1, I_0 \cdot I_1\}$
  $QM_{0,2} = \{I_0 \cdot I_1\}$
  $QM_{1,2} = \{\overline{I_0} \cdot I_1\}$
  $QM_{2,2} = \emptyset$

- $m = 2$.
  $QM(1) = \{I_1\}$
  $QM_{0,2} = \{I_0 \cdot I_1\}$
  $QM_{1,2} = \{\overline{I_0} \cdot I_1\}$
  $QM_{2,2} = \emptyset$

- $m = 1$.
  $QM(0) = \{I_1\}$
  $QM_{0,1} = \{I_1\}$
  $QM_{1,1} = \emptyset$

The method yields $O_1 = I_1$.   ♣

Example. The method applied to the carry flag

$$C_{out} = A_0 \cdot A_1 \cdot \overline{C_{in}} + \overline{A_0} \cdot A_1 \cdot C_{in} + A_0 \cdot \overline{A_1} \cdot C_{in} + A_0 \cdot A_1 \cdot C_{in}$$

of the full adder is as follows

- $m = 3$.
  $QM(3) = \{A_0 \cdot A_1 \cdot \overline{C_{in}}, \overline{A_0} \cdot A_1 \cdot C_{in}, A_0 \cdot \overline{A_1} \cdot C_{in}, A_0 \cdot A_1 \cdot C_{in}\}$
  $QM_{0,3} = \{A_0 \cdot A_1 \cdot C_{in}\}$
  $QM_{1,3} = \{\overline{A_0} \cdot A_1 \cdot C_{in}, A_0 \cdot \overline{A_1} \cdot C_{in}, A_0 \cdot A_1 \cdot \overline{C_{in}}\}$
  $QM_{2,3} = \emptyset$
  $QM_{3,3} = \emptyset$

- $m = 3$.

  $QM(2) = \{A_0 \cdot A_1 \cdot \overline{C_{in}}, \; A_1 \cdot C_{in}, \; A_0 \cdot \overline{A_1} \cdot C_{in}\}$

  $QM_{0,3} = \{A_0 \cdot A_1 \cdot C_{in}\}$

  $QM_{1,3} = \{\overline{A_0} \cdot A_1 \cdot C_{in}, \; A_0 \cdot \overline{A_1} \cdot C_{in}, \; A_0 \cdot A_1 \cdot \overline{C_{in}}\}$

  $QM_{2,3} = \emptyset$

  $QM_{3,3} = \emptyset$

- $m = 3$.

  $QM(2) = \{A_0 \cdot A_1 \cdot \overline{C_{in}}, \; A_1 \cdot C_{in}, \; A_0 \cdot C_{in}\}$

  $QM_{0,3} = \{A_0 \cdot A_1 \cdot C_{in}\}$

  $QM_{1,3} = \{\overline{A_0} \cdot A_1 \cdot C_{in}, \; A_0 \cdot \overline{A_1} \cdot C_{in}, \; A_0 \cdot A_1 \cdot \overline{C_{in}}\}$

  $QM_{2,3} = \emptyset$

  $QM_{3,3} = \emptyset$

- $m = 3$.

  $QM(2) = \{A_0 \cdot A_1, \; A_1 \cdot C_{in}, \; A_0 \cdot C_{in}\}$

  $QM_{0,3} = \{A_0 \cdot A_1 \cdot C_{in}\}$

  $QM_{1,3} = \{\overline{A_0} \cdot A_1 \cdot C_{in}, \; A_0 \cdot \overline{A_1} \cdot C_{in}, \; A_0 \cdot A_1 \cdot \overline{C_{in}}\}$

  $QM_{2,3} = \emptyset$

  $QM_{3,3} = \emptyset$

- $m = 2$.

  $QM(1) = \{A_0 \cdot A_1, \; A_1 \cdot C_{in}, \; A_0 \cdot C_{in}\}$

  $QM_{0,2} = QM(1)$

  $QM_{1,2} = \emptyset$

  $QM_{2,2} = \emptyset$

  $QM_{3,2} = \emptyset$

- $m = 1$.

  $QM(0) = \{A_0 \cdot A_1, \; A_1 \cdot C_{in}, \; A_0 \cdot C_{in}\}$

  $QM_{0,1} = \emptyset$

  $QM_{1,1} = \emptyset$

  $QM_{2,1} = \emptyset$

  $QM_{3,1} = \emptyset$

- $E = A_0 \cdot A_1 + A_1 \cdot C_{in} + A_0 \cdot C_{in}$

Thus we have reduced the expression to one consisting of only two types of operations and 5 operations in total. This is a large reduction compared to the original total of 11 operations. The example also illustrates that the process is long but simple enough to implement, making it a good application for a computing device. ♣

2.7 Example Programs

2.7.1 Efficient Set Operations Using Boolean Algebra

Let
$$U := \{\, o_0, o_1, \ldots, o_{n-1} \,\}$$

be the universal set of n objects. Any subset A of U can be represented with a sequence of bits
$$A := a_0 a_1 \ldots a_{n-1}$$

where $a_i = 1$ if $o_i \in A$ and $a_i = 0$ otherwise. For example, let $n = 8$ and consider the bitstring
$$10010111.$$

Then o_0, o_3, o_5, o_6, o_7 are in the set and o_1, o_2, o_4 are not in the set.

Now the set $A \cup B$ (union of A and B) corresponds to
$$A + B := (a_0 + b_0)(a_1 + b_1) \ldots (a_{n-1} + b_{n-1}),$$

and $A \cap B$ (intersection of A and B) corresponds to
$$A \cdot B := (a_0 \cdot b_0)(a_1 \cdot b_1) \ldots (a_{n-1} \cdot b_{n-1}).$$

The complement of A corresponds to
$$\overline{A} := \overline{a_0}\, \overline{a_1} \ldots \overline{a_{n-1}}.$$

For example if
$$
\begin{aligned}
A &= 11010100 \\
B &= 01101101
\end{aligned}
$$

then
$$
\begin{aligned}
A \cup B &= 11111101 \\
A \cap B &= 01000100 \\
\overline{A} &= 00101011.
\end{aligned}
$$

The following C++ program `bitset.cpp` implements these concepts. The class
BitSet implements all the bitwise operations introduced above. We could also use
the `bitset` which is part of the standard template library, which includes all the
methods needed to implement complement, intersection and union.

```cpp
// bitset.cpp

#include <iostream>
#include <string>

using namespace std;

class SetElementBase
{
 public:
    virtual void output(ostream&)=0;
};

template <class T>
class SetElement: public SetElementBase
{
 protected:
    T data;
 public:
    SetElement(T t) : data(t) {}
    virtual void output(ostream& o) { o << data; }
};

class BitSet
{
 protected:
    char *set;
    int len;
    SetElementBase **universe;
    static int byte(int);
    static char bit(int);
 public:
    BitSet(SetElementBase**,int,int*,int);
    BitSet(const BitSet&);
    BitSet &operator=(const BitSet&);
    BitSet operator+(const BitSet&) const;   // union
    BitSet operator*(const BitSet&) const;   // intersection
    BitSet operator~(void) const;            // complement
    void output(ostream&) const;
    ~BitSet();
};
```

```
int BitSet::byte(int n) { return n>>3; }
char BitSet::bit(int n) { return 1<<(n%8); }

// Create a BitSet with universe un of n elements,
// with m elements given by el
BitSet::BitSet(SetElementBase **un,int n,int *el,int m)
{
 int i;
 len = n;
 universe = un;
 set = new char[byte(len)+1];
 for(i=0;i<byte(len)+1;i++) set[i] = 0;
 if(m > 0)
 for(i=0;i<m;i++) set[byte(el[i])] |= bit(el[i]);
}


BitSet::BitSet(const BitSet &b)
{
 int i;
 len = b.len;
 universe = b.universe;
 set = new char[byte(len)+1];
 for(i=0;i<byte(len)+1;i++) set[i] = b.set[i];
}


BitSet &BitSet::operator=(const BitSet &b)
{
 if(this!=&b)
 {
  int i;
  delete[] set;
  len = b.len;
  universe = b.universe;
  set = new char[byte(len)+1];
  for(i=0;i<byte(len)+1;i++) set[i] = b.set[i];
 }
 return *this;
}

BitSet BitSet::operator+(const BitSet &b) const
{
 if(universe == b.universe)
 {
  int i;
  BitSet c(universe,len,NULL,0);
  for(i=0;i<byte(len)+1;i++) c.set[i] = set[i]|b.set[i];
  return c;
 }
```

```
  else return *this;
}

BitSet BitSet::operator*(const BitSet &b) const
{
 if(universe == b.universe)
 {
  int i;
  BitSet c(universe,len,NULL,0);
  for(i=0;i<byte(len)+1;i++) c.set[i] = set[i]&b.set[i];
  return c;
 }
 else return *this;
}

BitSet BitSet::operator~(void) const
{
 int i;
 BitSet b(universe,len,NULL,0);
 for(i=0;i<byte(len)+1;i++) b.set[i] = ~set[i];
 return b;
}

void BitSet::output(ostream &o) const
{
 int i,start = 0;
 o << "{";
 if((set[byte(0)]&bit(0)) !=0 ) universe[0]->output(o);
 for(i=0;i<len;i++)
  if((set[byte(i)]&bit(i)) != 0)
   { if(start) o << ", "; universe[i]->output(o); start=1; }
 o << "}";
}

BitSet::~BitSet() { delete[] set; }

ostream &operator<<(ostream& o,const BitSet &b)
{
 b.output(o);
 return o;
}

void main(void)
{
 SetElement<int> s1(5);
 SetElement<string> s2(string("element"));
 SetElement<double> s3(3.1415927);
 SetElement<int> s4(8);
```

```
SetElement<int> s5(16);
SetElement<int> s6(3);
SetElement<string> s7(string("string"));
SetElement<double> s8(2.7182818);
SetElement<int> s9(32);
SetElement<int> s10(64);
SetElementBase *universe[10]={&s1,&s2,&s3,&s4,&s5,
                              &s6,&s7,&s8,&s9,&s10};

cout << "Universe=" << (~BitSet(universe,10,NULL,0)) << endl;
cout << "Empty set=" << BitSet(universe,10,NULL,0) << endl;
int a[7] = {1,2,5,6,8,9};
int b[4] = {3,5,7,8};
BitSet A(universe,10,a,7);
BitSet B(universe,10,b,4);
cout << "A=" << A << endl;
cout << "B=" << B << endl;
cout << "~A=" << (~A) << endl;
cout << "A+B=" << (A+B) << endl;
cout << "A*B=" << (A*B) << endl;
}
```

The output of the bitset.cpp program is

```
Universe={55, element, 3.14159, 8, 16, 3, string, 2.71828, 32, 64}
Empty set={}
A={55, element, 3.14159, 3, string, 32, 64}
B={8, 3, 2.71828, 32}
~A={8, 16, 2.71828}
A+B={55, element, 3.14159, 8, 3, string, 2.71828, 32, 64}
A*B={3, 32}
```

2.7.2 Quine-McKluskey Implementation

The next C++ program illustrates the Quine-McKluskey method for the *full adder*. The algorithm is modified slightly to make the implementation easier. For an n-Boolean variable expression the program maintains $n + 1$ sets $S_0, \ldots, S_n$ where S_i contains product forms of exactly n variables. Initially only S_n is non-empty and contains all the product forms from the expression to be simplified. The program fills the set S_i by combining and simplifying two product forms from S_{i+1} which only differ in one literal (using previously discussed methods). Once all such product forms have been simplified the product forms used in simplification are removed from S_{i+1}. As input, the program takes an array of product forms (where product forms are arrays of characters) where the value 1 means the variable is present in the product form and 0 means the negation of the variable is present in the product form. Thus the arrays can be constructed directly from truth tables.

The program `quine.cpp` simplifies the carry and sum bits from the full adder. In the main function we consider the expressions for C_{out} and S for the full adder. We use an array of three `char` to represent a product form, a 1 indicates that the literal in the product form is not a negated variable and a 0 indicates that the literal is a negated variable. The variable is identified by the index in the array, for example the program uses index 0 for A_0, index 1 for A_1 and index 2 for C_{in}. These arrays (representing product forms) are placed in an array representing the final SOP form.

The function `complimentary` searches for complimentary literals in a set of product forms in order to perform simplification, `AddItem` is used to add elements to the sets maintained in the algorithm, `DeleteItem` removes elements from these sets (the simplification). The function `QuineRecursive` is the main implementation of the algorithm. It is implemented using recursion. This is not necessary but simplifies the implementation. The function `QuineMcKluskey` prepares the data for the `QuineRecursive` function.

The output of the program is

```
Cout=A0.A1+A1.Cin+A0.Cin
S=NOT(A0).A1.NOT(Cin)+A0.NOT(A1).NOT(Cin)
 +NOT(A0).NOT(A1).Cin+A0.A1.Cin
```

which is the same result obtained in earlier examples. The sum bit could not be simplified.

```
// quine.cpp

#include <iostream>

using namespace std;

struct QMelement
{
 int nvars,used;
 char *product;
 int *vars;
 QMelement *next;
};

int complementary(QMelement *p1,QMelement *p2)
{
 int sum = 0,i;
 if(p1->nvars != p2->nvars) return 0;
 for(i=0;i<p1->nvars;i++) sum += (p1->vars[i]!=p2->vars[i]);
 if(sum == 0)
  for(i=0;i<p1->nvars;i++)
   sum += (p1->product[p1->vars[i]]!=p2->product[p2->vars[i]]);
 else sum = 0;
 return (sum == 1);
}

void AddItem(QMelement* &list,char *product,int nvars,int *vars)
{
 int i;
 QMelement *item;
 if(list == (QMelement*)NULL)
 {
  list = new QMelement;
  list->nvars = nvars;
  list->next = (QMelement*)NULL;
  list->product = new char[nvars];
  list->vars = new int[nvars];
  list->used = 0;
  for(i=0;i<nvars;i++)
  {
   list->product[i] = product[i];
   list->vars[i] = vars[i];
  }
 }
 else
 {
  item = list;
  while(item->next != (QMelement*)NULL) item = item->next;
```

```
  item = (item->next = new QMelement);
  item->nvars = nvars;
  item->next = (QMelement*)NULL;
  item->product = new char[nvars];
  item->vars = new int[nvars];
  item->used = 0;
  for(i=0;i<nvars;i++)
  {
   item->product[i] = product[i];
   item->vars[i] = vars[i];
  }
 }
}


void DeleteItem(QMelement *&set,QMelement *item)
{
 QMelement *last = set;
 if(item == set) set = set->next;
 else
 {
  while(last->next != item) {last = last->next;}
  last->next = item->next;
 }
 delete[] item->product;
 delete[] item->vars;
 delete item;
}


void DeleteItems(QMelement *set)
{
 QMelement *item = set,*next;
 while(item != (QMelement*)NULL)
 {
  next = item->next;
  delete[] item->product;
  delete[] item->vars;
  delete item;
  item = next;
 }
}


void QuineRecursive(QMelement **sets,int index)
{
 if(index<1) return;
 if(sets[index] == (QMelement*)NULL) return;
 int i,j;
 QMelement *item1 = sets[index],*item2;
 while(item1 != (QMelement*)NULL)
```

```
{
 if(item1->next != (QMelement*)NULL)
 {
  item2 = item1->next;
  while(item2 != (QMelement*)NULL)
  {
   if(complementary(item1,item2))
   {
    char *product = new char[item1->nvars-1];
    int *vars = new int[item1->nvars-1];
    for(i=0,j=0;i<item1->nvars;i++)
     if(item1->product[i] == item2->product[i])
     {
      product[j] = item1->product[i];
      vars[j++] = i;
     }
    AddItem(sets[index-1],product,item1->nvars-1,vars);
    delete[] product;
    delete[] vars;
    item1->used = item2->used=1;
   }
   item2 = item2->next;
  }
 }
 item2 = item1;
 item1 = item1->next;
 if(item2->used) DeleteItem(sets[index],item2);
 }
 QuineRecursive(sets,index-1);
}

void QuineMcKluskey(char **sop,int nproducts,int nvars,char **names)
{
 int i,j,*vars = new int[nvars];
 QMelement **sets = new QMelement*[nvars+1];
 for(i=0;i<=nvars;i++)
  {sets[i] = (QMelement*)NULL; if(i < nvars) vars[i] = i;}
 for(i=0;i<nproducts;i++) AddItem(sets[nvars],sop[i],nvars,vars);
 QuineRecursive(sets,nvars);
 delete[] vars;
 for(i=0;i<=nvars;i++)
 {
  QMelement *item = sets[i];
  while(item != (QMelement*)NULL)
  {
   for(j=0;j<item->nvars;j++)
   {
    if(item->product[j] == 1) cout << names[item->vars[j]];
```

```
     else cout << "NOT(" << names[item->vars[j]] << ")";
    if(j != item->nvars-1) cout << ".";
   }
   item = item->next;
   if(item != (QMelement*)NULL) cout << "+";
   else if(i != nvars)
    if(sets[i+1] != (QMelement*)NULL) cout << "+";
  }
  DeleteItems(sets[i]);
 }
}


void main(void)
{
 //carry flag
 char c1[3]={1,1,0},c2[3]={0,1,1},c3[3]={1,0,1},c4[3]={1,1,1};
 //sum
 char s1[3]={0,1,0},s2[3]={1,0,0},s3[3]={0,0,1},s4[3]={1,1,1};
 char *Cout[4];
 char *S[4];
 char *names[3] = {"A0","A1","Cin"};
 Cout[0] = c1;
 Cout[1] = c2;
 Cout[2] = c3;
 Cout[3] = c4;
 S[0] = s1;
 S[1] = s2;
 S[2] = s3;
 S[3] = s4;
 cout << "Cout=";        QuineMcKluskey(Cout,4,3,names);
 cout << endl << "S="; QuineMcKluskey(S,4,3,names);
 cout << endl;
}
```

Chapter 3
Number Representation

3.1 Binary, Decimal and Hexadecimal Numbers

We are accustomed to using the *decimal number system*. For example the (decimal) number 34062 can be written as

$$34062 = 3 \cdot 10^4 + 4 \cdot 10^3 + 0 \cdot 10^2 + 6 \cdot 10^1 + 2 \cdot 10^0$$

where $10^1 = 10$ and $10^0 = 1$. In general, any positive integer can be represented in one and only one way in the form

$$a_0 \cdot 10^0 + a_1 \cdot 10^1 + a_2 \cdot 10^2 + \ldots + a_k \cdot 10^k$$

where $0 \le a_i \le 9$ for $0 \le i \le k$ and $a_k > 0$. This number is denoted

$$a_k a_{k-1} \cdots a_2 a_1 a_0$$

in standard decimal notation.

For any integer $r > 1$, every positive integer n can be represented uniquely in the form

$$a_0 \cdot r^0 + a_1 \cdot r^1 + a_2 \cdot r^2 + \cdots + a_m \cdot r^m$$

where $0 \le a_i \le r - 1$ for $0 \le i \le m$ and $a_m > 0$ and $r^0 = 1$. This can be proved by induction on n.

In particular, every positive integer can be represented in *binary notation*

$$a_0 \cdot 2^0 + a_1 \cdot 2^1 + a_2 \cdot 2^2 + \cdots + a_m \cdot 2^m$$

where $a_i \in \{0, 1\}$ for $0 \le i \le m$ and $a_m = 1$.

Obviously we have $1 = 1 \cdot r^0$. Further if

$$a = a_0 \cdot r^0 + a_1 \cdot r^1 + a_2 \cdot r^2 + \cdots + a_m \cdot r^m$$

where $0 \le a_i \le r - 1$ for $0 \le i \le m$ and $a_m > 0$ and $r^0 = 1$. Let k be the least integer in $\{0, 1, \ldots, m\}$ with $a_k < r - 1$. Either $k < m$ which gives

$$a + 1 = (a_k + 1)r^k + \cdots + a_m \cdot r^m,$$

or $k = 1$ which gives

$$a + 1 = r^{m+1}.$$

Example. The number 23 (in decimal notation) has the binary representation 10111, since $2^4 + 2^2 + 2 + 1 = 23$. The decimal number 101 has the binary representation 1100101, i.e. $2^6 + 2^5 + 2^2 + 1$.   ♣

A procedure for finding the binary representation of a number n is to find the highest power 2^m which is $\le n$, subtract 2^m from n, then find the highest power 2^j which is $\le n - 2^m$, etc.

Although the computer operates on binary-coded data, it is often more convenient for us to view this data in *hexadecimal* (base 16). There are three reasons for this:

1. Binary machine code is usually long and difficult to assimilate. Hexadecimal, like decimal, is much easier to read.

2. There is a direct correspondence between binary and hexadecimal. Thus, we can easily translate from hexadecimal to binary.

3. In today's CPU the length of the storage elements (called *registers*) are generally multiples of 8 bits (typically 32 or 64 bits). The general purpose registers are 32 bits long or 64 bits long. Thus it is convenient to show contents as multiples and fractions of 16 – hexadecimal. The storage sizes are 8 bits (a *byte*), 16 bits (a *word*), 32 bits (a *doubleword*), 64 bits (a *quadword*), and 80 bits (a *tenbyte*) – all multiples and fractions of 16.

Thus, although we think in decimal and the computer thinks in binary, hexadecimal is a number system that captures some of the important elements of both. In the remainder of this section we discuss the binary, decimal, and hexadecimal number systems and the methods for converting from one number system to another.

3.1.1 Conversion

In this section we describe the conversion from binary to hexadecimal, from hexadecimal to binary, binary to decimal, decimal to binary, decimal to hexadecimal, and hexadecimal to decimal.

Binary to Hexadecimal. To see the one-to-one correspondence between hexadecimal and binary, notice that if we use b to represent a bit and

$$b_n b_{n-1} \ldots b_2 b_1 b_0$$

is a binary number, then it has a value of

$$2^n b_n + \ldots + 2^8 b_8 + 2^7 b_7 + 2^6 b_6 + 2^5 b_5 + 2^4 b_4 + 2^3 b_3 + 2^2 b_2 + 2^1 b_1 + 2^0 b_0$$

or

$$\ldots + 256 b_8 + 128 b_7 + 64 b_6 + 32 b_5 + 16 b_4 + 8 b_3 + 4 b_2 + 2 b_1 + b_0$$

which can be written

$$\ldots + 1 b_8) \cdot 16^2 + (8 b_7 + 4 b_6 + 2 b_5 + 1 b_4) \cdot 16^1 + (8 b_3 + 4 b_2 + 2 b_1 + 1 b_0) \cdot 16^0.$$

Each of the sums in parentheses is a number between 0 (if all the b values are 0) and 15 (if all the b values are 1). These are exactly the digits in the hexadecimal number system. Thus to convert from binary to hexadecimal, we must gather up groups of 4 binary digits.

Example. Convert the following binary word to hexadecimal.

$$\underbrace{0010}_{2}\,\underbrace{1011}_{B}\,\underbrace{0011}_{3}\,\underbrace{1000}_{8}\,b$$

That is, 0010101100111000b = 2B38h.   ♣

The notation in assembly language is as follows. The letter **b** indicates that the number is in binary representation and the letter **h** indicates that the number is in hexadecimal representation. This is the notation used in assembly language. The letter **d** indicates a decimal number. The default value is decimal.

In C, C++, and Java decimal, octal and hexadecimal numbers are available. Hexadecimal numbers are indicated by **0x** ... in C, C++ and Java. For example the decimal number 91 would be expressed as **0x5B**, since

$$5 \cdot 16^1 + 11 \cdot 16^0 = 91.$$

Hexadecimal to Binary. To convert from hexadecimal to binary, we perform the opposite process from that used to convert from binary to hexadecimal. We must expand each hexadecimal digit to four binary digits.

Example. Convert the hexadecimal number D0h to binary. We find

$$\overbrace{1101}^{D}\,\overbrace{0000}^{0}\,b$$

That is, D0h = 11010000b.   ♣

Example. Convert FFh to binary.

$$\overbrace{1111}^{F}\,\overbrace{1111}^{F}\,b$$

Thus FFh = 11111111b.   ♣

Binary to Decimal. Write the binary sequence in its place-value summation form and then evaluate it.

Example.

$$10101010b = 1 \cdot 2^7 + 0 \cdot 2^6 + 1 \cdot 2^5 + 0 \cdot 2^4 + 1 \cdot 2^3 + 0 \cdot 2^2 + 1 \cdot 2^1 + 0 \cdot 2^0.$$

Thus

$$10101010b = 2^7 + 2^5 + 2^3 + 2^1 = 128 + 32 + 8 + 2 = 170d.$$

♣

Decimal to Binary. Divide the decimal number successively by 2: remainders are the coefficients of $2^0, 2^1, 2^2, \ldots$. We know that any non-negative integer a can be written in the form

$$a = a_0 + a_1 \cdot 2^1 + \ldots + a_m \cdot 2^m.$$

The technique follows simply from

$$\sum_{i=k}^{m} a_i 2^{i-k} = 2 \sum_{i=k+1}^{m} a_i 2^{i-k-1} + a_k.$$

The remainder after integer division by 2 gives a_k, and we continue until the division gives 0. The following example illustrates this.

Example. Convert $345d$ to binary.

$345/2 = \quad 172, \quad$ remainder 1; coefficient of 2^0 is 1

$172/2 = \quad 86, \quad$ remainder 0; coefficient of 2^1 is 0

$86/2 = \quad 43, \quad$ remainder 0; coefficient of 2^2 is 0

$43/2 = \quad 21, \quad$ remainder 1; coefficient of 2^3 is 1

$21/2 = \quad 10, \quad$ remainder 1; coefficient of 2^4 is 1

$10/2 = \quad 5, \quad$ remainder 0; coefficient of 2^5 is 0

$5/2 = \quad 2, \quad$ remainder 1; coefficient of 2^6 is 1

$2/2 = \quad 1, \quad$ remainder 0; coefficient of 2^7 is 0

$1/2 = \quad 0, \quad$ remainder 1; coefficient of 2^8 is 1

Thus, 345d = 101011001b.　　　♣

This method works because we want to find the coefficients $b_0, b_1, b_2, \ldots$ (which are 0 or 1) of $2^0, 2^1, 2^2, \ldots$ and so on. Thus, in the preceding example,

$$345 = b_{10}2^{10} + b_9 2^9 + b_8 2^8 + \cdots + b_1 2^1 + b_0 2^0$$

Dividing by 2,

$$345/2 = b_{10}2^9 + b_9 2^8 + \cdots + b_1 + (b_0/2)$$

Thus b_0 is the remainder on division by 2 and

$$(b_{10}2^9 + b_9 2^8 + \cdots + b_1)$$

is the quotient.

The following C++ program finds the binary representation of a non-negative integer. The operator % is used to calculate the remainder after integer division.

```
// remain.cpp

#include <iostream.h>

void main()
{
    int i;
    unsigned long N = 345;
    unsigned long array[32];
    for(i=0;i<32;i++)  {  array[i] = 0;  }
    for(i=0;i<32;i++)  {  array[i] = N%2;  N = N/2; }
    for(i=31;i>=0;i--) {  cout << array[i]; }
}
```

Decimal to Hexadecimal. Divide the decimal number successively by 16; remainders are the coefficients of $16^0, 16^1, 16^2, \ldots$

Example.

$$302/16 = 18, \quad \text{remainder } 14; \quad \text{coefficient of } 16^0 \text{ is E}$$
$$18/16 = 1, \quad \text{remainder } 2; \quad \text{coefficient of } 16^1 \text{ is 2}$$
$$1/16 = 0, \quad \text{remainder } 1; \quad \text{coefficient of } 16^2 \text{ is 1}$$

Therefore, 302d = 12Eh.   ♣

This works for the same reason that the method for decimal-to-binary conversion works. That is, division by 16 produces as a remainder the coefficient (h_0) of 16^0, and as a quotient the decimal number minus the quantity $(h_0 \cdot 16^0)$.

Hexadecimal to Decimal. Write the hexadecimal number in its place-value summation form and then evaluate.

Example.

$$CA14h = C \cdot 16^3 + A \cdot 16^2 + 1 \cdot 16^1 + 4 \cdot 16^0.$$

Thus

$$CA14h = 12 \cdot 4096 + 10 \cdot 256 + 1 \cdot 16 + 4 = 51732d.$$

                                                                    ♣

The following C++ program finds the hexadecimal representation of a non-negative integer.

```
// remain2.cpp

#include <iostream.h>

void main()
{
    int i;
    unsigned long N = 15947;
    unsigned char array[8];

    for(i=0;i<8;i++)
     array[i] = 0;

    for(i=0;i<8;i++)
    {
      array[i] = N%16;
      if(array[i]>9)
       array[i] += 'A'-10;
      else
       array[i] += '0';
      N = N/16;
    }

    for(i=7;i>=0;i--)
     cout << array[i];
}
```

The output is 00003E4B.

3.1.2   Arithmetic

The rules for addition of binary numbers are:

```
0 + 0 = 0
0 + 1 = 1
1 + 0 = 1
1 + 1 = (1) 0
```

where (1) denotes a carry of 1. Note that 10b is the binary equivalent of 2 decimal. Thus the sum $1 + 1$ requires two bits to represent it, namely 10, the binary form of the decimal number 2. This can be expressed as follows: one plus one yields a sum bit $s = 0$ and a carry bit $c = 1$. If we ignore the carry bit and restrict the sum to the single bit s, then we obtain $1 + 1 = 0$. This is a very useful special form of addition known as *modulo-2 addition*.

Doing arithmetic in the binary and hexadecimal number systems is best shown by examples and best learned by practice.

Example. Decimal Arithmetic

```
    45
+   57
-----
   102
```

Remember that $7 + 5$ is 2 with a 1 carry in decimal. $5 + 4+$ the carried 1 is 0 with a 1 carry.

Example. Binary Arithmetic

```
   1011
+  1001
------
  10100
```

Remember that $1 + 1$ is 0 with a 1 carry in binary.

Example. Binary Arithmetic

```
   1111
+  1111
-------
  11110
```

Example. Hexadecimal Arithmetic

```
  1A
+  5
----
  1F
```

♣

In decimal A + 5 is $10 + 5 = 15$. Thus $15d = Fh$.

Example. Hexadecimal Arithmetic

```
   FF
+   3
-----
  102
```

♣

F + 3 is 15 + 3 in decimal. Thus 18d = 12h. Thus we write down a 2, carry a 1.

Binary multiplication can be done by repeated addition. The following example shows how to multiply the two binary numbers 1001 (9 decimal) and 110 (6 decimal).

```
    1001 multiplicand
x    110 multiplier
------------------------------
    0000 first partial product
    1001 second partial product
    1001 third partial product
------------------------------
  110110 product   (54 decimal)
```

High speed multiplication techniques use addition and subtraction or uniform multiple shifts. *Binary divisions* can be performed by a series of subtractions and shifts.

3.1.3 Signed Integers

We can easily see how positive integers are stored. For example, 345 is stored as 101011001. This will not fit into a byte because it has more than 8 bits, but it fits into a word (2 consecutive bytes). A byte has 8 bits, a word has 16 bits and a double word has 32 bits.

Example. Show 65712d as a binary (a) byte, (b) word, (c) double word. We have

$$65712d = 10000000010110000b$$

Thus

(a) Does not fit in a byte (it is too large).

(b) Does not fit in a word (it is to large).

(c) 00000000000000010000000010110000

♣

The largest integer number which fits into a register with 32 bits is

$$2^{32} - 1 \equiv \sum_{k=0}^{31} 2^k \equiv 4294967295.$$

The largest integer number which fits into a register with 64 bits is

$$2^{64} - 1 \equiv \sum_{k=0}^{31} 2^k \equiv 18446744073709551615.$$

Storing negative integers presents a more difficult problem since the negative sign has to be represented (by a 0 or a 1) or some indication has to be made (in binary!) that the number is negative. There have been many interesting and ingenious ways invented to represent negative numbers in binary. We discuss three of these here:

1. Sign and magnitude

2. One's complement

3. Two's complement

Sign and Magnitude

The sign and magnitude representation is the simplest method to implement negative integers. Knuth [105] used sign and magnitude in his mythical MIX computer. In sign and magnitude representation of signed numbers, the leftmost (most significant) bit represents the sign:

$$0 \quad \text{for positive}$$

and

$$1 \quad \text{for negative.}$$

Example. The positive integer number 31 stored in a double word (32 bits) using sign and magnitude representations is

$$00000000000000000000000000011111b$$

Thus the negative integer -31 becomes

$$10000000000000000000000000011111b$$

♣

There are two drawbacks to sign and magnitude representation of signed numbers:

1. There are two representations of 0:

$$+0 = 00000000000000000000000000000000b$$

and

$$-0 = 10000000000000000000000000000000b.$$

Thus the CPU has to make two checks every time it tests for 0. Checks for 0 are done frequently, and it is inefficient to make two such checks.

2. Obviously,

$$a + (-b)$$

is not the same as

$$a - b.$$

What this means is that the logic designer must build separate circuits for subtracting; the adding circuit used for $a + b$ is not sufficient for calculating $a - b$.

Example. The following shows that

$$52 - 31$$

and

$$52 + (-31)$$

are not the same in sign and magnitude representation.

```
 52 =   00000000 00000000 00000000 00110100b
-31 = -00000000 00000000 00000000 00011111b
---------------------------------------------
 21 =   00000000 00000000 00000000 00010101b
```

On the other hand

```
52    =   00000000 00000000 00000000 00110100b
+ -31 =   10000000 00000000 00000000 00011111b
---------------------------------------------
21    =   10000000 00000000 00000000 01010011b
```

Thus this shows that the sign and magnitude representation is not useful for implementations on CPU's.

Furthermore $31 - 31$ gives

```
 31 =   00000000 00000000 00000000 00011111b
-31 = -00000000 00000000 00000000 00011111b
---------------------------------------------
  0 =   00000000 00000000 00000000 00000000b
```

and $31 + (-31)$ gives

```
31    =   00000000 00000000 00000000 00011111b
+ -31 =   10000000 00000000 00000000 00011111b
---------------------------------------------
 0    =   10000000 00000000 00000000 00111110b
```

♣

One's Complement

One's complement method of storing signed integers was used in computers more in the past than it is currently. Here we assume again that 32 bits are given (double word). In one's complement, the leftmost bit is still 0 if the integer is positive. For example,

$$00000000000000000000000000011111b$$

still represents +31 in binary. To represent the negative of this, however, we replace all 0's with 1's and all 1's with 0's. Thus

$$11111111111111111111111111100000b$$

represents −31. Note that the leftmost bit is again 1. Notice that in assembly language one starts counting from zero from the rightmost bit.

Example. Using a double word of storage, −1 is stored as

$$11111111111111111111111111111110b$$

since 1 is stored in binary as

$$00000000000000000000000000000001b$$

♣

Thus the second drawback to sign and magnitude representation has been eliminated. This means $a - b$ is the same as $a + (-b)$. Thus the circuit designer need only include an adder; it can also be used for subtraction by replacing all subtractions $a - b$ with $a + (-b)$.

The following example shows, however, that this adder must do a little more than just add.

Example. We show that $52 - 31$ and $52 + (-31)$ are the same in one's complement representation. For $52 - 31$ we have

```
 52 =   00000000 00000000 00000000 00110100b
-31 = -00000000 00000000 00000000 00011111b
---------------------------------------------
 21 =   00000000 00000000 00000000 00010101b
```

♣

Next we consider the one's complement. Since

```
31 = 00000000 00000000 00000000 00011111b
```

we find for the one's complement

-31 = 11111111 11111111 11111111 11100000b

Addition of the two terms $52 + (-31)$ yields

00000000 00000000 00000000 00010100b

plus an overflow bit. Addition of the overflow bit to the right-most bit yields

00000000 00000000 00000000 00010101b

which is 21 in binary representation.

The adder for one's complement arithmetic is more complicated; it must carry around any overflow bit in order to work correctly for subtraction. The first drawback is still with us, however. In one's complement, there are still two representations of 0

00000000 00000000 00000000 00000000b   positive 0

and

11111111 11111111 11111111 11111111b   negative 0

when viewed as a double word.

One's complement is implemented in C, C++ and Java with the ~ operator. The following program shows an application.

```
// complement.cpp

#include <iostream.h>

char *binary(unsigned int N)
{
 static char array[36];

 for(int i=34,j=27;i>=0;i--)
  if((i == j) && (i != 0)) { array[i]=' '; j-=9;}
  else { array[i] = N%2 + '0';   N = N/2; }
 array[35]='\0';
 return array;
}

void main()
{
 int a = 17; // binary 000000000 00000000 00000000 0010001
 cout << "a = " << a << endl << binary(a) << endl;
 int b = ~a; // binary 111111111 11111111 11111111 1101110
 cout << "~a = " << b << endl << binary(b) << endl;
}
```

Two's Complement

The *two's complement method* of storing signed integers is used in most present-day CPU's, including the 386, 486, Pentium and Alpha Dec. The two's complement is formed by

(1) forming the one's complement and then

(2) adding 1.

Example. Using two's complement and a double word (32 bits). Thus the decimal number 31 is stored as

$$00000000000000000000000000011111b.$$

Consequently -31 is stored as

$$11111111111111111111111111100001b.$$

♣

We can easily check that
$$a - b = a + (-b)$$
and that there is only one way to represent 0, i.e., $+0$ and -0 are stored the same, namely
$$00000000000000000000000000000000b.$$
The two's complement of a number is the true (one's) complement of the number plus 1.

With n bits we can represent numbers from

$$-2^{n-1} \quad \text{to} \quad 2^{n-1} - 1$$

in two's complement. If we have registers with 32 bits then we can store the integer numbers ($n = 32$)
$$-2147483648 \quad \text{to} \quad 2147483647.$$

Although taking the two's complement of a number is more difficult than taking its one's complement, addition of two's complement numbers is simpler than addition in one's complement or in signed-magnitude representations.

Next we consider some examples of addition in two's complement. We assume that 32 bits are given.

Addition of Two Positive Numbers

```
+3 = 00000000 00000000 00000000 00000011b
+4 = 00000000 00000000 00000000 00000100b
-------------------------------------------
+7 = 00000000 00000000 00000000 00000111b
```

Addition of Two Negative Numbers

```
-4 = 11111111 11111111 11111111 11111100b
-1 = 11111111 11111111 11111111 11111111b
-------------------------------------------
-5 = 11111111 11111111 11111111 11111011b
```

Addition of One positive and One Negative Number

```
-7 = 11111111 11111111 11111111 11111001b
+5 = 00000000 00000000 00000000 00000101b
-------------------------------------------
-2 = 11111111 11111111 11111111 11111110b
```

In two's complement, it is possible to add or subtract signed numbers, regardless of the sign. Using the usual rules of binary addition, the result comes out correctly, including the sign. The carry is ignored. This is a very significant advantage. If this were not the case, we would have to correct the result for sign every time, causing a much slower addition or subtraction time. For the sake of completeness, let us state that two's complement is simply the most convenient representation to use for microprocessors. All signed integers will be implicitly represented internally in two's complement notation.

The following Java program shows an implementation of two's complement.

```
// Twocomp.java

public class Twocomp
{
 public static void main(String[] args)
 {
  int r1 = 14;        // binary 1110

  // two's complement to find the negative number
  // of a given integer number
  // The operation ~ gives the one complement
  // and then we add 1 to find the two's complement
  int r2 = ~r1; r2++;
  System.out.println(r2); // => -14
 }
}
```

3.1.4 Overflow

If we do arithmetic operations with 32 bit registers *overflow* will occur in cases:

1. if we go beyond the range 0–4294967295 for the data type **unsigned long** in C and C++. This means we add numbers so that the sum is larger than 4294967295. Also negative numbers are out of range.

2. if we go out of the range −214783648 to 2147483647 (**long** in C and C++). This means if we add or subtract numbers which go beyond this range.

Example. Consider the sum

$$4294967295 + 1 = 4294967296$$

The number on the right-hand side is out of the range for a 32 bit register for the C and C++ data type **unsigned long**. Since

$$4294967295 = 11111111111111111111111111111111b$$

for unsigned long, the addition of 1 yields

$$00000000000000000000000000000000b$$

with one overflow bit. Thus the output is 0. ♣

Example. Consider the sum

$$+(-2147483648) + (-3) = -2147483651$$

The number on the right-hand side is out of range for a 32 bit register for long. Since

$$-2147483648 = 10000000000000000000000000000000b$$

and

$$-3 = 11111111111111111111111111111101b$$

we obtain

$$01111111111111111111111111111101b$$

Thus the output is 2147483651. ♣

Consider the following C++ program.

```
// overflow.cpp

#include <iostream.h>

int main()
{
    unsigned long a = 4294967295;
    unsigned long b = 1;
    unsigned long r1;
    r1 = a + b;
    cout << "r1 = " << r1 << endl;  // 0

    unsigned long c = 4294967295;
    unsigned long d = 2;
    unsigned long r2;
    r2 = c + d;
    cout << "r2 = " << r2 << endl;  // 1

    unsigned long e = 0;
    unsigned long f = -1;
    unsigned long r3 = e + f;
    r3 = e + f;
    cout << "r3 = " << r3 << endl; // 4294967295

    long g = -2147483648;
    long h = -3;
    long r4 = g + h;
    cout << "r4 = " << r4 << endl; // 2147483645

    unsigned long i = 0xFFFFFFFF; // hexadecimal number
    unsigned long j = 0x1;        // hexadecimal number
    unsigned long r5 = i + j;
    cout << "r5 = " << r5 << endl; // 0

    return 0;
}
```

The range of the data type **unsigned long** is 0 - 4294967295. The binary representation of 4294967295 is

11111111 11111111 11111111 11111111b

This is the largest number which fits into 32 bits. Thus if we add 1 to this binary number under the assumption that 32 bits are given we find

00000000 00000000 00000000 00000000b

with 1 overflow bit. The output of the C++ program ignores the overflow bit and displays the output 0.

Remark. Obviously the overflow flag of the CPU will be set.

Analogously we understand the other three outputs,

$$r2 = 1$$

$$r3 = 4294967295$$

and

$$r4 = 2147483645.$$

of the program.

Java has the signed data type **long**. The size is 64 bits. Thus the range is

-9223372036854775808 to 9223372036854775807.

3.1.5   Binary-Coded Decimal Form

The discussion so far has assumed that decimal numbers are translated into base-2 form for processing by digital circuits. An alternative approach is to encode the decimal digits into binary form, but maintain the base-10 positional notation in which all digits are weighted by powers of 10. Such numbers are called *binary-coded decimal numbers* or, if the context is clear, simply decimal numbers. We usually restrict the term binary-coded decimal, which is abbreviated *BCD*, to refer to the most widely used code of this sort.

An unsigned decimal number

$$N_{10} = d_{n-1}d_{n-2}\ldots d_1 d_0$$

is converted into the standard BCD form by mapping each digit d_i separately into a 4-bit binary number

$$B_i = b_{i3}b_{i2}b_{i1}b_{i0}$$

where $(B_i)_2 = (d_i)_{10}$. Thus, a 9 in N_{10} is mapped into 1001, an 8 into 1000, a 7 into 0111, and so on. For example, if $N_{10} = 7109_{10}$, then the decimal-to-BCD conversion process takes the form

$$N_{10} = \underbrace{7}_{0111}\,\underbrace{1}_{0001}\,\underbrace{0}_{0000}\,\underbrace{9}_{1001}$$

leading to

$$N_{\underline{10}} = 0111000100001001_{\underline{10}}$$

where the underlined subscript $\underline{10}$ is our notation for binary-coded decimal. This conversion process is, in fact, the same as that used for changing a hexadecimal number to binary. In this case, there are 10 digits instead of 16, so only 10 of the 16 possible 4-bit binary numbers are needed. Also, each 4-bit group must be assigned weight 10 rather than 16. For example, we get

$$N = 0111_2 \cdot 10^3 + 0001_2 \cdot 10^2 + 0000_2 \cdot 10^1 + 1001_2 \cdot 10^0$$

assuming N as an integer. The weight of an individual bit in $N_{\underline{10}}$ is of the form $j \cdot 10^i$, where j is 8, 4, 2, or 1. Standard BCD is therefore sometimes called 8421 decimal code.

Conversion from BCD to ordinary decimal form is achieved by replacing 4-bit groups with the equivalent decimal digit. For instance,

$$N'_{\underline{10}} = \underbrace{0010}_{2}\,\underbrace{1000}_{8}\,\underbrace{0100}_{4}\,\underbrace{1001}_{9}\,\underbrace{0000}_{0}\,\underbrace{0101}_{5}$$

implying that $N'_{10} = 284905_{10}$. Conversion between binary (base 2) and BCD requires the decimal-binary conversion procedure, in addition to the decimal digit-encoding procedure discussed above.

Not all the possible binary patterns correspond to BCD numbers. The six 4-bit patterns

$$\{\, 1010,\ 1011,\ 1100,\ 1101,\ 1110,\ 1111 \,\}$$

are not needed to represent decimal digits, and therefore cannot appear in the digit positions of a BCD number. These six digit patterns appear in BCD numbers only as the result of an error, such as failure of a hardware component or a programming mistake. The fact that some n-bit patterns are unused implies that a larger n is needed to represent a given range of numbers if BCD code is employed in place of binary. For example, to represent all integers from zero to a million requires 24 bits of BCD code, but only 20 bits of binary code. Note that $2^{20} > 10^6$. Therefore, BCD numbers have slightly greater storage requirements than binary numbers. Arithmetic operations on BCD numbers are more complicated than their binary counterparts. The main advantage of BCD code is that it eliminates most of the need for time-consuming base-10-to-base-2 and base-2-to-base-10 conversions. Digital computers are designed primarily to process binary numbers, but many have features to support BCD operations.

The following C++ program converts a number ($N = 15947$) to BCD form.

```
// bcd.cpp

#include <iostream.h>

void main()
{
    int i;
    unsigned long N = 15947;
    unsigned char array[4];
    unsigned char mask[2]={0x0F,0xF0};
    unsigned char shift[2]={0,4};
    for(i=0;i<4;i++)
     array[i] = 0;
    for(i=0;i<8;i++)
     { array[i/2] |= (N%10)<<shift[i%2]; N = N/10; }
    for(i=7;i>=0;i--)
     { cout << char(((array[i/2]&mask[i%2])>>shift[i%2])+'0'); }
}
```

3.2 Floating Point Representation

3.2.1 Introduction

We have seen how to store integer numbers in bit sequences, and in Chapter 2 an example illustrated a similar method for representing real numbers on a certain interval with a guaranteed accuracy. The interval was divided into equal subintervals according to the number of bit string combinations allowed. Certain real numbers can be converted without loss of accuracy. These are the numbers which are boundaries for the subintervals. The other real numbers will be encoded, with error, to one of these numbers. This is the *fixed point* method of storing real numbers. The maximum accuracy of the fractions is uniquely specified by the size of the sub intervals. Thus we can identify a fractional part and an integer part giving a fixed separation of bits representing the integer and fractional parts respectively. Traditionally this separation is indicated by a decimal point in the symbolic representation. *Floating point* number representations use unequal subintervals to represent real numbers. For example small numbers require better accuracy in calculations and thus are represented with smaller intervals. Larger numbers use larger intervals. To compromise we may require that the ratio of the error in representation to the number being represented be approximately constant.

Storing floating-point numbers presents a problem similar to that of storing signed integers. For integers, some indication of a positive or negative sign has to be represented. For floating-point instructions some method must be devised for showing where the decimal point should go. That is, we must distinguish between the fractional part to the right of the decimal point – called the mantissa – and the integer portion to the left of the decimal point. Different methods have been used in the past and different methods continue to be used by the various manufacturers of computers. There have been so many different ways of coding a floating-point number into binary that the Institute of Electrical and Electronics Engineers (IEEE) has proposed a standard format.

There are actually three formats – one that requires 32 bits, one that is used for 64 bits, and one for 80 bits. We describe the 32-bit format, called the *short real format*, here.

The table lists seven numeric data types showing the data format for each type. The table also shows the approximate range of normalized values that can be represented with each type. Denormal values are also supported in each of the real types, as required by IEEE Std 854.

Table: Numeric Data Types

Data Type	Bits	Significant Digits (Decimal)	Approximate Normalized Range (Decimal)
Word Integer	16	4	$-32768 \leq x \leq +32767$
Short Integer	32	9	$-2 \times 10^9 \leq x \leq +2 \times 10^9$
Long Integer	64	18	$-9 \times 10^{18} \leq x \leq +9 \times 10^{18}$
Packed Decimal	80	18	$-99\ldots99 \leq x \leq +99\ldots99(18 \text{ digits})$
Single Real	32	7	$1.18 \times 10^{-38} < |x| < 3.40 \times 10^{38}$
Double Real	64	15-16	$2.23 \times 10^{-308} < |x| < 1.79 \times 10^{308}$
Extended Real	80	19	$3.37 \times 10^{-4932} < |x| < 1.18 \times 10^{4932}$

All operands are stored in memory with the least significant digits starting at the initial (lowest) memory address. Numeric instructions access and store memory operands using only this initial address.

3.2.2   Representation

The first step to understanding how a binary fraction is stored using short real format is to *normalize* it. This is similar to putting a decimal point number into the familiar scientific notation in which we have a *sign*, an *exponent*, and a *mantissa*. To normalize a binary fraction, we write it so that the first 1 is just to the left of the binary point.

Example. Consider the binary number

$$0.000111101 = 0 \cdot \frac{1}{2} + 0 \cdot \frac{1}{2^2} + 0 \cdot \frac{1}{2^3} + 1 \cdot \frac{1}{2^4} + 1 \cdot \frac{1}{2^5} + 1 \cdot \frac{1}{2^6} + 1 \cdot \frac{1}{2^7} + 0 \cdot \frac{1}{2^8} + 1 \cdot \frac{1}{2^9}$$

Then the *normalized representation* is

$$1.11101 \cdot 2^{-4}$$

♣

The next step is to represent the important parts of the normalized fraction in 32 bits. The important parts are those that will allow us to recover the original number (and allow the computer to perform operations on it). These parts are the

1. Sign

2. Exponent (whose base is understood to be 2)

3. Mantissa

In the IEEE short real format, the sign is stored in the leftmost bit, the exponent is stored in the next 8-bits, after some alteration, and the mantissa is stored in the rightmost 23 bits, again after a minor adjustment.

1. To store the sign. 0 for positive, 1 for negative.

2. To store the exponent. Add 127 (1111111_2) to it. The number 127 is called a bias, and the resulting exponent is called a biased exponent. Biased exponents may range from 1 to 254, so that exponents range from -126 to $+128$.

3. To store the mantissa. Remove the leftmost 1 and store the rest of the fraction left-adjusted. This technique of not storing the first 1 before the binary point is a common way to store mantissas. It is called *hidden bit* storage. Computer circuitry knows that the 1 is really part of the mantissa.

Example. Find 0.0390625 (base 10) as it would be stored in short real format.

Step 1. Convert the fraction to binary

$$.0390625_{10} = .0000101_2$$

Step 2. Normalize the binary fraction.

$$.0000101 \text{ normalized is } 1.01 \cdot 2^{-5}$$

Step 3. Calculate the sign, the exponent, and the mantissa.

Sign: 0, since this is a positive number

Exponent: $-5 + 127 = 122$ (base 10) = 01111010 (base 2)

Mantissa: .01 left-adjusted into a field of width 23 is

$$.01000000000000000000000$$

Thus the entire number is represented by the bitstring

$$\underbrace{0}_{Sign} \; \underbrace{01111010}_{Exponent} \; \underbrace{01000000000000000000000}_{Fraction}$$

♣

The following C++ program implements this algorithm. The only difference is that the actual conversion to binary is delayed until after the normalization procedure. We use the above test example. For the output we find

```
0.0390625 (base 10) =
0 01111010 01000000000000000000000 (floating point base 2)
```

```
// float2bin.cpp

#include <iostream.h>
#include <math.h>

void normalize(float &f,char &e)
{
 e = 0;

 // numbers larger than 2 we reduce
 // down to 1 plus a fraction,
 // and a compensating exponent
 while(fabs(f) > 2)
 {
  f /= 2;
  e++;
 }

 // numbers smaller than 1 we promote
 // up to 1 plus a fraction,
 // and a compensating exponent
 while(fabs(f) < 1)
 {
  f *= 2;
  e--;
 }
}

void float2bin(float f,char *b)
{
 char e1;
 int e;
 int i;

 normalize(f,e1);

 // add the bias
 e = int(e1) + 127;

 // set the sign bit
 b[0] = (f < 0) ? '1':'0';

 f = fabs(f);

 // remove the leftmost 1 bit
 f -= 1;

 b[1]=b[10]=' ';
```

```
// convert the exponent
for(i=8;i>0;i--)
{
 b[i+1] = e%2 + '0';
 e/=2;
}

// convert the mantissa
for(i=1;i<24;i++)
{
 int bit = (f>=pow(2,-i));

 b[i+10] = (bit) ? '1':'0';
 if (bit) f -= pow(2,-i);
}

 b[34]='\0';
}

void main(void)
{
 char b[35];
 float f=0.0390625;

 float2bin(f,b);
 cout << f << " (base 10) = "
      << b << " (floating point base 2)"<<endl;
}
```

Example. What number is stored as

$$10111110111101000000000000000000 \ ?$$

We recover the parts as

$$1|01111101|11101000000000000000000$$

Sign: 1, so the number is negative

Exponent: $01111101_2 = 125_{10}, \quad 125 - 127 = -2$

Mantissa: affixing 1 to the left of

$$.11101000000000000000000$$

results in

$$1.11101000000000000000000$$

which is

$$1.11101_2 .$$

Multiplying by 2^{-2} (provided by the exponent) yields

$$.0111101_2 = \frac{1}{4} + \frac{1}{8} + \frac{1}{16} + \frac{1}{32} + \frac{1}{128} = -0.4765625_{10}$$

♣

Chapter 4
Logic Gates

4.1 Introduction

A digital electronic system uses a building-block approach. Many small operational units are interconnected to make up the overall system. The system's most basic unit is the gate circuit. These circuits have one output and one or more inputs. The most basic description of operation is given by the function table, which lists all possible combinations of inputs along with the resulting output in terms of voltage, high and low. Table 4.1(a) shows a function table for a 2-input circuit. This table indicates that if both inputs are low or both are high, the output will be low. If one input is high and the other is low, a high level will result on the output line. As we deal with logic design, it is appropriate to use 1s and 0s rather than voltage levels. Thus, we must choose a positive ($H = 1$, $L = 0$) or negative ($H = 0$, $L = 1$) logic scheme. Once this choice is made, we use the function table to generate a truth table. The function table describes inputs and outputs in terms of 1s and 0s rather than voltage levels. Function tables are used by manufacturers of logic gates to specify gate operation. The manufacturer conventionally defines gates in terms of positive logic.

Inputs		Output	Inputs		Output	Inputs		Output
A_1	A_2	X	A_1	A_2	X	A_1	A_2	X
L	L	L	0	0	0	1	1	1
L	H	H	0	1	1	1	0	0
H	L	H	1	0	1	0	1	0
H	H	L	1	1	0	0	0	1

L=low voltage level	Positive logic	Negative logic
H=high voltage level		
(a)	(b)	(c)

Table 4.1: Function Table and Truth Tables for a Logic Circuit

4.2 Gates

4.2.1 AND Gate

The AND gate has one output and two or more inputs. The output will equal 0 for all combinations of input values except when all inputs equal 1. When each input is 1, the output will also equal 1. Figure 4.1 shows the AND gate. Table 4.2 shows the function and positive logic truth tables. The AND gate will function as an OR gate for negative logic, but the gate is named for its positive logic function.

A_1	A_2	X
0	0	0
0	1	0
1	0	0
1	1	1

Table 4.2: Truth Table for the AND Gate

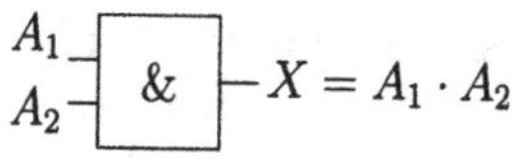

$$A_1 \quad \boxed{\&} \quad X = A_1 \cdot A_2$$
$$A_2$$

Figure 4.1: Symbol for 2-input AND Gate

The AND operation can be interpreted as the multiplication of a set of 1 bit numbers; a 0 among the input variables makes the result (product) 0; the product is 1 if and only if all the inputs are 1. For this reason the AND function is written as a product expression

$$X_{\text{AND}} := A_1 \cdot A_2$$

or

$$X_{\text{AND}} := A_1 \cdot \ldots \cdot A_n$$

if we have n inputs. Alternative AND symbols in common use are $\wedge$ and $\&$. The latter is the AND designator in the standard box symbol for an AND gate. As with multiplication the symbol $\cdot$ is sometimes omitted from AND expressions, so that $A_1 \cdot A_2$ reduces to $A_1 A_2$.

In CMOS the 4081 provides quad two-input AND gates.

4.2.2 OR Gate

The OR gate has one output and two or more inputs. If all inputs are equal to 0, the output will be equal to 0. The presence of a 1 bit leads to an output of 1. Table 4.3 describes this operation in terms of a truth table. The standard symbol for a 2-input OR gate is shown in Figure 4.2.

A_1	A_2	X
0	0	0
0	1	1
1	0	1
1	1	1

Table 4.3: Truth Table for the OR Gate

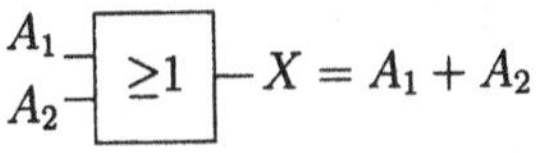

$$A_1 \quad \boxed{\geq 1} \quad X = A_1 + A_2$$
$$A_2$$

Figure 4.2: Symbol for 2-input OR Gate

The OR operation takes its name from the fact that the output X is 1 if and only if A_1 is 1 or A_2 is 1. In other words, the output X of an OR gate is 1 if and only if the number of 1s applied as input is one or greater.

We can have more than 2 input lines. The X is 1 if and only if A_1 is 1 or A_2 is 1 or $\ldots A_n$ is 1. This interpretation leads to the use of the symbol ≥ 1 in the OR box. By a somewhat weak analogy with numerical addition, the OR function is usually written as a sum expression

$$X_{OR}(A_1, A_2, \ldots, A_n) := A_1 + A_2 + \ldots + A_n.$$

Thus, + denotes OR in this context, and is read as "or" rather than plus. An alternative OR symbol is $\vee$.

In CMOS 4071 is a quad two-input gate and the CMOS 4072 is a two quad-input gate.

4.2.3 XOR Gate

If the exclusive OR gate (XOR gate) has two inputs, then the output gives a 1 when either input is 1, but not when both are 1. If the input is $A_1 = 0$ and $A_2 = 0$, then the output is 0. Table 4.4 shows the truth table for the XOR gate.

A_1	A_2	X
0	0	0
0	1	1
1	0	1
1	1	0

Table 4.4: Truth Table for the XOR Gate

The generalization of XOR to n input variables is most easily specified in terms of the parity of the number of 1's among the n input variables

$$X_{\text{XOR}}(A_1, A_2, \ldots, A_n) := \begin{cases} 1 & \text{if an odd number of inputs are 1} \\ 0 & \text{otherwise} \end{cases}$$

For this reason, XOR is also called the *odd-parity function*, and is the basis of error-handling circuits. This versatile function can also be interpreted as (numerical) summation modulo 2. Thus, another definition of XOR equivalent to the definition given above is

$$X_{\text{XOR}}(A_1, A_2, \ldots, A_n) := A_1 + A_2 + \ldots + A_n \ (\text{mod}2).$$

The XOR gate is a special gate and is widely employed in digital circuits that perform mathematical functions.

The symbol for the XOR gate are shown in the next figure.

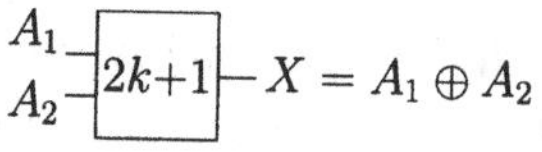

$$A_1 \underline{} \boxed{2k+1} \underline{} X = A_1 \oplus A_2$$

Figure 4.3: Symbol for 2-input XOR Gate

The use of the generic odd number $2k+1$ as the function designator in the standard box symbol reflects the fact that the output is 1 if and only if $2k + 1$ inputs are 1, for $k = 0, 1, 2, \ldots$. In logic expressions, the XOR operator is $\oplus$ which is read as *exclusive OR*, ring-sum, or sum modulo 2. Thus, we can write

$$X_{\text{XOR}} = A_1 \oplus A_2 \oplus \ldots \oplus A_n$$

The CMOS 4030 is a quad two-input exclusive OR gate.

4.2.4 NOT Gate (Inverter)

The *inverter* (also called NOT gate) performs the NOT or INVERT function and
has one output and one input. The output level is always opposite to the input level.
Thus an inverter converts a 0 to 1 and a 1 to 0, an operation known variously as
inversion, complementation, or the NOT function. NOT is denoted by an overbar
in functional expressions and by a small circle, the inversion symbol, in circuit
diagrams. We write

$$X_{\text{NOT}} = \overline{A}.$$

Table 4.5 shows the truth table for the inverter. Figure 4.4 shows the symbol for
the inverter.

A	X
0	1
1	0

Table 4.5: Truth Table for the NOT Gate

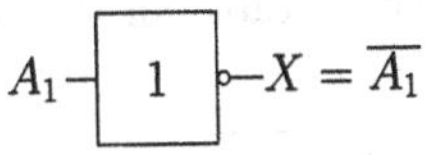

$$A_1 - \boxed{1} \!\!\circ - X = \overline{A_1}$$

Figure 4.4: Symbol for the NOT Gate

In CMOS the 4069 is a hex inverter. Each of the six inverters is a single stage.

The NOT gate can be combined with the AND, OR and XOR gate to provide the
NAND, NOR and XNOR gate.

4.2.5 NAND Gate

The *NAND gate* is an AND gate followed by an inverter. A NAND gate can have two or more inputs. Thus the NAND gate is formed by appending NOT to AND. The output will be 0 only when all inputs are 1. Its logic expression is

$$X = \overline{A_1 \cdot A_2 \cdot \ldots \cdot A_n}$$

which indicates that the inputs $A_1, A_2, \ldots, A_n$ are first ANDed and then the result inverted. Thus a NAND gate always produces an output that is the inverse (opposite) to an AND gate. The gate symbol is therefore formed by appending the graphic inversion symbol (a small circle) to the corresponding AND symbol.

A_1	A_2	X
0	0	1
0	1	1
1	0	1
1	1	0

Table 4.6: Truth Table for the NAND Gate

A_1 ─┐
      │ & ○─ $X = \overline{A_1 \cdot A_2}$
A_2 ─┘

Figure 4.5: Symbol for 2-input NAND Gate

Since both inverters and AND gates can be constructed from NAND gates, the NAND gate is seen to be a functionally complete set itself. The AND gate and inverter form a functionally complete set. This means that any logic function realized by logic gates can be realized with the AND and NOT functions. For example the XOR gate can be represented by

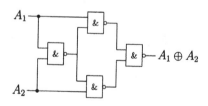

Figure 4.6: XOR Implemented With NAND Gates

In CMOS the 4011 provides a quad two-input NAND gate.

4.2.6 NOR Gate

The NOR gate is an OR gate followed by an inverter. The NOR gate can have two or more inputs. Thus the NOR gate combines the OR and NOT operations such that the output will be 0 when any input is 1. Its logical expression is

$$X = \overline{A_1 + A_2 + \ldots + A_n}$$

which indicates that $A_1, A_2, \ldots, A_n$ are first ORed and then the result is inverted. A NOR gate always gives an output that is the inverse of the OR gate. The gate is characterized by the tables and symbols of Table 4.7 and Figure 4.7.

A_1	A_2	X
0	0	1
0	1	0
1	0	0
1	1	0

Table 4.7: Truth Table for the NOR Gate

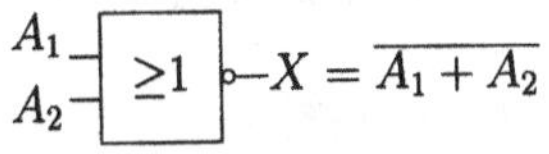

$$A_1 \quad A_2 \quad \boxed{\geq 1} \circ\!\!-X = \overline{A_1 + A_2}$$

Figure 4.7: Symbol for 2-input NOR Gate

All other gates can be constructed from NOR gates. For example, the XOR gate can be found as

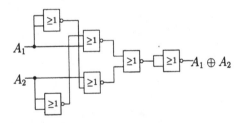

Figure 4.8: XOR Implemented With NOR Gates

In CMOS the 4001B is a quad two-input NOR gate.

4.2.7 XNOR Gate

The exclusive-NOR or XNOR gate produces a 1 output only when the inputs are at the same logic level. The exclusive-NOR gate is also known as the even-parity function for obvious reasons. The truth table is given in Table 4.8

A_1	A_2	X
0	0	1
0	1	0
1	0	0
1	1	1

Table 4.8: Truth Table for the XNOR Gate

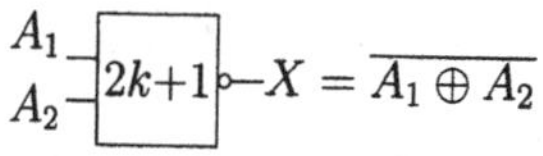

$$A_1 \atop A_2 \quad \boxed{2k+1} \!\!\!\!\circ\!\!-X = \overline{A_1 \oplus A_2}$$

Figure 4.9: Symbol for 2-input XNOR Gate

The XNOR gate is not a universal gate.

In CMOS the 4077 provides the quadruple exlusive-NOR gate.

4.3 Buffer

The *buffer* is an IC device that provides no change in logic at the output, but does provide a high input load impedance, and therefore good output drive capability. It works the same way as an emitter-follower circuit. The output of a MOS microprocessor, for example, has very poor drive capability when driving a TTL device. By inserting a buffer between the output of the MOS microprocessor and the input of the TTL device, we can solve the problem. The buffer provides an input load the processor can handle and an output drive that is TTL-compatible. The truth table and the symbol for a buffer are shown in Table 4.9 and Figure 4.10.

A	X
0	0
1	1

Table 4.9: Truth Table for the Buffer

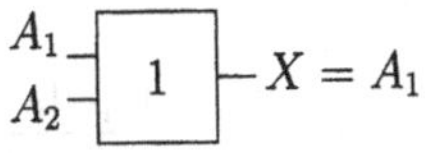

$$A_1 \boxed{\ 1\ } X = A_1$$
$$A_2$$

Figure 4.10: Symbol for the Buffer

As an example consider the buffering of MPU buses. The MPU, RAM and ROM are chips that are generally manufactured using CMOS technology. The decoders, gates, inverters, tri-state buffers, and output register are all TTL devices, usually LS-TTL to minimize power requirements and loading.

In CMOS the 4041B is a quadruple true/complement buffer which provides both an inverted active LOW output and a non-inverted active HIGH output (O) for each input (I).

4.4   Tri-State Logic

The development of bus organized computers led to the development of a type of logic circuitry that has three distinct output states. These devices, called *tri-state logic* (TSL) devices, have a third output condition in addition to the normal HIGH and LOW logic voltage levels. This third output condition is called the *high-impedance*, or *high-Z* state. Thus tri-state circuits have the high impedance state Z as a normal output variable in addition to the usual 0 and 1 values. We can convert any logic circuit C to tri-state form simply by inserting a switch S in its output line X. The control input signal E of S is called an enable signal if $E = 1$ makes $X = Y$, and $E = 0$ makes $X = Z$. Otherwise it is a disable signal When $E = 1$ the output X is said to be enabled, and assumes its normal 0-1 levels determined by C. The ENABLE input determines the output operation so that the output either acts as a normal TTL output (ENABLE=1) or as a high-Z output (ENABLE=0). In the enabled condition, the circuit behaves exactly as any logic buffer gate, producing an output voltage level equivalent to the input logic level. In the disabled high-Z state, the output terminal acts as if it were disconnected from the buffer; in other words, think of it as a virtual open circuit.

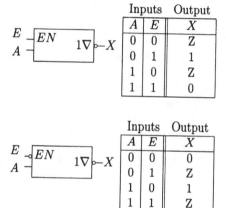

Inputs		Output
A	E	X
0	0	Z
0	1	1
1	0	Z
1	1	0

Inputs		Output
A	E	X
0	0	0
0	1	Z
1	0	1
1	1	Z

Figure 4.11: (a) A tri-state inverter with an enable line, (b) a tri-state buffer with a disable line

Tri-state buffers are often used in applications where several logic signals are to be connected to a common line called a bus. Many types of logic circuits are currently available with tri-state outputs. Other tri-state circuits include flip-flops, registers, memories, and almost all microprocessors and microprocessor interface chips. In CMOS the 400097 is a hex non-inverting buffer with 3-state outputs. The 3-state outputs are controlled by two enable inputs.

4.5 Feedback and Gates

Any logical circuit in which signal flow is unidirectional, a so-called feedforward circuit, has a finite memory span bounded by the maximum forward value of the combined propagation delays along any path from a primary output to a primary input. In order to construct a circuit with unbounded memory span from unidirectional logic elements, it is necessary to create a closed signal or *feedback loop*. Feedback is a basic property of sequential circuits (flip-flops and latches). The problem caused by feedback in a purely combinational logic circuit – that is, one with gate delay zero, is that it can lead to a logical inconsistency.

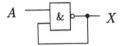

Figure 4.12: NAND Gate With Feedback

In Figure 4.12 there is a feedback loop from the output to the input of the NAND gate, implying the Boolean equation

$$X(t) = \overline{A(t) \cdot X(t)}$$

must be satisfied. This equation is satisfied by $A(t) = 0$ and $X(t) = 1$, since

$$\overline{0 \cdot 1} = \overline{0} = 1.$$

Consequently, the signal configuration is consistent and stable. If $A(t) = 1$ we obtain a logically inconsistent situation, since

$$\overline{A(t) \cdot X(t)} = 0.$$

Similarly, for $X(t) = 0$ we have a logically inconsistent situation, since $X(t)$ cannot be 0 and 1 at the same time.

The inconsistency present in this example disappears if the NAND gate has a nonzero propagation delay t_{pd}, which also makes a better model for the behaviour of a physical gate. Our equation changes to

$$X(t) = \overline{A(t - t_{pd}) \cdot X(t - t_{pd})}.$$

The output signal $A(t)$ is no longer a function of its present value. Instead, it depends on the past value $X(t - t_{pd})$, which can differ from $X(t)$. In particular, when $A(t - t_{pd}) = 1$, we can satisfy the equation with

$$X(t) = \overline{X(t - t_{pd})}.$$

Hence, if $A(t)$ changes from 0 to 1 at some time t, this change will cause $X(t)$ to change from 0 to 1 at some time $t + t_{pd}$. Owing to our equation, this second change will change $X(t)$ from 1 to 0 at $t + 2t_{pd}$, and so on. Hence, the value of $X(t)$ must change every t_{pd} time units. This type of regular and spontaneous changing, called oscillation, is an extreme form of unstable behaviour. However it is not logically inconsistent. This type of behaviour plays an important role in generating the *clock signal* that controls synchronous circuits. Spontaneous oscillation of the above kind involves narrow pulses of width t_{pd} that tend to be filtered out by the gate through which they pass. Consequently, such an oscillation usually dies out quickly.

Chapter 5
Combinational Circuits

5.1 Introduction

A combinational circuit consists of gates representing Boolean connectives; it is free of feedback loops. A combinational circuit has no state; its output depends solely on the momentary input values. Examples are the full adder, comparator, decoder and multiplexer. In reality, however, signal changes propagate through a sequence of gates with a finite speed. This is due to the capacitive loads of the amplifying transistors. Hence circuits have a certain propagation delay.

In this chapter we consider circuits such as adders, multipliers and comparators which can be used to build an arithmetic logic unit. Optimizations for some of these circuits are also considered.

Every Boolean function can be expressed in a normal form consisting of disjunctions of conjunctions, and it can therefore be implemented by two levels of gates only. Of considerable technical relevance are devices which represent two levels of gates in a general form. A specific function is selected by opening (or closing) connections between specific gates. This is called programming the device, and the device is a programmable logic device (PLD). The gates in a PLD are of the AND, OR and NOT types. Programming happens electrically under computer control. PLD's are highly attractive in order to reduce the number of discrete components in circuits. A specific form of a PLD is the read-only memory (ROM). Another example is programmable array logic (PAL) where the AND gates are programmable.

5.2 Decoder

In digital computers, binary codes are used to represent many different types of information, such as instructions, numerical data, memory addresses, and control commands. A code group that contains N bits can have 2^N different combinations, each of which represents a different piece of information. A logic circuit is required which can take the N-bit code as logic inputs and then generate an appropriate output signal to identify which of the 2^N different combinations is present. Such a circuit is called a *decoder*.

Thus a 1-out-of-n decoder is a circuit with n outputs and $N = \log_2 n = \operatorname{ld} n$ inputs, outputs X_j are numbered from 0 to $(n-1)$. An output goes to 1 when the input number A is identical to the number j of the relevant output. Figure 5.1 shows the truth table for a 1-out-of-4 decoder. The variables A_0 and A_1 represent the binary code of the decimal number m. The sum of the products (disjunctive normal form) of the recoding functions can be taken directly from the truth table. The circuit is also shown using AND and NOT gates. The functions are

$$X_0 = \overline{A_0} \cdot \overline{A_1}, \qquad X_1 = A_0 \cdot \overline{A_1}, \qquad X_2 = \overline{A_0} \cdot A_1, \qquad X_3 = A_0 \cdot A_1.$$

Most integrated-circuit decoders can decode 2-,3-, or 4-bit input codes.

In CMOS 4028 is a 4-bit BCD to 1-of-10 active high decoder.

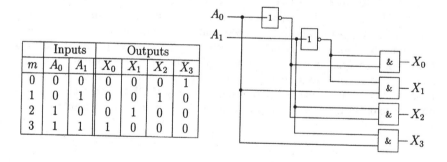

	Inputs		Outputs			
m	A_0	A_1	X_0	X_1	X_2	X_3
0	0	0	0	0	0	1
1	0	1	0	0	1	0
2	1	0	0	1	0	0
3	1	1	1	0	0	0

Figure 5.1: Truth Table and Circuit of a 1-out-of-4 Decoder

5.3 Encoder

A decoder takes an input code and activates the one corresponding output. An *encoder* performs the opposite operation; it generates a binary code corresponding to which input has been activated. A commonly used IC encoder is represented in Figure 5.2. It has eight active LOW inputs, which are kept normally high. When one of the inputs is driven to 0, the binary output code is generated corresponding to that input. For example, when input $\bar{I}_3 = 0$, the outputs will be $CBA = 011$, which is the binary equivalent of decimal 3. When $\bar{I}_6 = 0$, the outputs will be $CBA = 110$. For some encoders, if more than one input is made low the output would be garbage. For a *priority encoder*, the outputs would be the binary code for the highest-numbered input that is activated. For example, assume that the encoder of the Figure is a priority encoder and that inputs $\bar{I}_4$ and $\bar{I}_7$ are simultaneously made low. The output code will be $CBA = 111$ corresponding to $\bar{I}_7$. No matter how many inputs are activated, the code for the highest one will appear at the output.

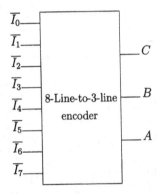

Figure 5.2: Typical IC Encoder

A simple 8-input encoder is given by

$$\begin{aligned}
A &= I_1 + I_3 + I_5 + I_7 \\
B &= I_2 + I_3 + I_6 + I_7 \\
C &= I_4 + I_5 + I_6 + I_7 \\
V &= I_0 + I_1 + I_2 + I_3 + I_4 + I_5 + I_6 + I_7
\end{aligned}$$

The output V is used to indicate when an input is 1 for the encoder; it differentiates between the 0 input I_0 and when no inputs are 1. The encoder is not a priority encoder, it performs a bitwise OR on all the inputs which are set to 1.

In CMOS the 4532 is an 8-input priority encoder with eight active HIGH priority inputs (I_0 to I_7), three active HIGH outputs (O_0 to O_2), an active HIGH enable input (E_{in}), an active HIGH enable output (E_{out}) and an active HIGH group select output (GS). Data is accepted on inputs I_0 to I_7. The binary code corresponding to the highest priority input (I_0 to I_7) which is HIGH, is generated on O_0 to O_2 if E_{in} is HIGH. Input I_7 is assigned the highest priority. GS is HIGH when one or more priority inputs and E_{in} are HIGH. E_{out} is HIGH when I_0 to I_7 are LOW and E_{in} is HIGH. E_{in}, when LOW, forces all outputs (O_0 to O_2, GS, E_{out}) LOW. The circuit is given below.

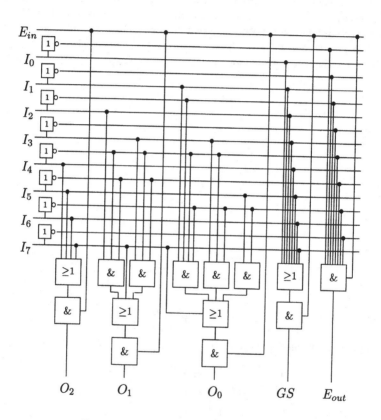

Figure 5.3: Circuit for the CMOS 4532

The logic equations are

$$O_2 = E_{in} \cdot (I_4 + I_5 + I_6 + I_7)$$

$$O_1 = E_{in} \cdot (I_2 \cdot \overline{I_4} \cdot \overline{I_5} + I_3 \cdot \overline{I_4} \cdot \overline{I_5} + I_7)$$

$$O_0 = E_{in} \cdot (I_1 \cdot \overline{I_2} \cdot \overline{I_4} \cdot \overline{I_6} + I_3 \cdot \overline{I_4} \cdot \overline{I_6} + I_5 \cdot \overline{I_6} + I_7)$$

$$E_{out} = E_{in} \cdot (\overline{I_0} \cdot \overline{I_1} \cdot \overline{I_2} \cdot \overline{I_3} \cdot \overline{I_4} \cdot \overline{I_5} \cdot \overline{I_6} \cdot \overline{I_7})$$

$$GS = E_{in} \cdot (I_0 + I_1 + I_2 + I_3 + I_4 + I_5 + I_6 + I_7).$$

Inputs									Outputs				
E_{in}	I_7	I_6	I_5	I_4	I_3	I_2	I_1	I_0	GS	O_2	O_1	O_0	E_{Out}
L	X	X	X	X	X	X	X	X	L	L	L	L	L
H	L	L	L	L	L	L	L	L	L	L	L	L	H
H	H	X	X	X	X	X	X	X	H	H	H	H	L
H	L	H	X	X	X	X	X	X	H	H	H	L	L
H	L	L	H	X	X	X	X	X	H	H	L	H	L
H	L	L	L	H	X	X	X	X	H	H	L	L	L
H	L	L	L	L	H	X	X	X	H	L	H	H	L
H	L	L	L	L	L	H	X	X	H	L	H	L	L
H	L	L	L	L	L	L	H	X	H	L	L	H	L
H	L	L	L	L	L	L	L	H	H	L	L	L	L

Table 5.1: Truth Table for the CMOS 4532

5.4 Demultiplexer

A demultiplexer can be used to distribute input information D to various outputs.
It represents an extension of the 1-out-of-n decoder. The addressed output does
not go to one, but assumes the value of the input variable D. Figure 5.4 shows
its implementation using AND and NOT gates. If we make D =const= 1, the
demultiplexer operates as a 1-out-of-n decoder.

The logic functions are

$$X_0 = D \cdot \overline{A_0} \cdot \overline{A_1}, \qquad X_1 = D \cdot A_0 \cdot \overline{A_1}, \qquad X_2 = D \cdot \overline{A_0} \cdot A_1, \qquad X_3 = D \cdot A_0 \cdot A_1$$

The following figure shows the basic mode of operation and the circuit.

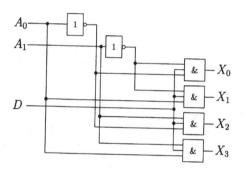

Figure 5.4: Demultiplexer Circuit

In CMOS the 4555 is a dual 1-of-4 decoder/demultiplexer. Each has two address
inputs (A_0 and A_1), an active LOW enable input ($\overline{E}$) and four mutually exclusive
outputs which are active HIGH (O_0 to O_3). When used as a decoder ($\overline{E}$ is HIGH),
then O_0 to O_3 is LOW. When used as a demultiplexer, the appropriate output is
selected by the information on A_0 and A_1 with $\overline{E}$ as data input. All unselected
outputs are LOW.

The truth table is given by Table 5.2.

Inputs			Outputs			
$\overline{E}$	A_0	A_1	O_0	O_1	O_2	O_3
L	L	L	H	L	L	L
L	H	L	L	H	L	L
L	L	H	L	L	H	L
L	H	H	L	L	L	H
H	X	X	L	L	L	L

Table 5.2: Truth Table for CMOS 4555

5.5 Multiplexer

A *multiplexer* or *data selector* is a logic circuit that accepts several data inputs and allows only one of them at a time to get through to the output. It is an extension of an encoder. The routing of the desired data input to the output is controlled by SELECT inputs (sometimes referred to as ADDRESS inputs). There are many IC multiplexers with various numbers of data inputs and select inputs. Thus the opposite of a demultiplexer is a multiplexer. The following figure shows the multiplexer circuit.

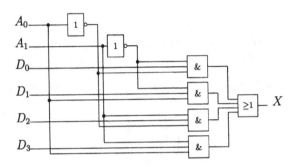

Figure 5.5: Multiplexer Circuit

The logic function is

$$X = \overline{A_0} \cdot \overline{A_1} \cdot D_0 + A_0 \cdot \overline{A_1} \cdot D_1 + \overline{A_0} \cdot A_1 \cdot D_2 + A_0 \cdot A_1 \cdot D_3.$$

In CMOS technology, a multiplexer can be implemented using both gates and analog switches (transmission gates). When analog switches are employed, signal transmission is bidirectional. In this case, therefore, the multiplexer is identical to the demultiplexer. The circuit is then known as an analog multiplexer/demultiplexer.

In CMOS the 4019 provides four multiplexing circuits with common select inputs (S_A, S_B). Each circuit contains two inputs (A_n, B_n) and one output (O_n). It may be used to select four bits of information from one of two sources. The A inputs are selected when S_A is HIGH, the B inputs are selected when S_B is HIGH. When S_A and S_B are HIGH, the output (O_n) is the logical OR of the A_n and B_n inputs $(O_n = A_n + B_n)$. When S_A and S_B are LOW, the output (O_n) is LOW independent of the multiplexer inputs.

5.6 Binary Adder

5.6.1 Binary Half Adder

Consider the task of adding two 1-bit numbers A_0 and A_1. Obviously,

$$0 + 0 = 0, \qquad 0 + 1 = 1, \qquad 1 + 0 = 1.$$

The sum 1+1 requires two bits to represent it, namely 10, the binary form of two(decimal). This can be expressed as follows: one plus one yields a sum bit $S = 0$ and a carry bit $C = 1$. If we ignore the carry bit and restrict the sum to the single bit s, then we obtain $1 + 1 = 0$. This is a very useful special form of addition known as modulo-2 addition.

The half adder circuit can be realized using an XOR gate and an AND gate. One output gives the sum of the two bits and the other gives the carry. In CMOS for the AND gate the 4081 can be used and for the XOR gate the 4030 can be used.

The circuit and the figure for the input and output is shown in the following figure.

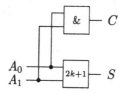

Figure 5.6: Half Adder Circuit

| Inputs || Outputs ||
A_0	A_1	S	C
0	0	0	0
0	1	1	0
1	0	1	0
1	1	0	1

Table 5.3: Half Adder Truth Table

The logic function for S and C are

$$S = A_0 \oplus A_1, \qquad C = A_0 \cdot A_1.$$

5.6.2 Binary Full Adder

A full adder adds three bits at a time, a necessary operation when two multi-bit binary numbers are added. The figure shows a full adder. It consistes of three AND gates, an OR gate and an XOR gate. The full adder computes the numerical sum of the three input bits in unsigned binary code. For example, when all three inputs are 1, the output is $Y_0 Y_1 = 11$, which is the binary representation of the decimal number three. One of the inputs is the carry bit from the previous addition.

We can construct logic expressions for these operations. The five gates (three AND gates, one OR gate and one XOR gate) used in the full adder given below lead to the logic equations

$$X_0 = A_0 \cdot A_1, \qquad X_1 = A_1 \cdot A_2, \qquad X_2 = A_0 \cdot A_2,$$

$$Y_0 = X_0 + X_1 + X_2, \qquad Y_1 = A_0 \oplus A_1 \oplus A_2.$$

Note that the $+$ is the logical OR operation, and $\cdot$ is the AND operation and $\oplus$ is the XOR operation.

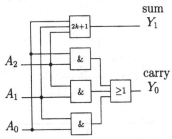

Figure 5.7: Full Adder Circuit

Inputs			Outputs	
A_0	A_1	A_2	Y_0	Y_1
0	0	0	0	0
0	0	1	0	1
0	1	0	0	1
0	1	1	1	0
1	0	0	0	1
1	0	1	1	0
1	1	0	1	0
1	1	1	1	1

Table 5.4: Full Adder Truth Table

5.6.3   Binary Four-Bit Adder

To add 3+3 (decimal) (i.e. 11+11 in binary), two full adders in parallel are required, as shown in the figure. Actually the first full adder, FA_1, need only be a half adder since it handles just two bits. The output is 6 (decimal) or 110 in binary.

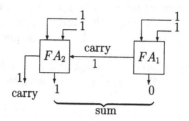

Figure 5.8: Two Full Adders in Parallel

To add two four-bit numbers, four adders are connected in parallel as shown for the addition of binary 1110 (14 decimal) and 0111 (7 decimal) to give the sum 10101 (21 in decimal). By joining more full adders to the left of the system, numbers with more bits can be added.

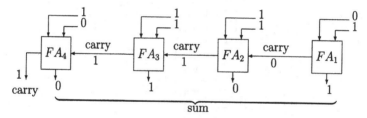

Figure 5.9: Four Bit Adder Consisting of Four Adders

In CMOS the 4008 is a 4-bit binary full adder with two 4-bit data inputs, a carry input, four sum outputs, and a carry output. The IC uses full look-ahead across 4-bits to generate the carry output. This minimizes the necessity for extensive look-ahead and carry cascading circuits.

5.6.4 Faster Addition

The 4-bit adder calculates a carry bit three times before the final carry value can be passed to another 4-bit adder. This is known as *carry cascading*. A speed increase is obviously possible if the final carry value can be calculated immediatly from the input values (to reduce the carry cascading). Let A_i, B_i and C_{i-1} ($i = 0, 1, 2$) denote the inputs for each stage i of a 3-bit adder. The equations for the carry bits are given by

$$C_{-1} := C_{in}$$

$$C_i = A_i \cdot B_i + A_i \cdot C_{i-1} + B_i \cdot C_{i-1} \qquad i = 0, 1, 2.$$

Thus for C_{out} we find

$$C_2 = A_2 \cdot B_2 + A_2 \cdot A_1 \cdot B_1 + A_1 \cdot B_2 \cdot B_1 +$$
$$A_2 \cdot A_1 \cdot A_0 \cdot B_0 + A_2 \cdot A_0 \cdot B_1 \cdot B_0 + A_1 \cdot A_0 \cdot B_2 \cdot B_0 + A_0 \cdot B_2 \cdot B_1 \cdot B_0 +$$
$$A_2 \cdot A_1 \cdot A_0 \cdot C_{in} + A_2 \cdot A_1 \cdot B_0 \cdot C_{in} + A_2 \cdot A_0 \cdot B_1 \cdot C_{in} + A_1 \cdot A_0 \cdot B_2 \cdot C_{in} +$$
$$A_2 \cdot B_1 \cdot B_0 \cdot C_{in} + A_1 \cdot B_2 \cdot B_0 \cdot C_{in} + A_0 \cdot B_2 \cdot B_1 \cdot C_{in} + B_2 \cdot B_1 \cdot B_0 \cdot C_{in}.$$

Each full adder requires 2 levels of gates to calculate the carry bit. Thus we have reduced the carry computation from 6 levels of gates to 2 levels of gates. The circuit for the computation is given below.

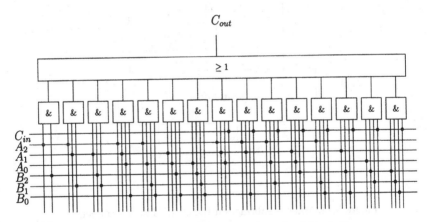

Figure 5.10: Circuit for the Carry Bit of a 3-bit Adder

Similary the calculation for the carry bit of a 4-bit adder will be reduced from 8 levels of gates to 2 levels of gates.

5.7   Binary Subtraction

Binary subtraction can be built in the same way as binary addition. However, it is possible to perform binary subtraction using binary addition which can simplify circuits designed to perform binary arithmetic. In chapter 3 a number of ways to represent negative integers in binary format were introduced. The two's complement method made it possible to add signed integers using standard binary addition. Thus it makes sense to use the two's complement format to enable subtraction using existing methods.

To subtract B from A where B and A are 4-bit numbers we use

$$A - B = A + (-B)$$

where the negation is *two's complement.*

The circuit is as follows

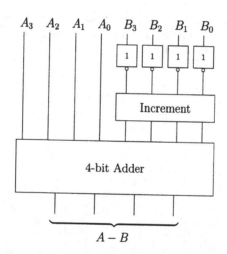

Figure 5.11: Binary Subtraction Using the Two's Complement

5.8   Binary Multiplication

5.8.1   Unsigned Integer Multiplication

Binary multiplication can be implemented with simple addition. This is very slow. Some improvement can be obtained by using the distributive properties of binary arithmetic. Suppose

$$A = \sum_{i=0}^{n} a_i 2^i \quad \text{and} \quad B = \sum_{i=0}^{n} b_i 2^i$$

with $a_i, b_i \in \{0, 1\}$ are to be multiplied. Using distributivity we find

$$A \cdot B = \sum_{j=0}^{n} b_j \left(\sum_{i=0}^{n} a_i 2^{i+j} \right).$$

Multiplying by a power of 2 can be implemented as a simple shift operation. The following algorithm summarizes the technique.

1. $j := 0, result := 0$

2. if b_j is 1 add A to $result$

3. shift left A

4. increment j

5. if $j \leq n$ goto (2)

The algorithm requires a number of steps. To keep the circuits simple, control mechanisms introduced in Chapter 7 will be needed. The alternative is a more complicated implementation consisting of calculating all the shifted values of A first and then adding each input where the input lines of B are 1. The shift left operation is simple. To perform addition controlled by the b_j inputs it suffices to examine for each output line the input value I, the calculated output O (in this case for addition) and b_j. Thus the final output is given by

$$b_j \cdot O + \overline{b_j} \cdot I.$$

To ensure that the product of two n-bit numbers can be represented the output may be extended to $2n$ bits.

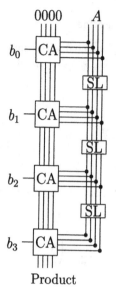

Product

CA- Controlled 4-bit Adder

SL - Logical 4-bit Shift Left

Figure 5.12: Unsigned 4-bit Multiplication

The *Russian peasant method* [104] uses the same technique for multiplication, with a small change to simplify implementation. It is a practical method for multiplication by hand, since it involves only the operations of doubling, halving and adding. Western visitors to Russia in the nineteenth century found the method in wide use there, from which the method derives its name.

1. $j := 0, result := 0$

2. if b_0 is 1 add A to *result*

3. shift left A

4. shift right B

5. increment j

6. if $j \leq n$ goto (2)

5.8.2 Fast Multiplication

Instead of using an iterative algorithm, multiplication can be performed in parallel by determining each output bit from the input bits only (at the same time). Consider the following multiplication of two bits $A_1 A_0$ and $B_1 B_0$.

$$
\begin{array}{cccc}
 & B_1 & & B_0 \\
 & A_1 & & A_0 \\
\hline
 & A_0 \cdot B_1 & & A_0 \cdot B_0 \\
A_1 \cdot B_1 & A_1 \cdot B_0 & & \\
\hline
P_3 & P_2 & P_1 & P_0
\end{array}
$$

The product is given by

$$
\begin{aligned}
P_0 &= A_0 \cdot B_0 \\
P_1 &= (A_0 \cdot B_1) \oplus (A_1 \cdot B_0) \\
P_2 &= (A_1 \cdot B_1) \oplus (A_1 \cdot A_0 \cdot B_1 \cdot B_0) \\
P_3 &= A_1 \cdot A_0 \cdot B_1 \cdot B_0
\end{aligned}
$$

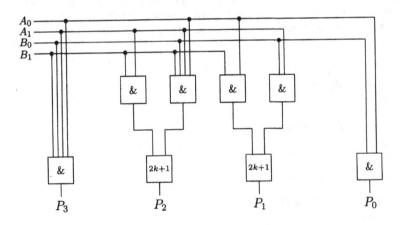

Figure 5.13: 2-bit Fast Unsigned Multiplication

5.8.3 Signed Integer Multiplication

Two signed integers can be multiplied by multiplying their absolute values and setting the appropriate sign afterwards. For two's complement numbers *Booth's algorithm* is used. It is an efficient algorithm for multiplying signed integers and optimizes the calculation in the following way: suppose an N-bit unsigned integer has a block of consecutive 1's in its binary representation. If the first 1 is at bit k and the last 1 is at bit n then the decimal representation for this block is $2^{n+1} - 2^k$. This follows from the fact that

$$\sum_{k=0}^{n} 2^k = 2^{n+1} - 1.$$

This can be extended for two's complement numbers. The two's complement (negation) of the same number gives $2^N - 2^n + 2^{k+1}$. The contribution of 2^N is an overflow and does not influence the operation. Thus addition and subtraction can be performed whenever a 1 and 0 are adjacent in the bit representations.

For Booth's algorithm we introduce an extra bit Q_{-1} which is used to determine the boundaries of blocks of 0's or 1's. The final product is in AQ. A and Q are n-bit registers. The arithmetic shift right (SHR) operation shifts all bits one position right, and leaves the highest order bit (sign bit) at its previous value.

1. $A := 0$, $Q_{-1} := 0$
   $M :=$ Multiplicand
   $Q :=$ Multiplier
   $C := n$

2. If $Q_0 = Q_{-1}$ goto (5)

3. If $Q_0 Q_{-1} = 01$ $A := A + M$

4. If $Q_0 Q_{-1} = 10$ $A := A - M$

5. Arithmetic SHR A, Q, Q_{-1}
   $C = C - 1$

6. If C is nonzero goto (2)

5.9 Binary Division

Binary division is modelled on the long division process, for example

1. $A := 0$
   $M :=$ Divisor, $Q :=$ Dividend, $C := n$

2. SHL AQ

3. $A := A - M$

4. if $(A < 0)$ goto (6)

5. $Q_0 := 1$, goto (7)

6. $Q_0 := 0$, $A := A + M$

7. increment C

8. if $(C > 0)$ goto (2)

Two's complement division:

1. Load the divisor into the M register and the dividend into the AQ registers. The dividend must be expressed as a $2n$ two's complement number. Thus, for example, the 4-bit number 0111 becomes 00000111, and 1001 becomes 11111001.

2. Shift left AQ 1 bit position

3. If M and A have the same signs, perform $A := A - M$; otherwise, $A := A + M$.

4. The above operation is successful if the sign of A is the same before and after the enumeration.

   (a) If the operation is successful or $(A = 0$ AND $Q = 0)$ then set $Q_0 = 1$.
   (b) If the operation is unsuccessful and $(A \neq 0$ OR $Q \neq 0)$, then set $Q_0 = 0$ and restore the previous value of A.

5. Repeat steps (2) through (4) as many times as there are bit positions in Q.

6. The remainder is in A. If the signs of the divisor and dividend are the same, the the quotient is in Q; otherwise the correct quotient is the two's complement of Q.

5.10   Magnitude Comparator

One of the basic operations in computation is comparing two integer numbers. Let a and b be two integers. We have to consider three cases $a > b$, $a = b$ and $a < b$. For combinational circuits the comparison is done bit by bit. For example, 5 in binary is 0101b and 3 in binary is 0011b. First we compare the most significant bits of 5 and 3. Both are 0. Thus we move to the next bit (from left to right). For 5 the bit is set, namely 1 and for 3 we have 0. Thus $5 > 3$.

In CMOS the 4585 is a *four-bit magnitude comparator* which compares two 4-bit words (A and B), whether they are 'less than', 'equal to' or 'greater than'. Each word has four parallel inputs (A_0 to A_3) and (B_0 to B_3); A_3 and B_3 being the most significant inputs. Three outputs are provided. A greater than B ($O_{A>B}$), A less than B ($O_{A<B}$) and A equal to B ($O_{A=B}$). Three expander inputs ($I_{A>B}$, $I_{A<B}$ and $I_{A=B}$) allow cascading of the devices without external gates. For proper comparison operation the expander inputs to the least significant position must be connected as follows:

$$I_{A=B} = I_{A>B} = HIGH, \qquad I_{A<B} = LOW$$

For words greater than 4-bits, units can be cascaded by connecting output $O_{A<B}$ and $O_{A=B}$ to the corresponding inputs of the next significant comparator (input $I_{A>B}$ is connected to a HIGH). Operation is not restricted to binary codes, the devices will work with any monotonic code. Table 5.5 displays the truth table for the CMOS 4585. The following notation is used. H=HIGH state (the more positive voltage), L=LOW state (the less positive voltage), X = state is immaterial. The upper 11 lines describe the normal operation under all conditions that will occur in a single device or in a serial expansion scheme. The lower 2 lines describe the operation under abnormal conditions on the cascading inputs. These conditions occur when the parallel expansion technique is used. The circuit consists of 8 XNOR gates and one NAND gate.

In CMOS the 74LV688 is an 8-bit magnitude comparator. It takes two 8-bit numbers provided by the inputs P_0 to P_7 and Q_0 to Q_7. The output is

$$\overline{P = Q}.$$

Table 5.6 shows the function table for the CMOS 74LV688 and Figure 5.14 the logic diagram for the CMOS 74LV688.

Comparing inputs				Cascading inputs			Outputs		
A_3, B_3	A_2, B_2	A_1, B_1	A_0, B_0	$I_{A>B}$	$I_{A<B}$	$I_{A=B}$	$O_{A>B}$	$O_{A<B}$	$O_{A=B}$
$A_3 > B_3$	X	X	X	H	X	X	H	L	L
$A_3 < B_3$	X	X	X	X	X	X	L	H	L
$A_3 = B_3$	$A_2 > B_2$	X	X	H	X	X	H	L	L
$A_3 = B_3$	$A_2 < B_2$	X	X	X	X	X	L	H	L
$A_3 = B_3$	$A_2 = B_2$	$A_1 > B_1$	X	H	X	X	H	L	L
$A_3 = B_3$	$A_2 = B_2$	$A_1 < B_1$	X	X	X	X	L	H	L
$A_3 = B_3$	$A_2 = B_2$	$A_1 = B_1$	$A_0 > B_0$	H	X	X	H	L	L
$A_3 = B_3$	$A_2 = B_2$	$A_1 = B_1$	$A_0 < B_0$	X	X	X	L	H	L
$A_3 = B_3$	$A_2 = B_2$	$A_1 = B_1$	$A_0 = B_0$	X	L	H	L	L	H
$A_3 = B_3$	$A_2 = B_2$	$A_1 = B_1$	$A_0 = B_0$	H	L	L	H	L	L
$A_3 = B_3$	$A_2 = B_2$	$A_1 = B_1$	$A_0 = B_0$	X	H	L	L	H	L
$A_3 = B_3$	$A_2 = B_2$	$A_1 = B_1$	$A_0 = B_0$	X	L	H	L	H	H
$A_3 = B_3$	$A_2 = B_2$	$A_1 = B_1$	$A_0 = B_0$	L	L	L	L	L	L

Table 5.5: Truth Table for the CMOS 4585

Inputs		Output
Data	Enable	
P_n, Q_n	E	$\overline{P = Q}$
$P = Q$	L	L
X	H	H
$P > Q$	L	H
$P < Q$	L	H

Table 5.6: Function Table for CMOS 74LV688

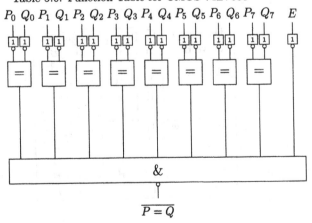

$P_0\ Q_0\ P_1\ Q_1\ P_2\ Q_2\ P_3\ Q_3\ P_4\ Q_4\ P_5\ Q_5\ P_6\ Q_6\ P_7\ Q_7\quad E$

&

$\overline{P = Q}$

Figure 5.14: Logic Diagram for the CMOS 74LV688

5.11   4-Bit ALU

The *Arithmetic logic unit* (*ALU*) is responsible for the arithmetic and logic operations discussed so far. Typically some input lines are used to select the required functionality. The ALU combined with memory (such as registers) and control logic is essentially all that is required for a classical computer. In CMOS the MC14581B is a 4-bit ALU capable of providing 16 functions of two Boolean variables and 16 binary arithmetic operations on two 4-bit words. The level of the mode control input determines whether the output function is logic or arithmetic. The desired logic function is selected by applying the appropriate binary word to the select inputs (S_0 thru S_3) with the mode control input HIGH, while the desired arithmetic operation is selected by applying a LOW to the mode control input, the required level to carry in, and the appropriate word to the select inputs. The word inputs and function outputs can be operated on with either active high or active low data. The arithmetic functions interpret the input words as two's complement numbers. As noted in the table, when C_n is opposite to the given value, the result is the given arithmetic function plus 1 because C_n is interpreted as a carry bit.

Carry propagate (P) and carry generate (G) outputs are provided to allow a full look-ahead carry scheme for fast simultaneous carry generation for the four bits in the package. Fast arithmetic operations on long words are obtainable by using the MC14582B as a second order look ahead block. An inverted ripple carry input (C_n) and a ripple carry output (C_{n+4}) are included for ripple through operation.

When the device is in the subtract mode (LHHL), comparison of two 4-bit words present at the A and B inputs is provided using the $A = B$ output. It assumes a high-level state when indicating equality. Also, when the ALU is in the subtract mode the C_{n+4} output can be used to indicate relative magitude as shown in this table

Data level	C_n	C_{n+4}	Magnitude
Active High	H	H	$A \leq B$
	H	H	$A \leq B$
	L	H	$A < B$
	H	L	$A > B$
	L	L	$A \geq B$
Active Low	H	H	$A \leq B$
	L	L	$A \leq B$
	H	L	$A < B$
	L	H	$A > B$
	H	H	$A \geq B$

The truth table is as follows.

Function Select				Inputs/Outputs Active Low		Inputs/Outputs Active High	
S_3	S_2	S_1	S_0	Logic Function (MC=H)	Arithmetic* Function (MC=L, C_n=L)	Logic Function (MC=H)	Arithmetic* Function (MC=L, $\overline{C_n}$=H)
L	L	L	L	$\overline{A}$	A minus 1	$\overline{A}$	A
L	L	L	H	$\overline{A \cdot B}$	$A \cdot B$ minus 1	$\overline{A+B}$	$A+B$
L	L	H	L	$\overline{A}+B$	$A \cdot \overline{B}$ minus 1	$\overline{A} \cdot B$	$A + \overline{B}$
L	L	H	H	Logic 1	minus 1	Logic 0	minus 1
L	H	L	L	$\overline{A+B}$	A plus $(A+\overline{B})$	$\overline{A} \cdot B$	A plus $A \cdot \overline{B}$
L	H	L	H	$\overline{B}$	$A \cdot B$ plus $(A+\overline{B})$	B	$(A+B)$ plus $A \cdot \overline{B}$
L	H	H	L	$\overline{A \oplus B}$	A minus B minus 1	$A \oplus B$	A minus B minus 1
L	H	H	H	$A+\overline{B}$	$A+\overline{B}$	$A \cdot \overline{B}$	$A \cdot \overline{B}$ minus 1
H	L	L	L	$\overline{A} \cdot B$	A plus $(A+B)$	$\overline{A}+B$	A plus $A \cdot B$
H	L	L	H	$A \oplus B$	A plus B	$\overline{A \oplus B}$	A plus B
H	L	H	L	B	$A \cdot \overline{B}$ plus $(A+B)$	B	$(A+\overline{B})$ plus $A \cdot B$
H	L	H	H	$A+B$	A plus B	$A \cdot B$	$A \cdot B$ minus 1
H	H	L	L	Logic 0	A plus A	Logic 1	A plus A
H	H	L	H	$A \cdot \overline{B}$	$A \cdot B$ plus A	$A+\overline{B}$	$(A+B)$ plus A
H	H	H	L	$A \cdot B$	$A \cdot \overline{B}$ plus A	$A+B$	$(A+\overline{B})$ plus A
H	H	H	H	A	A	A	A minus 1

The * indicates that the inputs are expressed in two's complement form. For arithmetic functions with C_n in the opposite state, the resulting function is as shown plus 1.

Thus, for active high inputs, the basic logic functions are achieved with the following selections and $MC = H$.

S_3	S_2	S_1	S_0	Logic Function
L	L	L	L	NOT
H	L	H	H	AND
H	H	H	L	OR
L	H	L	L	NAND
L	L	L	H	NOR
L	H	H	L	XOR

The basic arithmetic operations of addition, subtraction, increment and decrement are provided when $MC = L$.

S_3	S_2	S_1	S_0	C_n	Arithmetic Function
H	L	L	H	L	Addition
L	H	H	L	H	Subtraction
L	L	L	L	L	Increment
H	H	H	H	L	Decrement

5.12   Read Only Memory (ROM)

Another frequently encountered purely combinational circuit is the read-only memory. Its structure is such that any Boolean function of n variables, where n is the number of inputs, can be generated. Since complicated functions are usually not perceived as functions, but rather as individual values corresponding to the possible combinations of input values, the device is called a memory rather than a function generator.

A ROM essentially consists of a decoder of the binary encoded input number, called the *address*, an array of OR gates, and a set of output drivers. The decoder yields a selector signal for each input value, addressing each cell.

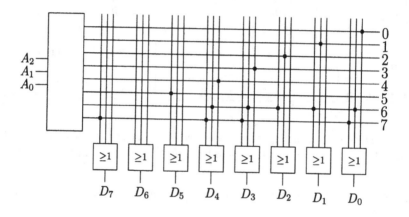

Figure 5.15: Logic Diagram for a ROM

As an example when the combination $A_2 = 1$, $A_1 = 1$ and $A_0 = 0$ the decoder selects line 6 which means that the output lines give the bit string

$$D_7 D_6 D_5 D_4 D_3 D_2 D_1 D_0 = 00011111$$

which is the binary representation of 31. The ROM can be used to speed up certain tasks. For example it can store multiplication tables. Of course ROMs can also be used to store identification strings or any other data.

5.13 Combinational Programmable Logic Devices

In the previous section the ROM was introduced. To make circuit design easier more flexible approaches are available. Many hardware implementations implement the SOP form (disjunctive normal form) of an expression directly. The circuit designed determines the connections to make for a given configuration of gates (for example selecting the inputs to the OR gates in the ROM configuration).

A programmable gate is one where the inputs to the gate can be selected from a given set of inputs (for example from other gates). If no inputs are selected we assume that all inputs are 0. We introduce a new notation to simplify circuit representation. A cross $\times$ indicates a programmable connection to a gate, in other words a connection which can be removed (for example a fuse that can be burnt open). A dot $\bullet$ indicates a fixed connection. The following figure shows an AND gate with programmable inputs (the inputs from A_0 and A_2 can still be removed) and an OR gate with two fixed inputs.

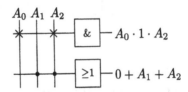

Figure 5.16: Input Representation for Programmable Gates

In the following examples we use two inputs, four AND gates and one OR gate for the output. In general, for n inputs and m outputs, 2^n AND gates and m OR gates are required. One way to implement a programmable AND gate is to have an AND gate with $2n$ inputs (for an input and its inverse) and to set the input to 1 whenever an input is not connected. Similarly for the OR gate an input can be set to 0. A special case is when no input is selected, the output of the gate must be zero (as if the gate is not present). In this case we set all inputs to the gate to 0. In this way gates with a fixed number of inputs can be used as programmable gates. For each architecture we show the circuit before programming and after programming the XOR operation.

PROM stands for programmable read only memory. These devices consist of a number of fixed AND gates (fixed in input) and programmable OR gates. The AND gates are over all possible inputs. For an n variable system there are 2^n AND gates. All connections are initially closed, the unwanted connections are then burnt by applying a voltage to the appropriate inputs. Once a PROM is programmed it cannot be reprogrammed. The *EPROM* or *erasable* PROM can be erased (all connections are closed). The *EEPROM* is an *electrically erasable* PROM.

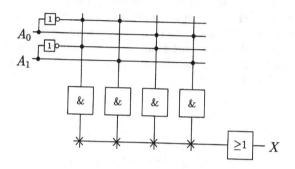

Figure 5.17: PROM Device Architecture

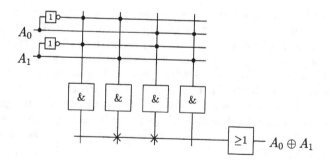

Figure 5.18: PROM Implementation of XOR

PAL stands for programmable array logic. A number of programmable AND gates feed fixed OR gates in these devices. The AND gates represent the product forms of the desired expression's SOP form. Specific AND gates are dedicated to specific OR gates.

GAL stands for generic array logic. They are used to emulate PALs. Different types of PALs can then be replaced with a single device type (the GAL device).

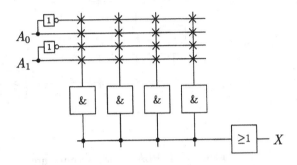

Figure 5.19: PAL Device Architecture

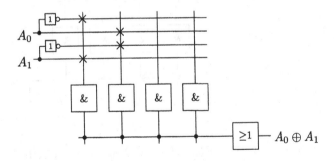

Figure 5.20: PAL Implementation of XOR

PLA stands for programmable logic array. These devices provide the greatest programming flexibility through the use of programmable AND gates feeding programmable OR gates. Any AND gate can feed any OR gate.

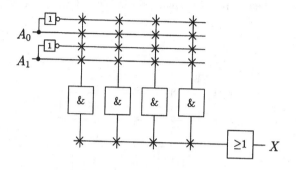

Figure 5.21: PLA Device Architecture

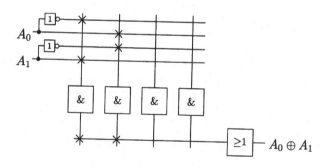

Figure 5.22: PLA Implementation of XOR

5.14 Programmable Gate Arrays

A *programmable gate array (PGA)* consists of an array (or matrix) of cells. Each cell's function and connections can be selected. Two kinds of *field* programmable gate arrays (FPGA) are commonly used. The first kind consists of cells which are loadable and erasable as a whole like an EPROM (i.e. all the cells are erased and written at the same time). The second type consists of cells which are not modifiable but the connections between them are. For example, suppose a cell can implement any of the 16 possible Boolean function of two variables. The cell can be programmed using a multiplexer to select the inputs used and a second multiplexer to determine the output. Of course, a multiplexer to select only the output function (where all 16 functions are implemented separately) is also possible. In the following figure the input multiplexers are assumed to be fixed (until reconfigured) and selection signals are not shown

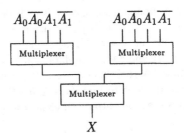

Figure 5.23: Example of a Combinational FPGA Cell

To design a circuit using FPGA the cell functions must be specified as well as the connections between cells. Determining which connections must be closed is called *routing*. For example a FPGA may have a grid pattern where the output can be connected to four adjacent cells. Each outward going arrow is a duplicate of the function output. Each input arrow can be configured to be closed or open.

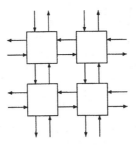

Figure 5.24: Grid Pattern for PGA Design

5.15   VHDL

VHDL [151] is a standardized language that is not tied to any single tool vendor
or hardware manufacturer. It is a complete programming language with built-in
mechanisms to handle and synchronize parallel processes, and also supports abstract
data types and high level modelling. The IEEE adopted VHDL as a standard in
1987.

VHDL was initially intended to describe digital electronics systems. It can be used
to model existing hardware for verification and testing, and also for synthesis.

The following code example is a VHDL description of a 4-bit equality comparator.
The data types `bit` and `bit_vector` and are basic data type in VHDL. The assignment operator is `<=` and the comparison operator is `=`. A comment is preceded
by `--`, everything following on the same line is part of the comment. The `entity`
declaration describes the communication mechanism. It describes pins (`port`) used
for input and output. In the example below a and b represent 2 four bit inputs.
Thus `a(0)` refers to the first bit of a and `b(3)` refers to the last bit of b. We refer to the logic values as '0' and '1'. The data type `bit_vector` uses literals in
double quotes, for example `"0010"`. The `architecture` declaration describes the
functionality. In this case the architecture `dataflow` which is associated with the
entity `eqcomp4`. It ensures that the output pin `equals` is always defined.

```
-- eqcomp4.vhd

-- eqcomp4 is a four bit equality comparator
entity eqcomp4 is
  port (a, b  : in  bit_vector(3 downto 0);
        equals: out bit);  -- equals is active high
end eqcomp4;

architecture dataflow of eqcomp4 is
begin
  equals <= '1' when (a = b) else '0';
end dataflow
```

Chapter 6
Latches and Registers

6.1 Introduction

The combinational circuits introduced so far perform a function and, except for the ROM, do not provide any memory. The ROM provides a static memory, the content is predetermined. A system providing dynamic memory is required to store data which cannot be predetermined. Any two state (bistable) system which can be dynamically controlled will provide this function. Many systems acting in parallel can provide the required data width for operations provided by combinational circuits. The bistable systems are called *latches* and the parallel combination *registers*. Combinational circuits are free of loops. In this chapter we examine circuits with feedback loops. This is what allows them to store information. Propagation delays are important in the analysis of these circuits.

The following chapter introduces mechanisms for an external source of timing. The timing system helps describe the logic functions of these circuits under specific conditions.

In this chapter we discuss the SR latch and JK latch which use two inputs. One sets the logical value of the latch to 0 and the other sets the logical value of the latch to 1. The D latch has only one input and remembers the logical value of the input for one time interval. The D register and JK register use D latches and JK latches respectively, to provide the same logical action as the latches, but with different physical characteristics.

6.2 SR Latch

This circuit has two inputs labelled S (Set) and R (Reset), and two outputs Q and $\overline{Q}$, and consists of two NOR gates connected in a feedback arrangement.

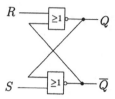

Figure 6.1: Logic Diagram for the SR Latch

The following table summarizes the characteristics of the operation of the latch. The S and R should not both be set at the same time as this gives an undetermined value for Q.

S_t	R_t	Q_t	Q_{t+1}
0	0	0	0
0	0	1	1
0	1	0	0
0	1	1	0
1	0	0	1
1	0	1	1
1	1	0	-
1	1	1	-

Table 6.1: Characteristic Table for the SR Latch

The circuit is stable when $S = R = 0$ ($Q_{t+1} = Q_t$). The output is time dependent and there is a delay from the time that one of S or R are set to one and the time when the circuit is stable again. If $S = 0$ and $R = 1$ the system is reset. If $S = 1$ and $R = 0$ the system is set. The logical equation for Q_{t+1} (if S_t and R_t are not 1 at the same time) is

$$Q_{t+1} = S_t + \overline{R_t} \cdot Q_t.$$

6.3   D Latch

The input $S = R = 1$ must be avoided when using the SR Latch. The D latch overcomes this by using only a single D input. The ouput is always the same as the last D input.

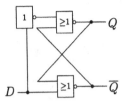

Figure 6.2: Logic Diagram for the D Latch

The D latch is sometimes called the data latch because it stores 1 bit of information. It is also called the delay latch because it delays the output of the 0 or 1 (in an environment where a CLOCK input is provided, the delay is one clock cycle). The characteristic table for the D latch is as follows:

D	Q_{t+1}
0	0
1	1

Table 6.2: Characteristic Table for the D Latch

The latch described above is called *transparent* since the output Q is the same as the input D. An extra input can be introduced to indicate when to set the output identical to the given input.

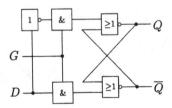

Figure 6.3: Logic Diagram for the D Latch with Enable

The G input is called an enable input.

6.4 JK Latch

The JK latch takes two inputs. Unlike the SR latch all input combinations are valid. The J input performs the set function while the K input performs the reset function. When $J = K = 1$ the *toggle* function is performed (the outputs are inverted).

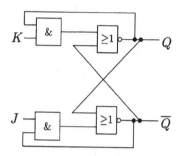

Figure 6.4: Logic Diagram for the JK Latch

The characteristic table describes the functionality of the circuit.

J_t	K_t	Q_t	Q_{t+1}
0	0	0	0
0	0	1	1
0	1	0	0
0	1	1	0
1	0	0	1
1	0	1	1
1	1	0	1
1	1	1	0

Table 6.3: Characteristic Table for the JK Latch

The logic equation for Q_{t+1} is

$$Q_{t+1} = J_t \cdot \overline{K_t} + J_t \cdot \overline{Q_t} + \overline{K_t} \cdot Q_t.$$

6.5   D Register

The transparency of a D latch can be undesirable. It may be preferable to accept
an input value upon the rising edge of a control signal, and retain the stored value
before and after the transition. Latches are level-sensitive whereas registers are
edge-sensitive. An example implementation of this is the *master-slave* latch pair. It
consists of two latches connected in series with each enable input the inverse of the
other. This separates the storage of the input D and the output Q. The boxes with
the symbols D, G and Q represent D latches.

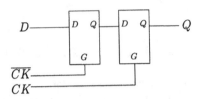

Figure 6.5: D Register Using Two D Latches

This is called an edge-triggered D register. Typically CK is a periodical signal. The
signals CK and $\overline{CK}$ must never be active at the same time to guarantee nontrans-
parency. The output of a transparent D latch can be used for this purpose. The
following figure shows a D-Master-Slave register.

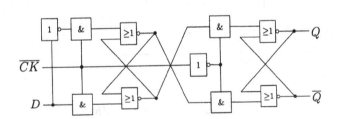

Figure 6.6: Logic Diagram for the D Register

6.6   JK Register

Similar to the D register, the principle for the JK register is based on the JK latch. A master-slave configuration can again be used to implement this register.

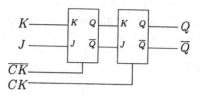

Figure 6.7: JK Register Using Two JK Latches

Each JK latch has two additional AND gates for each input where the appropriate CK or $\overline{CK}$ is the second input to each AND gate. This construction has the same purpose as the G input in a D latch. A variation on the register is to use the Q and $\overline{Q}$ feedback loops directly to the first input and not for each latch. The following figure shows a JK-Master-Slave register.

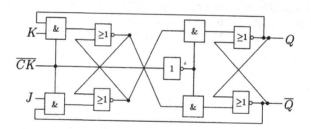

Figure 6.8: Logic Diagram for the JK Register

Chapter 7
Synchronous Circuits

7.1 Introduction

Circuits that react immediately to the stimulus of the input are called *asynchronous*. This term is a combination of the greek words meaning "without regard to time". In digital systems it is important that outputs change at precise points in time. Circuits that operate in this manner are called *synchronous*. Digital circuits often use time reference signals called *clocks*. A clock signal is nothing more than a square wave that has a precise known period. The clock will be the timing reference that synchronizes all circuit activity and tells the device when it should execute its function. Thus the clock signal is the signal that causes things to happen at regularly spaced intervals. In particular, operations in the system are made to take place at times when the clock signal is making a transition from 0 to 1 or from 1 to 0. These transitions are pointed out in the figure. The 0-to-1 transition is called the rising edge or positive-going edge of the clock signal. The synchronous action of the clock signal is the result of using clocked latches, which are designed to change states on either (but not both) the rising edge or the falling edge of the clock signal. In other words, the clocked latches will change states at the appropriate clock transition and will rest between successive clock pulses. The frequency of the clock pulses is generally determined by how long it takes the latches and gates to respond to the level changes by the clock pulse, that is, the propagation delays of the various logic circuits.

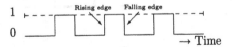

Figure 7.1: Example Clock Signal

Many ways of designing and controlling latches have evolved over the years. They differ not only in their logic design but also how they use the clock signal. Let us consider a latch. During the period $t_1 : t_2$ when the clock is enabled $C = 1$, any change made to the data signal may enter the latch immediately. After some propagation delay, these changes affect the latch's data output Q (and also $\overline{Q}$) during the period $t_3 : t_4$. Thus, ignoring the brief and somewhat uncertain transition periods when the data and clock signals are actually changing values, the latch responds to all input changes that occur when C is at the inactive 1 level. For this reason latches are said to be level sensitive or *level-triggered*.

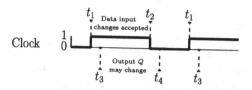

Figure 7.2: Level Sensitive Latch

To obtain latch behavior, we must ensure that the period $t_1 : t_2$ (when input data changes are accepted) and the period $t_3 : t_4$ (when the output data changes) do not overlap. One way a latch can meet this requirement is by accepting input changes when $C = 1$, and changing its output when $C = 0$. This pulse mode of operation was used in some early designs for bistables. The clocking method most commonly used in modern latch design is *edge triggering*, in which a transition or edge of the clock signal C causes the actions required in $t_1 : t_2$ and $t_3 : t_4$ to take place, as shown in the figure.

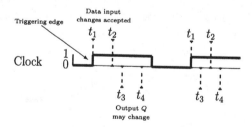

Figure 7.3: Edge Triggered Latch

7.2 Shift Registers

Shift registers are classed as sequential logic circuits, and as such they are constructed from latches. Thus shift registers are chains of latches which allow data applied to the input to be advanced by one latch with each clock pulse. After passing through the chain, the data is available at the output with a delay but otherwise is unchanged. Shift registers are used as temporary memories and for shifting data to the left or right. Shift registers are also used for changing serial to parallel data or parallel to serial data.

Identification of shift registers may be made by noting how data is loaded into and read from the storage unit. In the following figure we have a register 8 bits wide. The registers are classified as:

1. Serial-in serial-out

2. Serial-in parallel-out

3. Parallel-in serial-out

4. Parallel-in Parallel-out

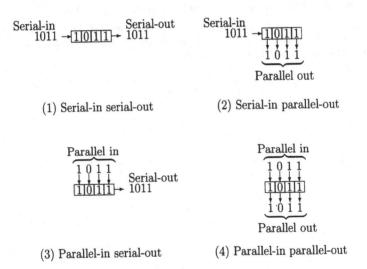

(1) Serial-in serial-out   (2) Serial-in parallel-out

(3) Parallel-in serial-out   (4) Parallel-in parallel-out

Figure 7.4: Types of Shift Registers

A simple four-bit shift register is displayed in the following figure. It uses four D-latches. Data bits (0s and 1s) are fed into the D input of latch 1. This input is labelled as the serial data input. The clear input will reset all four D latches to 0 when activated by a LOW. A pulse at the clock input will shift the data from the serial-data input to the position A (Q of latch 1). The indicators (A, B, C, D) across the top of the figure show the contents of the register. This register can be classified as a serial-in parallel-out unit if data is read from the parallel inputs (A, B, C, D) across the top.

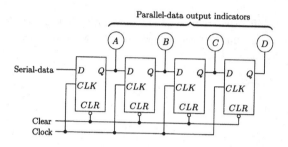

Figure 7.5: Logic Diagram of a 4-bit Serial-Load Shift-Right Register

In CMOS the 4014B is a fully synchronous edge-triggered 8-bit static shift register with eight synchronous parallel inputs, a synchronous serial data input, a synchronous parallel enable, a LOW to HIGH edge-triggered clock input and buffered parallel outputs from the last three stages.

7.3 Binary Counter

Latches can be connected in various arrangements to function as binary counters
that count input clock pulses. A wide variety of counters are available as standard
integrated-circuit packages and it is seldom necessary to construct a counter from
individual latches. We review the external operating characteristics of currently
available IC counters.

Next we discuss the basic counter operation. The Figure 7.6 shows the schematic
representation of a 4-bit counter. This counter contains four latches, one per bit,
with outputs labelled A, B, C, and D. Two inputs are shown, the clock pulse
input, CP, and $Reset$. The counter operates such that the states of the four latches
represent a binary number equal to the number of pulses that have been applied to
the CP input. The diagram shows the sequence which the latch outputs follow as
pulses are applied. The A output represents the LSB (least significant bit) and D
is the MSB (most significant bit) of the binary count. For example, after the fifth
input pulse, the outputs $DCBA = 0101$, which is the binary equivalent of 5. The
CP input has a small circle and triangle to indicate that the latches in the counter
change states on the negative going edge of the clock pulses. Counters that trigger
on positive-going edges are also available and they do not have the circle on the CP
input.

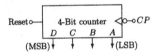

Figure 7.6: Four-Bit Binary Counter

	D	C	B	A	
	0	0	0	0	Before 1st input pulse
16 Different	0	0	0	1	After 1st input pulse
possible	0	0	1	0	After 2nd input pulse
states	0	0	1	1	After 3rd input pulse
	⋮	⋮	⋮	⋮	⋮
	1	1	1	1	After 15th input pulse
Sequence	0	0	0	0	After 16th input pulse
repeats	0	0	0	1	After 17th input pulse
	⋮	⋮	⋮	⋮	⋮

Table 7.1: Counting Sequence

In general, a counter with N latches can count from 0 up to $2^N - 1$, for a total
of 2^N different states. The total number of different states is called the counter's
MOD number. The counter in the Figure is a MOD-16 counter. A counter with N
latches would be a MOD-2^N counter. Some IC counters are designed so that the
user can vary the counting sequence through the appropriate external additional
logic connections. These are usually referred to as *variable-MOD counters*.

In addition to counting pulses, all counters can perform *frequency division*. This is
illustrated in the following figure for the 4-bit, MOD-16 counter. The state of the
A output is seen to change at a rate exactly $\frac{1}{2}$ that of the CP input. The C output
is $\frac{1}{2}$ that of the A output and $\frac{1}{4}$ of the CP input. The C output is $\frac{1}{2}$ the frequency
of the B output and $\frac{1}{8}$ the input frequency, and the D output is $\frac{1}{2}$ the frequency of
the C output and $\frac{1}{16}$ the input frequency. In general, the waveform out of the MSB
latch of a counter will divide the input frequency by the MOD number.

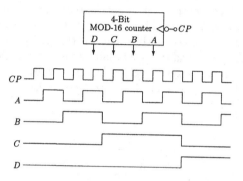

Figure 7.7: Counter Waveforms Showing Frequency Division

The counters described above can count up from zero to some maximum count and
then reset to zero. There are several IC counters that can count in either direction
and are called up/down counters. The following figure shows the two basic up/down
counter arrangements. The counter in this Figure has a single CP input that is
used for both count-up and count-down operations. The UP/DOWN input is used
to control the counting direction. One logic level applied to this input causes the
counter to count up from 0000 to 1111 as pulses are applied to CP. The other logic
level applied to UP/DOWN causes the counter to count down from 1111 to 0000
as pulses are applied to CP. The second counter does not use and UP/DOWN
control input. Instead, it uses separate clock inputs CP_U and CP_D for counting up
and down, respectively. Pulses applied to CP_U cause the counter to count up, and
pulses applied to CP_D cause the counter to count down. Only one CP input can
be pulsed at one time, or erratic operations will occur.

In CMOS the 4516 is an edge triggered synchronous up/down 4-bit binary counter
with a clock input and an up/down count control input.

Figure 7.8: Representation of Two Types of Up/Down Counters

An example of a mod-4 ripple counter implemented with clocked JK latches is given below. The JK latch was chosen for the simple toggle capability.

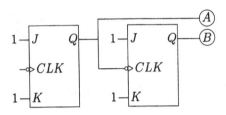

Figure 7.9: 2-bit Binary Ripple Counter

The CLK input of the second latch is driven by the output of the first latch. The CLK input of the first latch is driven by an external clock signal. Every second toggle action of the first latch will cause the the second latch to toggle. The output A is the least significant bit and B is the most significant bit of the binary counter.

This ripple action of one latch depending on the output of the previous latch can necessitate potentially large clock cycles, due to propagation delays. To avoid this lag, latches can be updated in parallel. The latches are driven by the same external clock at their CLK inputs. This is illustrated below in another mod-4 counter.

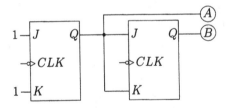

Figure 7.10: 2-bit Binary Parallel Counter

VHDL can also be used to simulate a synchronous circuit. For example, consider

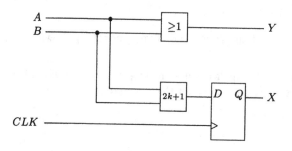

Figure 7.11: Synchronous Circuit Specified in VHDL Program

The VHDL program is given below.

```
-- simple.vhd

entity simple is
  port(A, B, CLK: in  bit;
       X, Y     : out bit);
end simple;

architecture break_out of simple is
begin
  Y <= A or B;

  p1: process begin
    wait until CLK = '1';
    X <= A xor B;
  end process;
end break_out;
```

7.4 Example Program

For the PIC16F8X processor, the rotate instruction can be used to do fast multiplication. For example if we want to multiply two 8 bit numbers and store the result in two 8 bit registers (HBYTE and LBYTE), we apply the instruction rotate right f through carry, where f is an 8 bit register. For the PIC16F8X, RRF is the "rotate right f through carry" instruction. The following diagram illustrates the operation.

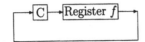

The instruction BTFSC is the "bit test f and skip if clear" instruction. If the tested bit of f is 0 the next instruction is skipped. Thus if the BTFSC is executed with the operands STATUS and 0, the carry flag (STATUS register bit 0) is tested to determine if the next instruction is executed. The instruction DECFSZ is the "decrement f and skip if zero" instruction. The value of register f is decremented, and if the result is zero the next instruction is skipped.

```
; multiply.asm
;****************************************************************
; Multiplies two 8 bit numbers
; 00000011 (decimal 3)
; and
; 01100101 (decimal 101)
; and stores the result (16 bits)
; 00000001 00101111 (decimal 303)
; in LBYTE and HBYTE
; LBYTE: 00101111
; HBYTE: 00000001
;
; RRL rotate right f through carry
; The contents of register f are rotated
; one bit to the right through the Carry Flag.
;
;****************************************************************
;
PROCESSOR 16f84
INCLUDE "p16f84.inc"
;
; Variable Declarations
LBYTE EQU       H'11'   ; variable at address 0x11 in SRAM
HBYTE EQU       H'12'   ; variable at address 0x12 in SRAM
COUNT EQU       H'13'   ; variable at address 0x13 in SRAM
NOA   EQU       H'20'   ; first number at address 0x20 in SRAM
NOB   EQU       H'21'   ; second number at address 0x21 in SRAM
;
```

```
        ORG  H'00'
;
Start
      BSF STATUS, RP0
      MOVLW  B'11111111'
      MOVWF PORTA
      MOVLW  B'00000000'
      MOVWF  PORTB
      BCF STATUS, RP0

      CLRF LBYTE
      CLRF HBYTE
      MOVLW 8
      MOVWF COUNT

      MOVLW B'00000011'
      MOVWF NOA
      MOVLW B'01100101'
      MOVWF NOB

      MOVF NOB, W
      BCF    STATUS, 0

LOOP
      RRF NOA
      BTFSC STATUS, 0
      ADDWF HBYTE
      RRF HBYTE
      RRF LBYTE
      DECFSZ COUNT, F
      GOTO LOOP

      MOVF HBYTE, 0
      MOVWF PORTB
Stop    GOTO    Stop
;
END
```

Chapter 8

Recursion

8.1 Introduction

Recursion is a fundamental concept in mathematics and computer science. It is a useful tool for simplifying solutions to problems. A recursive solution is possible if a problem can be solved using the solution of a simpler problem of the same type and a solution to the simplest of problems of the same type is known. A recursive solution to a problem consists of

- A solution to a simplest problem of the same type (*base* problem or *stopping condition*)

- A method to solve the problem if the solution to a simpler problem of the same type is known

Let us now list some recursive structures. One of the most important recursive structures are strings. The string manipulation functions can be implemented using recursion, for example to find the length of a string and reverse a string. The linear linked list is a recursive structure; it has a head followed by a linked list. An example implementation of a recursive linked list is given later in the next chapter, it allows lists to be copied, compared, searched and items to be inserted and deleted recursively. Another structure which is recursive is the binary tree.

In mathematics, recursion is the name given to the technique of defining a function in terms of itself. Any recursive definition must have an explicit definition for some value or values of the argument(s), otherwise the definition is circular. Recursion can also occur in another form if a process is defined in terms of subprocesses, one of which is identical to the main process.

For example, consider the double integral

$$\int\limits_{a}^{b}\int\limits_{c}^{d} f(x,y)dx\, dy.$$

One method of evaluation is to write the double integral as a repeated integral

$$\int\limits_{a}^{b}\left(\int\limits_{c}^{d} f(x,y)dy\right) dx.$$

The evaluation of the outer integral requires us to know the value of the integrand at selected points, and calculation of the integrand requires the evaluation of an integral, so that the subprocess is the same as the main process.

Example. The set $\mathbf{N}_0^2$ is bijective with $\mathbf{N}_0$. To see this we write the elements of $\mathbf{N}_0^2$ in a table.

$$
\begin{array}{ccccc}
(0,0) & (0,1) & (0,2) & (0,3) & \ldots \\
(1,0) & (1,1) & (1,2) & (1,3) & \ldots \\
(2,0) & (2,1) & (2,2) & (2,3) & \ldots \\
(3,0) & (3,1) & (3,2) & (3,3) & \ldots \\
\vdots & \vdots & \vdots & \vdots & \ddots
\end{array}
$$

We now write down the elements of this table by moving along the diagonals which go from north-east to south-west, that is, we write them in the sequence

$$(0,0),\ (0,1),\ (1,0),\ (0,2),\ (1,1),\ (2,0),\ (0,3),\ (1,2),\ \ldots$$

Since there are $(k+1)$ pairs (r,s) with $r+s=k$, we see that the pair (m,n) occurs in the position

$$1+2+\ldots+(m+n)+m = \frac{(m+n)(m+n+1)}{2}+m.$$

Hence we have a bijection $f : \mathbf{N}_0^2 \to \mathbf{N}_0$ given by

$$f(m,n) = \frac{1}{2}(m+n)(m+n+1)+m.$$

We have two functions g and h from $\mathbf{N}_0 \to \mathbf{N}_0$ such that $f^{-1}(r) = (g(r), h(r))$. They are given by the following formulas. Find $s \in \mathbf{N}_0$ so that

$$\frac{1}{2}s(s+1) \leq r < \frac{1}{2}(s+1)(s+2).$$

Let m be

$$r - \frac{1}{2}s(s+1).$$

Then $m \leq s$, and $g(r) = m$, and $h(r) = s - m$.

We can use f to obtain bijections

$$f_k : \mathbf{N}_0^k \to \mathbf{N}_0$$

for all k using recursion. We define f_1 to be the identity, and we define f_3 by

$$f_3(x_1, x_2, x_3) = f(x_1, f(x_2, x_3)).$$

If f_k has been defined, then f_{k+1} is defined by

$$f_{k+1}(x_1, x_2, \ldots, x_{k+1}) := f(x_1, f_k(x_2, \ldots, x_{k+1})).$$

The inverse of f_k has as its components composites of g and h. For instance

$$f_3^{-1}(r) = (g(r), g(h(r)), h(h(r))).$$

♣

Example. Let $n = 0, 1, \ldots$ and $f(n) = 2^n$. Then we can find the recursive definition as follows

$$f(n+1) = 2^{n+1} = 2 \cdot 2^n = 2f(n)$$

Thus $f(n+1) = 2f(n)$ where $f(0) = 1$.

♣

Example. Another typical example of recursion is the *Fibonacci sequence* given by

$$F_{n+2} = F_{n+1} + F_n \qquad n = 0, 1, 2, \ldots$$

where $F_0 = F_1 = 1$.

♣

Example. The *Bessel functions* $J_n(x)$ are solutions of the linear second order differential equation

$$x^2 \frac{d^2 y}{dx^2} + x \frac{dy}{dx} + (x^2 - n^2)y = 0 \qquad n = 0, 1, 2, \ldots.$$

A recurrence formula for Bessel functions is given by

$$J_{n+2}(x) = \frac{2(n+1)}{x} J_{n+1}(x) - J_n(x) \qquad n = 0, 1, 2, \ldots$$

where

$$J_0(x) = \sum_{j=0}^{\infty} (-1)^j \frac{x^{2j}}{\prod_{k=1}^{j} (2k)^2}$$

♣

Example. Given a first order differential equation

$$\frac{dy}{dx} = f(x, y(x)), \qquad y(x_0) = y_0$$

where f is an analytic function of x and y. Formal integration yields

$$y(x) = y_0 + \int_{x_0}^{x} f(s, y(s)) ds.$$

Thus a recursive definition for an approximation of $y(x)$ is given by

$$y_{n+1}(x) = y_0 + \int_{x_0}^{x} f(s, y_n(s)) ds.$$

This is known as *Picard's method*.

As an example we consider

$$\frac{dx}{dy} = x + y, \qquad x_0 = 0, \; y(x_0) = 1.$$

The approximation at each step is given by

$$y_{n+1}(x) = 1 + \int_0^x (s + y_n(s))ds.$$

The method yields a polynomial approximation after each step. Thus

$$y_0(x) \; = \; 1$$

$$y_1(x) \; = \; \frac{x^2}{2} + x + 1$$

$$y_2(x) \; = \; \frac{x^3}{6} + x^2 + x + 1$$

$$y_3(x) \; = \; \frac{x^4}{24} + \frac{x^3}{3} + x^2 + x + 1$$

$$y_4(x) \; = \; \frac{x^5}{120} + \frac{x^4}{12} + \frac{x^3}{3} + x^2 + x + 1$$

$$y_5(x) \; = \; \frac{x^6}{720} + \frac{x^5}{60} + \frac{x^4}{12} + \frac{x^3}{3} + x^2 + x + 1$$

♣

8.2   Example Programs

Example. The Towers of Hanoi problem illustrates the benefits of recursion very
well. The problem has an easy recursive solution. The problem is as follows

- There are 3 pegs A, B and C.

- There are n discs of different sizes.

- Initially all n discs are on peg A with the largest disc at the bottom and discs
  decrease in size towards the top of the pile. If disc 1 is above disc 2 then disc
  1 is smaller than disc 2.

- Only one disc may be moved at a time. A disc must be moved from the top
  of a pile on one peg to the top of a pile on another peg. A larger disc may not
  be placed on a smaller one.

- The task is to move all the discs from peg A to peg B.

If $n=1$ we can move the disc from A to B. If $n=2$ we can move a disc from A to C,
A to B, C to B. This is the inspiration for the solution to the general problem. If
$n > 2$ move the pile of $n - 1$ discs from A to C, move the disc on A to B and move
the pile on peg C to peg B.

```cpp
// hanoi.cpp

#include <iostream>

using namespace std;

void hanoi(unsigned long n,char A,char B,char C)
{
 if(n==1)
  cout << A << " -> " << B << endl;
 else
 {
  hanoi(n-1,A,C,B);
  cout << A << " -> " << B << endl;
  hanoi(n-1,C,B,A);
 }
}

void main(void)
{
 cout << "Tower of Hanoi with 1 disc:" << endl;
 hanoi(1,'A','B','C');
 cout << "Tower of Hanoi with 2 discs:" << endl;
 hanoi(2,'A','B','C');
 cout << "Tower of Hanoi with 3 discs:" << endl;
```

```
hanoi(3,'A','B','C');
cout << "Tower of Hanoi with 4 discs:" << endl;
hanoi(4,'A','B','C');
}
```

The output of the program is

```
Tower of Hanoi with 1 disc:
A -> B
Tower of Hanoi with 2 discs:
A -> C
A -> B
C -> B
Tower of Hanoi with 3 discs:
A -> B
A -> C
B -> C
A -> B
C -> A
C -> B
A -> B
Tower of Hanoi with 4 discs:
A -> C
A -> B
C -> B
A -> C
B -> A
B -> C
A -> C
A -> B
C -> B
C -> A
B -> A
C -> B
A -> C
A -> B
C -> B
```

Example. The number of multiplications involved in calculating an integer power of
a number can be reduced significantly with a recursive solution. Using the identity

$$a^n \equiv \begin{cases} a^{\frac{n}{2}} a^{\frac{n}{2}} & n \text{ even} \\ a a^{\frac{n}{2}} a^{\frac{n}{2}} & n \text{ odd} \end{cases}$$

with $a \in \mathbf{R}$ and $n \in \mathbf{N}$ and $\frac{n}{2}$ is calculated using integer division (i.e. $\lfloor \frac{n}{2} \rfloor$). The
program **power.cpp** implements the solution.

```
// power.cpp

#include <iostream>
#include <iomanip>

using namespace std;

double power(double a,unsigned int n)
{
  double power_ndiv2;

  if(n == 0) return 1.0;

  power_ndiv2 = power(a,n/2);
  if(n )
   return a*power_ndiv2*power_ndiv2;
  return power_ndiv2*power_ndiv2;
}

void main(void)
{
  cout << "13.4^0=" << power(3.4,0) << endl;
  cout << "2^24=" << setprecision(9) << power(2,24) << endl;
  cout << "3.1415^7=" << setprecision(9) << power(3.1415,7) << endl;
}
```

The output of the program is

```
13.4^0=1
2^24=16777216
3.1415^7=3019.66975
```

Example. If R is a relation on a set A and S is a sequence of elements from A then the sequence S can be sorted. For R to be a relation we require

1. $R \subseteq A \times A$.
   We view the statement $(a, b) \in R$ with $a, b \in A$ as a proposition. We also write $(a, b) \in R$ as aRb

2. aRb and bRc implies aRc

A fast sorting method would be to place the elements of S in a tree as they occur in the sequence and traverse the tree to find the sorted sequence. Another fast sorting algorithm called *quicksort* is implemented using recursion. The algorithm first partitions the sequence around an element s_i such that all elements on the left of s_i have the property $s_j R s_i$ and all elements to the right of s_i do not. The next step is to sort each of the partitions, and we use use quicksort to do this (i.e. recursively). The program qsort.cpp uses the function **partition** to partition the sequence at each step of the qsort algorithm. This is the most important part of the algorithm. The function takes an element of the array and rearranges the array such that all elements before are less than the given element and all elements after are greater than the given element.

```
// qsort.cpp

#include <iostream>
#include <string>

using namespace std;

// general definition of ordering R(t1,t2)
// returns >0 if t2 R t1, <=0 otherwise
template <class T>
void partition(T *array,int n,int (*R)(T,T),int &p)
{
  // partition around the first element of the array
  // any element could have been used.

  // p is the index of the element around which the
  // partition is made
  // pe (declared below) points to the element after
  // the second partition
  int i = n-1, pe = 1;
  T temp1,temp2;

  p=0;
  while(i > 0)
  {
   if(R(array[p],array[pe]) > 0)
   {
    temp1 = array[p]; temp2 = array[p+1];
```

```
      array[p++] = array[pe];// put element in first partition
      array[p] = temp1;        // move element around which partition
                               // is made, one element right
    if(pe-p > 0)               // if the second partition is not empty
    array[pe] = temp2;         // move second partition one element right
   }
  pe++;
  i--;
 }
}


template <class T>
void qsort(T *array,int n,int (*R)(T,T))
{
 int pelement;
 if(n <= 1) return;

 partition(array,n,R,pelement);
 qsort(array,pelement,R);
 qsort(array+pelement+1,n-pelement-1,R);
}


int less_int(int n1,int n2) { return n1>n2; }
int less_string(string n1,string n2) { return (n1>n2); }

void main(void)
{
 int test1[9] = {1,5,3,7,2,9,4,6,8};
 string test2[6] = {"orange","grape","apple","pear","banana","peach"};
 int i;

 qsort<int>(test1,9,less_int);
 qsort<string>(test2,6,less_string);
 for(i=0;i<9;i++) cout << test1[i] << " ";
 cout << endl;
 for(i=0;i<6;i++) cout << test2[i] << "   ";
 cout << endl;
}
```

The output of the program is

```
1  2  3  4  5  6  7  8  9
apple  banana  grape  orange  peach  pear
```

Example. The Ackermann function $f : \mathbf{N_0} \times \mathbf{N_0} \to \mathbf{N}$ is defined as

$$f(n,m) := \begin{cases} m+1 & \text{if } n = 0 \\ f(n-1,1) & \text{if } m = 0 \\ f(n-1,f(n,m-1)) & \text{otherwise} \end{cases}$$

Thus f is defined recursively. Ackermann's function f is a total function; each pair of numbers (n,m) yields a value $f(n,m)$ of the function. We see that $f(n,m)$ depends only on the values of $f(r,p)$ with $r < n$ and $p < m$.

```cpp
// acker.cpp
#include <iostream>

using namespace std;

unsigned long ackermann(unsigned long n,unsigned long m)
{
  if(n==0) return m+1;
  if(m==0) return ackermann(n-1,1);
  return ackermann(n-1,ackermann(n,m-1));
}

void main(void)
{
  cout<<"f(1,1)="<<ackermann(1,1)<<"   "
      <<"f(2,1)="<<ackermann(2,1)<<"   "
      <<"f(3,1)="<<ackermann(3,1)<<endl;
  cout<<"f(1,2)="<<ackermann(1,2)<<"   "
      <<"f(2,2)="<<ackermann(2,2)<<"   "
      <<"f(3,2)="<<ackermann(3,2)<<endl;
  cout<<"f(1,3)="<<ackermann(1,3)<<"   "
      <<"f(2,3)="<<ackermann(2,3)<<"   "
      <<"f(3,3)="<<ackermann(3,3)<<endl;
}
```

The output of the program is

```
f(1,1)=3   f(2,1)=5   f(3,1)=13
f(1,2)=4   f(2,2)=7   f(3,2)=29
f(1,3)=5   f(2,3)=9   f(3,3)=61
```

Example. The *logistic map* $f : [0, 1] \to [0, 1]$ is given by

$$f(x) = 4x(1 - x)$$

The map can be written as a difference equation

$$x_{t+1} = 4x_t(1 - x_t) \qquad t = 0, 1, 2, \ldots$$

where $x_0 \in [0, 1]$ is the initial value. Thus we can implement the function to compute x_t recursively. Of course it makes more sense to implement the function using iteration [164].

```cpp
// logistic.cpp

#include <iostream>

using namespace std;

double logistic(unsigned int t,double x0)
{
 double x;
 if(t==0) return x0;
 x=logistic(t-1,x0);
 return 4.0*x*(1.0-x);
}

void main(void)
{
 cout << "x100 = " << logistic(100,0.3899)
      << " when x0=0.3899" << endl;
 cout << "x500 = " << logistic(500,0.5)
      << " when x0=0.5" << endl;
 cout << "x10000 = " << logistic(10000,0.89881)
      << " when x0=0.89881" << endl;
}
```

The output of the program is

```
x100 = 0.744501 when x0=0.3899
x500 = 0 when x0=0.5
x10000 = 0.311571 when x0=0.89881
```

Example. *Horner's rule* is used to reduce the number of multiplications in evaluating a polynomial. Consider, for example the polynomial

$$P_5(x) = a_5 x^5 + a_4 x^4 + a_3 x^3 + a_2 x^2 + a_1 x^1 + a_0$$

where x, a_5, a_4, a_3, a_2, a_1 and a_0 are given numbers. Finding P_5 would involve $5 + 4 + 3 + 2 + 1 = 15$ multiplications and 5 additions. Rewriting this in the form (Horner's rule)

$$P_5 = ((((a_5 x + a_4)x + a_3)x + a_2)x + a_1)x + a_0$$

reduces the number of multiplications to five and we still have five additions. In general, let

$$P_n(x) = a_n x^n + a_{n-1} x^{n-1} + \ldots + a_1 x + a_0$$

which can be rewritten as

$$P_n(x) = (\ldots ((a_n x + a_{n-1})x + \ldots a_1)x + a_0.$$

Then we have n multiplications and n additions. The next program shows a non-recursive implementation of Horner's rule in C++.

```cpp
// horner1.cpp

#include <iostream>

using namespace std;

template <class T>
T P(T x,const T *a,int n)
{
  T s = a[n];
  while(--n >= 0) s = s*x + a[n];
  return s;
}

void main(void)
{
  const double a[5] = { 1.0,0.5,0.0,-18.0,3.0 };

  cout << "P(x) = 3x^4-18x^3+x/2+1" << endl;
  cout << "P(0.0) = " << P(0.0,a,4) << endl;
  cout << "P(-1.0) = " << P(-1.0,a,4) << endl;
  cout << "P(5.0) = " << P(5.0,a,4) << endl;
}
```

A recursive implementation of Horner's rule is given in the next program.

```cpp
// horner2.cpp

#include <iostream>

using namespace std;

template <class T>
T P(T x,const T *a,int n)
{
 if(n==0)
  return a[0];
 return a[0]+x*P(x,a+1,n-1);
}

void main(void)
{
 const double a[5] = { 1.0,0.5,0.0,-18.0,3.0 };

 cout << "P(x) = 3x^4-18x^3+x/2+1" << endl;
 cout << "P(0.0) = " << P(0.0,a,4) << endl;
 cout << "P(-1.0) = " << P(-1.0,a,4) << endl;
 cout << "P(5.0) = " << P(5.0,a,4) << endl;
}
```

The output of both programs is

```
P(x) = 3x^4-18x^3+x/2+1
P(0.0) = 1
P(-1.0) = 21.5
P(5.0) = -371.5
```

Example. The following Java program is used to construct a graphic pattern called a *Hilbert curve*. Each curve H_i consists of four half-sized copies of H_{i-1} with a different orientation. The Hilbert curve is the limit of this construction process, i.e H_∞. Thus we can implement the methods A(), B(), C() and D() to draw the four copies for each step in the construction of the Hilbert curve using recursion. Lines are drawn to connect the four copies. For example, the first three steps in constructing the Hilbert curve are given below.

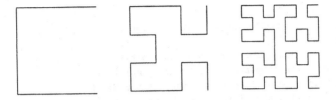

Figure 8.1: First 3 Steps in the Construction of the Hilbert Curve

```java
// Hilbert.java

import java.awt.*;
import java.awt.event.*;

public class Hilbert extends Frame
        implements WindowListener, ActionListener
{
 public Hilbert()
 {
  addWindowListener(this);
  drawButton.addActionListener(this);
  setTitle("Hilbert");
  Panel parameterPanel = new Panel();
  parameterPanel.setLayout(new GridLayout(2,1));
  Panel nStepsPanel = new Panel();
  nStepsPanel.add(new Label("no of steps = "));
  nStepsPanel.add(nStepsField);
  Panel buttonPanel = new Panel();
  buttonPanel.add(drawButton);
  parameterPanel.add(nStepsPanel);
  parameterPanel.add(buttonPanel);
  add("North",parameterPanel);
  add("Center",hilbertCurve);
  setSize(400,400); setVisible(true);
 }
```

```
    public static void main(String[] args)
    { new Hilbert(); }

    public void actionPerformed(ActionEvent action)
    {
      if(action.getSource() == drawButton)
      hilbertCurve.setSteps(Integer.parseInt(nStepsField.getText()));
      System.out.println(Integer.parseInt(nStepsField.getText()));
    }

    public void windowClosing(WindowEvent event)
    { System.exit(0); }
    public void windowClosed(WindowEvent event){}
    public void windowOpened(WindowEvent event){}
    public void windowDeiconified(WindowEvent event){}
    public void windowIconified(WindowEvent event){}
    public void windowActivated(WindowEvent event){}
    public void windowDeactivated(WindowEvent event){}

    TextField nStepsField = new TextField("5",5);
    Button drawButton = new Button("Draw");
    HilbertCurve hilbertCurve = new HilbertCurve();
}

class HilbertCurve extends Canvas
{
  private int x, y, h, n, len;

  public HilbertCurve() { n = 5; }

  public void A()
  {
    if(n > 0)
    {
      Graphics g = getGraphics(); n--;
      D(); g.drawLine(x, y, x-h, y); x-=h;
      A(); g.drawLine(x, y, x, y-h); y-=h;
      A(); g.drawLine(x, y, x+h, y); x+=h;
      B(); n++;
    }
  }
  public void B()
  {
    if(n > 0)
    {
      Graphics g = getGraphics(); n--;
      C(); g.drawLine(x, y, x, y+h); y+=h;
      B(); g.drawLine(x, y, x+h, y); x+=h;
```

```
  B(); g.drawLine(x, y, x, y-h); y-=h;
  A(); n++;
 }
}
public void C()
{
 if(n > 0)
 {
  Graphics g = getGraphics(); n--;
  B(); g.drawLine(x, y, x+h, y); x+=h;
  C(); g.drawLine(x, y, x, y+h); y+=h;
  C(); g.drawLine(x, y, x-h, y); x-=h;
  D(); n++;
 }
}
public void D()
{
 if(n > 0)
 {
  Graphics g = getGraphics(); n--;
  A(); g.drawLine(x, y, x, y-h); y-=h;
  D(); g.drawLine(x, y, x-h, y); x-=h;
  D(); g.drawLine(x, y, x, y+h); y+=h;
  C(); n++;
 }
}

public void paint(Graphics g)
{
 Dimension size = getSize();
 h = 4*Math.min(size.width,size.height)/5;
 x = size.width/2+h/2;
 y = size.height/2+h/2;

 for(int i=len=1;i<n;i++) len = 2*len+1;
 h/=len; A();
}

public void setSteps(int nSteps)
{ n = nSteps; repaint(); }
}
```

8.3   Mutual Recursion

If the solution to a problem relies on the solution to another problem which in turn relies on a simpler problem of the first type, a recursive solution can be implemented. Mutual recursion refers to the recursive dependence of one solution on another. This concept can be extended for more than two problems.

The *Jacobi elliptic functions* can be defined as inverse of the elliptic integral of first kind [53]. Thus, if we write

$$x(\phi, k) = \int_0^\phi \frac{ds}{\sqrt{1 - k^2 \sin^2 s}} \tag{8.1}$$

where $k \in [0, 1]$ we then define the following functions

$$\text{sn}(x, k) := \sin(\phi), \qquad \text{cn}(x, k) := \cos(\phi), \qquad \text{dn}(x, k) := \sqrt{1 - k^2 \sin^2 \phi}. \tag{8.2}$$

For $k = 0$ we obtain

$$\text{sn}(x, 0) \equiv \sin(x), \qquad \text{cn}(x, 0) \equiv \cos(x), \qquad \text{dn}(x, 0) \equiv 1 \tag{8.3}$$

and for $k = 1$ we have

$$\text{sn}(x, 1) \equiv \tanh(x), \qquad \text{cn}(x, 1) \equiv \text{dn}(x, 1) \equiv \frac{2}{e^x + e^{-x}}. \tag{8.4}$$

We have the following identities

$$\text{sn}(x, k) \equiv \frac{2\text{sn}(x/2, k)\text{cn}(x/2, k)\text{dn}(x/2, k)}{1 - k^2\text{sn}^4(x/2, k)} \tag{8.5}$$

$$\text{cn}(x, k) \equiv \frac{1 - 2\text{sn}^2(x/2, k) + k^2\text{sn}^4(x/2, k)}{1 - k^2\text{sn}^4(x/2, k)} \tag{8.6}$$

$$\text{dn}(x, k) \equiv \frac{1 - 2k^2\text{sn}^2(x/2, k) + k^2\text{sn}^4(x/2, k)}{1 - k^2\text{sn}^4(x/2, k)}. \tag{8.7}$$

The expansions of the Jacobi elliptic functions in powers of x up to order 3 are given by

$$\mathrm{sn}(x,k) = x - (1+k^2)\frac{x^3}{3!} + \dots \tag{8.8}$$

$$\mathrm{cn}(x,k) = 1 - \frac{x^2}{2!} + \dots \tag{8.9}$$

$$\mathrm{dn}(x,k) = 1 - k^2\frac{x^2}{2!} + \dots \tag{8.10}$$

For x sufficiently small these will be good approximations.

We can now use the identities (8.5)–(8.7) and the expansions (8.8)–(8.10) to implement the Jacobi elliptic functions using one recursive call. The recursive call in scdn uses half of the provided parameter x. In other words the absolute value of the parameter passed in the recursive call is always smaller (by $\frac{1}{2}$). This guarantees that for fixed $\epsilon > 0$ the parameter $|x|$ will satisfy $|x| < \epsilon$ after a finite number of recursive calls. At this point a result is returned immediately using the polynomial approximation (8.8)–(8.10). This ensures that the algorithm will complete successfully. The recursive call is possible due to the identities for the sn, cn and dn functions given in (8.5)–(8.7). Since the identities depend on all three functions sn, cn and dn we can calculate all three at each step instead of repeating calculations for each of sn, cn and dn [81]. Lastly some optimization was done to reduce the number of multiplications used in the double angle formulas. We also use the fact that the denominator of all three identities is the same.

The advantage of this approach is that all three Jacobi elliptic functions are found with one function call. Furthermore the cases $k = 0$ and $k = 1$ include the sine, cosine, tanh and sech functions. Obviously, for these special cases faster routines are available. Elliptic functions belong to the class of doubly periodic functions in which $2K$ plays a similar role to π in the theory of circular functions, where $K = F(1, k)$ is the complete elliptic integral of first kind. We have the identities

$$\mathrm{sn}(x \pm 2K, k) \equiv -\mathrm{sn}(x, k), \quad \mathrm{cn}(x \pm 2K, k) \equiv -\mathrm{cn}(x, k), \quad \mathrm{dn}(x \pm 2K, k) \equiv \mathrm{dn}(x, k).$$

To reduce the argument of the Jacobi elliptic functions we can also apply these identities.

The recursion method described above can be implemented using C++ as follows. The arguments to the function scdn are

- x, the first argument to sn, cn and dn.

- k2, the square of the second argument to sn, cn and dn.

- eps, the upper bound on the argument x for application of the Taylor expansion approximation.

- s, a variable for the value of $sn(x, k)$.

- c, a variable for the value of $cn(x, k)$.

- d, a variable for the value of $dn(x, k)$.

Using the implementation we calculate the sine, cosine, identity, hyperbolic tan, hyperbolic sec and hyperbolic cosec functions for the value 3.14159.

```
// jacobi.cpp

#include <iostream.h>
#include <math.h>

// forward declaration
void scdn(double,double,double,double&,double&,double&);

void main(void)
{
 double x, k, k2, eps;
 x = 3.14159;
 eps = 0.01;

 double res1,res2,res3;

 cout << "x = " << x << endl;

 // sin,cos,1 of x
 k = 0.0;
 k2 = k*k;
 scdn(x,k2,eps,res1,res2,res3);
 cout << "sin(x) = " << res1 << endl;
 cout << "cos(x) = " << res2 << endl;
 cout << "1(x) = " << res3 << endl;

 // tanh,sech,sech of x
```

```
  k = 1.0;
  k2 = k*k;
  scdn(x,k2,eps,res1,res2,res3);
  cout << "tanh(x) = " << res1 << endl;
  cout << "sech(x) = " << res2 << endl;
  cout << "sech(x) = " << res3 << endl;
}

void scdn(double x,double k2,double eps,double &s,double &c,double &d)
{
  if(fabs(x) < eps)
  {
  double x2 = x*x/2.0;
  s = x*(1.0 - (1.0 + k2)*x2/3.0);
  c = 1.0 - x2;
  d = 1.0 - k2*x2;
  }
  else
  {
  double sh,ch,dh;

  scdn(x/2.0,k2,eps,sh,ch,dh);  // recursive call

  double sh2 = sh*sh;
  double sh4 = k2*sh2*sh2;
  double denom = 1.0 - sh4;

  s = 2.0*sh*ch*dh/denom;
  c = (1.0 - 2.0*sh2+sh4)/denom;
  d = (1.0 - 2.0*k2*sh2+sh4)/denom;
  }
}
```

8.4 Wavelets and Recursion

The *discrete wavelet transform* (or DWT) is an orthogonal function which can be applied to a finite group of data. Functionally, it is very much like the discrete Fourier transform, in that the transforming function is orthogonal, a signal passed twice through the transformation is unchanged, and the input signal is assumed to be a set of discrete-time samples. Both transforms are convolutions. Whereas the basis function of the Fourier transform is sinusoidal, the wavelet basis is a set of functions which are defined by a recursive difference equation

$$\phi(x) = \sum_{k=0}^{M-1} c_k \phi(2x - k)$$

where the range of the summation is determined by the specified number of nonzero coefficients M. The number of the coefficients is not arbitrary and is determined by constraints of orthogonality and normalization. Owing to the periodic boundary condition we have

$$c_k :\equiv c_{k+nM}$$

where $n \in \mathbf{N}$. Generally, the area under the wavelet curve over all space should be unity, i.e.

$$\int_{\mathbf{R}} \phi(x)dx = 1.$$

It follows that

$$\sum_{k=0}^{M-1} c_k = 2.$$

In the Hilbert space $L_2(\mathbf{R})$, the function ϕ is orthogonal to its translations; i.e.

$$\int_{\mathbf{R}} \phi(x)\phi(x - k)dx = 0, \qquad k \neq 0.$$

What is desired is a function ψ which is also orthogonal to its dilations, or scales, i.e.,

$$\int_{\mathbf{R}} \psi(x)\psi(2x - k)dx = 0.$$

Such a function ψ does exist and is given by (the so-called associated wavelet function)

$$\psi(x) = \sum_{k=1} (-1)^k c_{1-k} \phi(2x - k)$$

which is dependent on the solution of ϕ. Normalization requires that

$$\sum_k c_k c_{k-2m} = 2\delta_{0m}$$

which means that the above sum is zero for all m not equal to zero, and that the sum of the squares of all coefficients is two. Another equation which can be derived from the above conditions is

$$\sum_k (-1)^k c_{1-k} c_{k-2m} = 0.$$

A way to solve for ϕ is to construct a matrix of coefficient values. This is a square $M \times M$ matrix where M is the number of nonzero coefficients. The matrix is designated L with entries

$$L_{ij} = c_{2i-j}.$$

This matrix has an eigenvalue equal to 1, and its corresponding (normalized) eigenvector contains, as its components, the value of the function ϕ at integer values of x. Once these values are known, all other values of the function ϕ can be generated by applying the recursion equation to get values at half-integer x, quarter-integer x, and so on down to the desired dilation. This determines the accuracy of the function approximation.

An example for ψ is the Haar function

$$\psi(x) := \begin{cases} 1 & 0 \le x < \frac{1}{2} \\ -1 & \frac{1}{2} \le x < 1 \\ 0 & \text{otherwise} \end{cases}$$

and ϕ is given by

$$\phi(x) = \begin{cases} 1 & 0 \le x < 1 \\ 0 & \text{otherwise} \end{cases}.$$

The functions

$$\psi_{m,n}(x) := 2^{-\frac{m}{2}} \psi(2^{-m} x - n), \quad m, n \in \mathbf{Z}$$

form a basis in the Hilbert space $L_2(\mathbf{R})$.

This class of wavelet functions is constrained, by definition, to be zero outside of a small interval. This is what makes the wavelet transform able to operate on a finite set of data, a property which is formally called *compact support*. The recursion relation ensures that a wavelet function ϕ is non-differentiable everywhere. The following table lists coefficients for three wavelet transforms. The pyramid algorithm

Wavelet	c_0	c_1	c_2	c_3	c_4	c_5
Haar	1.0	1.0				
Daubechies-4	$\frac{1}{4}(1+\sqrt{3})$	$\frac{1}{4}(3+\sqrt{3})$	$\frac{1}{4}(3-\sqrt{3})$	$\frac{1}{4}(1-\sqrt{3})$		
Daubechies-6	0.332671	0.806891	0.459877	-0.135011	-0.085441	0.035226

Table 8.1: Coefficients for Three Wavelet Functions

operates on a finite set on N input data, where N is a power of two; this value will be referred to as the input block size. These data are passed through two convolution functions, each of which creates an output stream that is half the length of the original input. These convolution functions are filters, one half of the output is produced by the "low-pass" filter

$$a_i = \frac{1}{2} \sum_{j=0}^{N-1} c_{2i-j+1} f_j, \qquad i = 0, 1, \ldots, \frac{N}{2} - 1$$

and the other half is produced by the "high-pass" filter function

$$b_i = \frac{1}{2} \sum_{j=0}^{N-1} (-1)^j c_{j-2i} f_j, \qquad i = 0, 1, \ldots, \frac{N}{2} - 1$$

where N is the input block size, c_j are the coefficients, f is the input function, and a and b are the output functions. In the case of the lattice filter, the low- and high-pass outputs are usually referred to as the odd and even outputs, respectively. In many situations, the odd or low-pass output contains most of the information content of the original input signal. The even, or high-pass output contains the difference between the true input and the value of the reconstructed input if it were to be reconstructed from only the information given in the odd output. In general, higher order wavelets (i.e. those with more nonzero coefficients) tend to put more information into the odd output, and less into the even output. If the average amplitude of the even output is low enough, then the even half of the signal may be discarded without greatly affecting the quality of the reconstructed signal. An important step in wavelet-based data compression is finding wavelet functions which cause the even terms to be nearly zero.

The Haar wavelet represents a simple interpolation scheme. After passing these data through the filter functions, the output of the low-pass filter consists of the average of every two samples, and the output of the high-pass filter consists of the difference of every two samples. The high-pass filter contains less information than the low pass output. If the signal is reconstructed by an inverse low-pass filter of the form

$$f_j^L = \sum_{i=0}^{N/2-1} c_{2i-j+1} a_i, \qquad j = 0, 1, \ldots, N-1$$

then the result is a duplication of each entry from the low-pass filter output. This is a wavelet reconstruction with $2\times$ data compression. Since the perfect reconstruction is a sum of the inverse low-pass and inverse high-pass filters, the output of the inverse high-pass filter can be calculated. This is the result of the inverse high-pass filter function

$$f_j^H = \sum_{i=0}^{N/2-1} (-1)^j c_{j-1-2i} b_i, \qquad j = 0, 1, \ldots, N-1.$$

The perfectly reconstructed signal is

$$f = f^L + f^H$$

where each f is the vector with elements f_j. Using other coefficients and other orders of wavelets yields similar results, except that the outputs are not exactly averages and differences, as in the case using the Haar coefficients.

The following C++ program implements the Haar wavelet transform.

```
// wavelet.cpp

#include <iostream.h>
#include <math.h>

void main()
{
  const double pi = 3.14159;
  int n = 16;  // n must be a power of 2
  double* f = new double[n];

  // input signal
  int k;
  for(k=0; k < n; k++)
    f[k] = sin(2.0*pi*(k+1)/n);
```

```cpp
double* c = new double[n];
for(k=0; k < n; k++)
 c[k] = 0.0;

c[0] = 1.0; c[1] = 1.0;

double* a = new double[n/2];
for(k=0; k < n/2; k++)
 a[k] = 0.0;
double* b = new double[n/2];
for(k=0; k < n/2; k++)
 b[k] = 0.0;

int i, j;
for(i=0; i < n/2; i++)
{
 for(j=0; j < n; j++)
 {
  if(2*i-j+1 < 0) a[i] += c[2*i-j+1+n]*f[j];
  else a[i] += c[2*i-j+1]*f[j];
 }
 a[i] = 0.5*a[i];
}

for(i=0; i < n/2; i++)
{
 for(j=0; j < n; j++)
 {
  if(j-2*i < 0) b[i] += pow(-1.0,j)*c[j-2*i+n]*f[j];
  else b[i] += pow(-1.0,j)*c[j-2*i]*f[j];
 }
 b[i] = 0.5*b[i];
}

for(k=0; k < n/2; k++)
cout << "a[" << k << "] = " << a[k] << endl;
for(k=0; k < n/2; k++)
cout << "b[" << k << "] = " << b[k] << endl;

//inverse
double* fL = new double[n];
double* fH = new double[n];

for(j=0; j < n; j++)
 fL[j] = 0.0;
for(j=0; j < n; j++)
 fH[j] = 0.0;
```

```cpp
for(j=0; j < n; j++)
{
 for(i=0; i < n/2; i++)
 {
  if(2*i-j+1 < 0) fL[j] += c[2*i-j+1+n]*a[i];
  else fL[j] += c[2*i-j+1]*a[i];
 }
}

for(k=0; k < n; k++)
 cout << "fL[" << k << "] = " << fL[k] << endl;

for(j=0; j < n; j++)
{
 for(i=0; i < n/2; i++)
 {
  if(j-1-2*i < 0) fH[j] += pow(-1.0,j)*c[j-1-2*i+n]*b[i];
  else fH[j] += pow(-1.0,j)*c[j-1-2*i]*b[i];
 }
}

for(k=0; k < n; k++)
 cout << "fH[" << k << "] = " << fH[k] << endl;

// input signal reconstructed
double* g = new double[n];
for(k=0; k < n; k++)
 g[k] = fL[k] + fH[k];

for(k=0; k < n; k++)
 cout << "g[" << k << "] = " << g[k] << endl;
}
```

8.5   Primitive Recursive Functions

Let N_0 be the natural numbers including 0, i.e. $\{0, 1, 2, \ldots\}$. For a function to be computable, there must be an algorithm or procedure for computing it. So in a formal definition of this class of functions we must replace the intuitive, semantic ideas with precise descriptions of the functions [48, 63, 77, 97, 115].

To begin, we take as variable the letters $n, x_1, x_2, \ldots$. We write $\mathbf{x}$ for $(x_1, \ldots, x_k)$.

Next we list the basic, incontrovertibly computable functions which we use as building blocks for all others.

$$
\begin{array}{lll}
\text{zero function } Z : \mathbf{N}_0 \to \mathbf{N}_0 & Z(n) = 0 \text{ for all } n \\
\text{successor function } S : \mathbf{N}_0 \to \mathbf{N}_0 & S(n) = n + 1 \\
\text{projection functions } P_k^i : \mathbf{N}_0^k \to \mathbf{N}_0 & P_k^i(x_1, \ldots, x_k) = x_i \text{ for } 1 \le i \le k
\end{array}
$$

These functions are called *initial functions*. We sometimes call the projections pick-out functions, and P_1^1 the identity function, written $id(x) = x$.

Next, we specify the ways we allow new functions to be defined from ones we already have.

Composition. If g is a function of m-variables and $h_1, \ldots, h_m$ are functions of k variables, which are already defined, then composition yields the function

$$f(\mathbf{x}) = g(h_1(\mathbf{x}), \ldots, h_m(\mathbf{x})).$$

Primitive recursion. For functions of one variable the schema is

$$
\begin{aligned}
f(0) &= d \\
f(n+1) &= h(f(n), n)
\end{aligned}
$$

where d is a number and h is a function already defined.

For functions of two or more variables, if g and h are already defined then f is given by primitive recursion on h with basis g as

$$
\begin{aligned}
f(0, \mathbf{x}) &= d \\
f(n+1, \mathbf{x}) &= h(f(n, \mathbf{x}), n, \mathbf{x})
\end{aligned}
$$

The reason we allow both n and $\mathbf{x}$ as well as $f(n, \mathbf{x})$ to appear in h is that we may wish to keep track of both the stage we are at and the input.

The *primitive recursive functions* are exactly those which are either an initial function or can be obtained from the initial functions by a finite number of applications of the basic operations.

Example. The sum $x_1 + x_2$ is primitive recursive defined by

$$sum(0, x_2) = P_1^1(x_2)$$
$$sum(x_1 + 1, x_2) = S(sum(x_1, x_2))$$

Example. The product $x_1 x_2$ is primitive recursive defined by

$$product(0, x_2) = 0$$
$$product(x_1 + 1, x_2) = sum(product(x_1, x_2), x_2)$$

Example. The *predecessor* of x is primitive recursive defined by

$$pred(0) = 0$$
$$pred(x + 1) = x$$

Example. The function

$$minus(x_1, x_2) = \begin{cases} x_1 - x_2 & x_1 \geq x_2 \\ 0 & \text{otherwise} \end{cases}$$

is primitive recursive.

$$minus(0, x_2) = 0$$
$$minus(x_1 + 1, x_2) = S(minus(x_1, x_2))$$

Example. We can now describe the mod function

$$mod(n, m) = \begin{cases} n & n < m \\ mod(n - m, m) & \text{otherwise} \end{cases}$$

The definition is as follows.

$$mod(0, m) = 0$$
$$mod(n + 1, m) = k(mod(n, m), n, m)$$

$$k(p, n, m) = minus(S(p), product(m, minus(S(p), m)))$$

The function k is a composition of primitive recursive functions (some of which are compositions themselves) and is primitive recursive. Thus $mod(n, m)$ is primitive recursive. ♣

The Ackermann function f is not primitive recursive. It is obviously not an initial function. Nor can it be defined for the case $f(0, m)$ as a number independent of m. Thus the only way the Ackermann function can be primitive recursive is if it is a composition of primitive recursive functions. Since for $m \neq 0$ and $n \neq 0$ f relies on an evaluation of f itself it cannot be a composition of primitive recursive functions.

Definition. A set C of total functions from $\mathbf{N}_0^n$ to $\mathbf{N}_0$ (for all n) is called *primitive recursively closed* if it satisfies the following conditions.

1. all initial functions are in C

2. C is closed under primitive recursion, if f comes from g and h by primitive recursion and $g, h \in C$ then $f \in C$

3. C is closed under composition, if f comes from g and $h_1, \ldots, h_r$ by composition $(f(\mathbf{x}) = g(h_1(\mathbf{x}), \ldots, h_r(\mathbf{x})))$ and $g, h_1, \ldots, h_r \in C$ then $f \in C$.

Theorem. The set of all primitive recursive functions is primitive recursively closed. [48, 63, 97]

Theorem. Any primitive recursively closed set contains every primitive recursive function. [48, 63, 97]

Definition. The definition of μ-recursive is

- The primitive recursive functions are μ-recursive.

- If
$$f : \mathbf{N}_0^{n+1} \to \mathbf{N}_0$$
is μ-recursive then
$$\mu f : \mathbf{N}_0^n \to \mathbf{N}_0$$
defined by
$$\mu f := \min\{\ x \mid f(x, \mathbf{y}) = 0 \text{ with } f(u, \mathbf{y}) \text{ defined for } u \leq x\ \}$$
is μ-recursive. The function μf will be undefined if no such x exists.

- Functions defined by composition and primitive recursion of μ-recursive functions are also μ-recursive.

In other words μ acts as a minimalization operator in the sense that it maps to the minimum x such that $f(x, \mathbf{y}) = 0$.

The Ackermann function is μ-recursive.

8.6 Backtracking

A common approach to finding a solution, when no simple solution algorithm is avaialable, is trial and error. Suppose a configuration is built up by a number of well defined steps and then tested to see if it is a solution. If it is not a solution we return to an earlier step and try a different option. *Backtracking* is the technique of returning to an earlier stage in a solution process to make a different choice in the attempt to find solutions.

Example. 8-Queens problem. A chessboard is 8 columns wide and 8 rows high. The 8 queens problem requires us to place 8 queens on the chessboard so that no queen is attacking another. A queen attacks another if it is on the same row, column or diagonal as the other queen. An example solution is

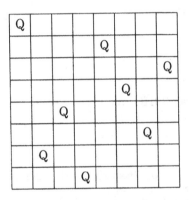

Figure 8.2: A Solution to the 8-Queens Problem

The recursive solution is to place a queen on a position which is not attacked, row by row. If there is no position available for a queen, the algorithm return to the previous row and moves the queen to the next position which is not attacked. To check every possible placement of the 8 queens includes configurations which are obviously incorrect and will also take a long time. This algorithm uses a technique called *pruning* to reduce the number of configurations to test. We can form a tree according to the square we place each queen in. The root of the tree corresponds to an empty board. The branches from the root correspond to each possible placement of a queen. By rejecting certain options earlier, the corresponding branches and entire sub-trees are removed from consideration.

```
// queens.cpp

#include <iostream>

using namespace std;

const char QUEEN = 'Q';
const char SPACE = '#';

void printboard(char board[8][8])
{
 int i,j;
 for(i=0;i<8;i++)
 {
  for(j=0;j<8;j++)
   cout << board[i][j];
  cout << endl;
 }
 cout << endl;
}

int attacking(char board[8][8],int row,int col)
{
 int i;
 for(i=0;i<8;i++)
 {
  if((board[row][i]==QUEEN)||(board[i][col]==QUEEN)) return 1;
  if((i+row-col>=0)&&(i+row-col<8))
   if(board[i+row-col][i]==QUEEN)
    return 1;
  if((col+row-i>=0)&&(col+row-i<8))
   if(board[col+row-i][i]==QUEEN)
    return 1;
 }
 return 0;
}

void queens(char board[8][8],int row)
{
 int i;
 if(row<0) return;
 if(row>7) { printboard(board);return; }
 for(i=0;i<8;i++)
  if(!attacking(board,row,i))
  {
   board[row][i]=QUEEN;
   queens(board,row+1);
   board[row][i]=SPACE;
```

```
  }
}

void main(void)
{
 char board[8][8];
 int i,j;

 for(i=0;i<8;i++)
  for(j=0;j<8;j++)
    board[i][j]=SPACE;
 queens(board,0);
}
```

The program output is

```
Q#######
####Q###
#######Q
#####Q##
##Q#####
######Q#
#Q######
###Q####

Q#######
#####Q##
#######Q
##Q#####
######Q#
###Q####
#Q######
####Q###
```

⋮

8.7   Stacks and Recursion Mechanisms

8.7.1   Recursion Using Stacks

Most implementations of recursion rely on a *stack* which is a *last-in first-out* (*LIFO*) storage mechanism. The stack has two operations, namely to *push* data onto the top of a stack and *pop* data off the top of the stack. For a recursive function call all the local data must be preserved. This is done by pushing the local data onto a stack. After the recursive function call has completed, the local data is popped off the stack into registers or other local storage. This stack can also be used to return the result of a function call. Next we introduce two routines called CALL and RETURN to implement the recursion. CALL has three arguments.

1. The address of the function to enter.

2. The address of the first register of workspace to be preserved.

3. The number of registers to be preserved.

The CALL routine pushes the specified registers and return address onto the stack, and then transfers control to the specified function.
RETURN has two arguments.

1. The address of the first register of the work-space to be restored.

2. The number of registers to be restored.

The RETURN routine restores the specified registers from the stack and then returns control to the return address on the stack. In shortened form

```
CALL(function,registers) : PUSH currentaddress
                           PUSH registers
                           GOTO function
RETURN(registers)         : POP into registers
                           POP into returnaddress
                           GOTO returnaddress
```

A simple recursive function would then be

```
function : IF (basecase) ... RETURN(registers)
           ...
           CALL(function,registers)
           ...
           RETURN(registers)
```

8.7.2   Stack Free Recursion

It is sometimes possible to convert a recursive function to an iterative one. The stack is very important in implementing recursion. It allows the recursive function to return to a previous state which it has stored. This is also possible if the function to be computed has an inverse. The inverse allows an iterative routine to return to previous values of the function evaluation.

Example. The Fibonacci sequence defined recursively by

$$F_{n+2} = F_{n+1} + F_n \qquad n = 0, 1, 2, \ldots$$

where $F_0 = F_1 = 1$ can be implemented iteratively by simply storing the previous two function evaluations. In this case the inverse is given by

$$F_n = F_{n+2} - F_{n+1} \qquad n = 0, 1, 2, \ldots$$

but it is not necessary to use this information.

```cpp
// fibonacci.cpp

#include <iostream>

using namespace std;

void main(void)
{
  int i;
  unsigned long F0 = 1,F1 = 1;
  unsigned long temp;

  for(i=0;i<10;i++)
  {
   cout << F0 << " ";
   temp = F1;
   F1 = F0 + F1;
   F0 = temp;
  }
  cout << endl;
}
```

To remove the dependence on a stack, the changes made to variables by a recursive call must be reversible. This excludes some variables such as those used exclusively for return values. If we consider the towers of Hanoi problem, the algorithm just swaps variables and decrements a variable. This can obviously be reversed. Swapping two variables is reversed by the same action and the decrement is reversed by an increment. The changes must be made immediately before the recursive call and reversed immediately after the recursive call.

```cpp
// hanoi2.cpp

#include <iostream>

using namespace std;

char A,B,C;
unsigned long n;

void hanoi()
{
  if(n==1) cout << A << " -> " << B << endl;
  else
  {
    n--; C = B^C; B = B^C; C = B^C;     // swap B and C
    hanoi();
    C = B^C; B = B^C; C = B^C;          // swap B and C back
    cout << A << " -> " << B << endl;
    C = A^C; A = A^C; C = A^C;          // swap A and C
    hanoi();
    n++; C = A^C; A = A^C; C = A^C;     // swap A and C back
  }
}

void main(void)
{
  A = 'A'; B = 'B'; C = 'C'; n = 1;
  cout << "Tower of Hanoi with 1 disc:" << endl;
  hanoi();
  A = 'A'; B = 'B'; C = 'C'; n = 2;
  cout << "Tower of Hanoi with 2 discs:" << endl;
  hanoi();
  A = 'A'; B = 'B'; C = 'C'; n = 3;
  cout << "Tower of Hanoi with 3 discs:" << endl;
  hanoi();
}
```

Chapter 9

Abstract Data Types

9.1 Introduction

Programming languages such as C++ and Java have built in data types (so-called *basic data types* or *primitive data types*) such as integers that represent information and have operations that can be performed on them (such as multiplication and addition). For example the built in basic data types in C++ are short, int, long, float, double and char.

An *abstract data type* (ADT) consists of data and the operations which can be performed on it. Generally the data is represented with standard data types of the language in which it is implemented but can also include other abstract data types. The operations defined on the ADT provide access to the information and manipulation of the data without knowing the implementation of the ADT. The abstract data type is implemented using constructors, data fields and methods (functions). *Information hiding* is when ADT data is inaccessible (no operation can retrieve the data). *Encapsulation* refers to the hiding of inner details (such as implementation). In C++ the concepts of public, private and protected data fields and methods are important in the implementation of an ADT. Public members of an ADT are always accesible. Private members are only accesible by the ADT itself and protected members are only accesible by the ADT and any derived ADT's. A derived ADT may override the accessibility of members by forcing all inherited members to a specified level if they are more accessible.

For example the Standard Template Library [2] in C++ introduces many ADT's such as Vector, list, stack, queue and set. Standard C++ now includes the abstract data type string. Symbolic C++ [169] includes the template classes Rational, Complex, Quaternion, Vector, Matrix, Polynomial and Sum. Operations such as addition and multiplication, determinant, trace and inverse of matrices are included in the matrix class. An instance of Matrix could then be used in the same way that integers are used without knowing the internal differences.

In the following sections various useful ADT's will be introduced.

9.2 Linked List

The linked list is a useful data structure that can dynamically grow according to data storage requirements. This is done by viewing data as consisting of a unit of data and a link to more units of data.

A linked list is useful in the implementations of dynamic arrays, stacks, strings and sets. The linked list is the basic ADT in some languages, for example LISP. LISP stands for List Processing. All the program instructions in LISP operate on lists.

Linked lists are most useful in environments with dynamic memory allocation. With dynamic memory allocation dynamic arrays can grow and shrink with less cost than in a static memory allocation environment. Linked lists are also useful to manage dynamic memory environments. Diagrammatically a linear linked list can be viewed as follows.

Figure 9.1: Diagrammatic Representation of a Linked List

The list consists of data elements. Each data element has an associated link to the next item in the list. The last item in the list has no link. In C++ we can implement this by using a null pointer. The first element of the list is called the head, the last element is called the tail.

Extensions to the ADT include double-linked lists where links exist for the next data and the previous data allowing easier access to data earlier in the list, and sorted linked lists. We use a template class definition so that the linked list can store any kind of data without having to change or reimplement any functionality.

The following class is a C++ implementation of the ADT list. It has methods for creating and destroying a list, copying one list to another (assignment operator), adding items to and removing items from the list (additem, insertitem, removeitem), merging lists (operators for addition), iteration (first, next, last, position, data) and indexing elements (operator[]).

```
// list.h

#ifndef LIST_HEADER
#define LIST_HEADER

using namespace std;

template <class T>
struct listitem { T data; listitem *next; };

template <class T>
class list
{
 protected:
        listitem<T> *head;
        listitem<T> *current;
        int size;
 public:
        list();
        list(const list&);
        ~list();
        list &operator=(const list&);
        void additem(T);
        void insertitem(T);
        int insertitem(T,int);
        void removeitem(void);
        int removeitem(int);
        list operator+(const list&) const;
        list &operator+=(const list&);
        T operator[](int);
        T data(void);
        int next(void);
        int first(void);
        int last(void);
        int position(int);
        int getsize(void);
};

template <class T>
list<T>::list() { head=current=NULL; size=0; }

template <class T>
list<T>::list(const list &l)
{
 listitem<T> *li=l.head;
 head=current=NULL;
 size=l.size;
 if(li!=NULL)
```

```
 {
  head=new listitem<T>;
  head->data=li->data;
  head->next=NULL;
  current=head;
  li=li->next;
 }
 while(li!=NULL)
 {
  current->next=new listitem<T>;
  current=current->next;
  current->data=li->data;
  current->next=NULL;
  li=li->next;
 }
 current=head;
}


template <class T>
list<T>::~list()
{
 while(head!=NULL)
 {
  current=head;
  head=head->next;
  delete current;
 }
}

template <class T>
list<T> &list<T>::operator=(const list &l)
{
 listitem<T> *li=l.head;
 if(this==&l) return *this;

 size=l.size;
 while(head!=NULL)
 {
  current=head;
  head=head->next;
  delete current;
 }

head=current=NULL;
if(li!=NULL)
{
 head=new listitem<T>;
```

```
 head->data=li->data;
 head->next=NULL;
 current=head;
 li=li->next;
}
while(li!=NULL)
{
 current->next=new listitem<T>;
 current=current->next;
 current->data=li->data;
 current->next=NULL;
 li=li->next;
}
 current=head;
 return *this;
}

template <class T>
void list<T>::additem(T t)
{
 listitem<T> *li=head;
 if(head==NULL)
 {
  head=new listitem<T>;
  head->data=t;
  head->next=NULL;
  current=head;
 }
 else
 {
  while(li->next!=NULL) li=li->next;
  li->next=new listitem<T>;
  li=li->next;
  li->data=t;
  li->next=NULL;
 }
 size++;
}

template <class T>
void list<T>::insertitem(T t)
{
 listitem<T> *li;
 if(head==NULL)
 {
  head=new listitem<T>;
  head->data=t;
  head->next=NULL;
```

```
  current=head;
 }
 else
 {
  li=current->next;
  current->next=new listitem<T>;
  current->next->data=t;
  current->next->next=li;
 }
 size++;
}


template <class T>
int list<T>::insertitem(T t,int i)
{
 int j=0;
 listitem<T> *li1, *li2;
 li1=head;
 while((j<i)&&(li1->next!=NULL)) { li1=li1->next;j++; }
 if(j==i)
 {
  li2=li1->next;
  li1->next=new listitem<T>;
  li1=li1->next;
  li1->data=t;
  li1->next=li2;
  size++;
  return 1;
 }
 return 0;
}


template <class T>
void list<T>::removeitem(void)
{
 listitem<T> *li=head;
 if(head==NULL) return;
 if(head==current)
 {
  delete head;
  head=current=NULL;
  size--;
 }
 if(current!=NULL)
 {
  while(li->next!=current) li=li->next;
  li->next=current->next;
  delete current;
```

```
  current=li;
  size--;
 }
}

template <class T>
int list<T>::removeitem(int i)
{
 int j=0;
 listitem<T> *li1, *li2;
 li1=head;
 if(head==NULL) return 0;
 while((j<i-1)&&(li1->next!=NULL)) { li1=li1->next;j++; }
 if(j==i-1)
 {
  li2=li1->next;
  li1->next=li1->next->next;
  if(li2!=NULL) delete li2;
  size--;
  return 1;
 }
 return 0;
}

template <class T>
list<T> list<T>::operator+(const list &l) const
{
 list<T> l2(*this);
 l2+=l;
 return l2;
}

template <class T>
list<T> &list<T>::operator+=(const list &l)
{
 listitem<T> *li=l.head;
 while(li!=NULL) { additem(li->data); li=li->next; }
 return *this;
}

template <class T>
T list<T>::operator[](int i)
{
 int j=0;
 listitem<T> *li=head;
 if(li==NULL) return T();
 while((j<i)&&(li->next!=NULL)) { li=li->next; j++; }
 if(j==i) return li->data;
```

```
  return T();
 }

 template <class T>
 T list<T>::data(void)
 {
  if(current==NULL) return T();
  return current->data;
 }

 template <class T>
 int list<T>::next(void)
 {
  if(current->next==NULL) return 0;
  current=current->next;
  return 1;
 }

 template <class T>
 int list<T>::first(void)
 {
  current=head;
  if(head==NULL) return 0;
  return 1;
 }

 template <class T>
 int list<T>::last(void)
 {
  current=head;
  if(head==NULL) return 0;
  while(current->next!=NULL) current=current->next;
  return 1;
 }

 template <class T>
 int list<T>::position(int i)
 {
  int res;
  res=first();
  while(res&&(i>0)) { next();i--; }
  return res;
 }

 template <class T>
 int list<T>::getsize(void) { return size; }

 #endif
```

Now the ADT is used in an example program to illustrate the available operations.

```cpp
// listeg.cpp

#include <iostream>
#include "list.h"

using namespace std;

void main(void)
{
 int i;
 list<int> l1;
 l1.additem(1);
 l1.additem(2);
 l1.additem(3);
 l1.additem(5);
 l1.additem(8);
 list<int> l2(l1),l3;
 l3=l2;
 l1.next();
 l1.insertitem(13); l1.insertitem(21,5);
 l1.first();
 cout<<"l1: ";
 do cout<<l1.data()<<" ";
 while(l1.next());
 cout<<endl;
 cout<<"l2: ";
 do cout<<l2.data()<<" ";
 while(l2.next());
 cout<<endl;
 cout<<"l3: ";
 do cout<<l3.data()<<" ";
 while(l3.next());
 cout<<endl;
 list<int> l4=l1+l2;
 cout<<"l4: ";
 do cout<<l4.data()<<" ";
 while(l4.next());
 cout<<endl;
 l4+=l3;
 l4.first();
 cout<<"l4: ";
 do cout<<l4.data()<<" ";
 while(l4.next());
 cout<<endl;
 l4.first();
 cout<<"The first item of l4 is "<<l4.data()<<endl;
```

```
l4.last();
cout<<"The last item of l4 is "<<l4.data()<<endl;
cout<<"The fourth item of l4 is "<<l4[3]<<endl;
l4.position(3);
l4.removeitem();
l4.removeitem(7);
l4.first();
cout<<"l4: ";
do
 cout<<l4.data()<<" ";
while(l4.next());
cout<<endl;
cout<<"Size of l1 is "<<l1.getsize()<<endl;
cout<<"Size of l2 is "<<l2.getsize()<<endl;
cout<<"Size of l3 is "<<l3.getsize()<<endl;
cout<<"Size of l4 is "<<l4.getsize()<<endl;
}
```

The program output is:

```
l1: 1 2 13 3 5 8 21
l2: 1 2 3 5 8
l3: 1 2 3 5 8
l4: 1 2 13 3 5 8 21 1 2 3 5 8
l4: 1 2 13 3 5 8 21 1 2 3 5 8 1 2 3 5 8
The first item of l4 is 1
The last item of l4 is 8
The fourth item of l4 is 3
l4: 1 2 13 5 8 21 1 3 5 8 1 2 3 5 8
Size of l1 is 7
Size of l2 is 5
Size of l3 is 5
Size of l4 is 15
```

The linked list can also be viewed as a recursive structure with the first element followed by a linked list. This view can make the implementation of many methods easier.

Suppose an item is to be inserted into the sorted list. The item either comes before the head of the list (which can be easily implemented) or after the head in which case the item actually has to be inserted in a list with the second element of the list as the head. Similarly to delete an item either the head must be removed or the item must be deleted from a list with the second element of the list as the head.

The benefits of the recursive structure is demonstrated with a function **Reverse** which reverses the list. This is as simple as removing the head of the list, reversing the rest of the list and then adding the head to the end of the list. The assignment and equality operators also demonstrate the benefits of the structure with simple implementations. Assignment first deletes the current list (if the list is not being assigned to itself). The head is copied first and then the rest of the list can be copied. Equality is determined by testing the equality of the heads of two lists. On success the equality of the rest of the two lists are tested. A linear search also has a simple implementation. First the head of the list is examined to see if it contains the data. If this is not the case, the rest of the list is searched.

In each of the above cases the rest of the list is a list itself and the method can be applied recursively. Usually the simplest case for each recursive method is for the empty list.

Special care must be taken when destroying the list. Simply deleting the head of the list will create a memory leak. The remaining list must be destroyed before the head can be destroyed. In the implementation the head data is part of the class data and so the memory leak is avoided.

The data members of the class are few since most of the data management is done by the recursive structure. The data member **head** stores the data for the node in the linked list. Since a linked list can be empty, a node with the data member **empty** set to one represents an empty list. A pointer **tail** provides access to the rest of the list.

```
// rlist.h

#ifndef RLIST_HEADER
#define RLIST_HEADER

#include <assert.h>

using namespace std;

template <class T>
class RList
{
 public:
   RList();
   RList(const RList &);
   ~RList();
   RList &operator = (const RList&);
   int operator == (const RList&);
   void Insert(const T&);
   int Search(const T&);
   int Delete(const T&);
   T Head(void);
   RList *Tail(void);
   int Empty(void);
   RList *Reverse(RList*);
 private:
   T head;
   RList* tail;
   int empty;
};

template <class T>
RList<T>::RList() { empty = 1; }

template <class T>
RList<T>::RList(const RList<T> &RL)
{
 empty = RL.empty;
 head = RL.head;
 tail = new RList<T>(*RL.tail);
}

template <class T>
RList<T>::~RList() { if(!empty) delete tail; }

template <class T>
RList<T> &RList<T>::operator=(const RList<T> &RL)
{
```

```
 if(this == &RL) return;
 if(!empty) delete tail;
 empty = RL.empty;
 head = RL.head;
 tail = new RList<T>(*RL.tail);
}

template <class T>
int RList<T>::operator==(const RList<T> &RL)
{
 if(empty&&RL.empty) return 1;
 if(this == &Rl) return 1;
 if(head != RL.head) return 0;
 return (*tail == *RL.tail);
}

template <class T>
void RList<T>::Insert(const T &toInsert)
{
 if(empty)
 {
  head = toInsert;
  tail = new RList<T>;
  empty=0;
 }
  else tail->Insert(toInsert);
}

template <class T>
int RList<T>::Search(const T &toSearch)
{
 if(empty) return 0;
 else if(head == toSearch) return 1;
 else return tail -> Search(toSearch);
}

template <class T>
int RList<T>::Delete(const T &toDelete)
{
 if(empty) return 0;
 else if(head==toDelete)
 {
  head = tail -> head;
  empty = tail -> empty;
  tail -> Delete(tail -> head);
  if (tail -> empty) delete tail;
  return 1;
 }
```

```
 else return tail -> Delete(toDelete);
}


template <class T>
T RList<T>::Head(void)
{
 assert(!empty);
 return head;
}


template <class T>
RList<T> *RList<T>::Tail(void)
{
 assert(!empty);
 return tail;
}


template <class T>
int RList<T>::Empty(void) { return empty; }

template <class T>
RList<T> *RList<T>::Reverse(RList<T> *RL)
{
 if(RL->Empty())
 {
  RList<T> *temp;
  temp = new RList<T>;
  return temp;
 }
 else
 {
  RList<T> *R;
  R = Reverse(RL->Tail());
  (*R).Insert(RL->Head());
  return R;
 }
}


#endif
```

Now the ADT is used in an example program to illustrate the available operations.

```cpp
// rlisteg.cpp

#include <iostream>
#include "rlist.h"

using namespace std;

int main(void)
{
 RList<int> L;
 int i;
 for(i=1; i<=8; i++)
  L.Insert(i);

 RList<int>* LX = &L;
 cout << "The initial list is: " << endl;

 while(!LX -> Empty())
 {
  cout << LX -> Head() << ' ';
  LX = LX -> Tail();
 }
 cout << endl << endl;

 RList<int>* R = L.Reverse(&L);
 RList<int>* LP = R;

 while(!LP -> Empty())
 {
  cout << LP -> Head() << ' ';
  LP = LP -> Tail();
 }
 cout << endl << endl;

 cout << "what happened to the initial list: "<< endl;
 LP = &L;
 while(!LP -> Empty())
 {
  cout << LP -> Head() << ' ';
  LP = LP -> Tail();
 }
 cout << endl;

 cout << "remove some items: "<< endl;
 L.Delete(1);
 L.Delete(4);
```

```
L.Delete(8);
LP = &L;
while(!LP -> Empty())
{
 cout << LP -> Head() << ' ';
 LP = LP -> Tail();
}
cout << endl;

cout << "is 3 in the list: " << L.Search(3) << endl;
cout << "is 4 in the list: " << L.Search(4) << endl;

return 0;
}
```

The program output is:

```
The initial list is:
1 2 3 4 5 6 7 8

8 7 6 5 4 3 2 1

what happened to the initial list:
1 2 3 4 5 6 7 8
remove some items:
2 3 5 6 7
is 3 in the list: 1
is 4 in the list: 0
```

9.3 Stack

The *stack* is a *LIFO* (last in first out structure). The last value stored (and not yet retrieved) is the only value that can be retrieved. The traditional analogy is a stack of plates where only the top plate can be removed, and a plate can only be placed on top of the stack. Due to the dynamic nature of a stack the implementation is based on the linked list.

Since we have already created a list ADT which can grow or shrink in size as needed we can reduce the amount of work needed to implement the stack ADT. The list enables access to any element in the structure, the stack can be viewed as a restricted list with access to only the tail. The operation of putting data on the stack is referred to as "pushing" data onto the stack, and the operation of retrieving data from the stack is referred to as "popping" data off the stack. The stack is an important structure for implementing recursion.

Diagrammatically a stack can be viewed as follows.

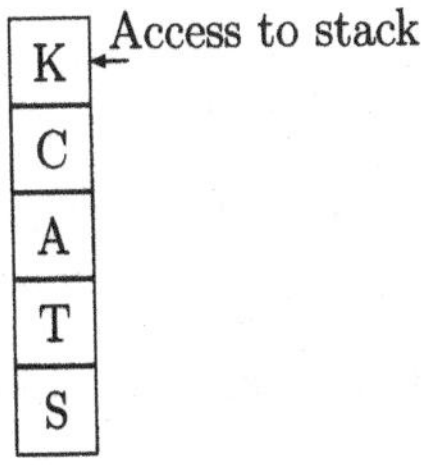

Figure 9.2: Diagrammatic Representation of a Stack

We implement the stack as a class in C++. The class has methods for creating a stack using an empty list (the constructor), copying one stack to another (the assignment operator), pushing data onto the stack (**push** which simply adds the data to the end of the list) and popping data off the stack (**pop** which simply removes the last element of the list). No destructor is needed since the list destructor is called automatically.

```
// stack.h

#ifndef STACK_HEADER
#define STACK_HEADER
#include "list.h"
using namespace std;

template <class T>
class stack
{
 protected:
         list<T> stacklist;
 public:
         stack();
         stack(const stack&);
         stack &operator=(const stack&);
         void push(T);
         T pop(void);
         int getsize(void);
};

template <class T>
stack<T>::stack() : stacklist() {}

template <class T>
stack<T>::stack(const stack &s) : stacklist(s.stacklist) {}

template <class T>
stack<T> &stack<T>::operator=(const stack & s)
{ stacklist=s.stacklist; return *this; }

template <class T>
void stack<T>::push(T t) { stacklist.additem(t); }

template <class T>
T stack<T>::pop(void)
{
 T data;
 stacklist.last();
 data=stacklist.data();
 stacklist.removeitem();
 return data;
}

template <class T>
int stack<T>::getsize() { return stacklist.getsize(); }

#endif
```

Now the ADT is used in an example program to illustrate the available operations.

```cpp
// stackeg.cpp

#include <iostream>
#include "stack.h"

using namespace std;

void main(void)
{
 int i;
 stack<int> s1;
 s1.push(1);
 s1.push(2);
 s1.push(3);
 s1.push(5);
 s1.push(7);
 s1.push(11);
 stack<int> s2(s1);
 stack<int> s3;
 s3=s1;
 stack<int> s4;
 cout<<"Size of s1 is "<<s1.getsize()<<endl;
 cout<<"s1: ";
 while(s1.getsize()>0) {cout<<(i=s1.pop())<<" ";s4.push(i);}
 cout<<endl<<"s2: ";
 while(s2.getsize()>0) cout<<s2.pop()<<" ";
 cout<<endl<<"s3: ";
 while(s3.getsize()>0) cout<<s3.pop()<<" ";
 cout<<endl<<"s4: ";
 while(s4.getsize()>0) cout<<s4.pop()<<" ";
 cout<<endl;
}
```

The program output is:

```
Size of s1 is 6
s1: 11 7 5 3 2 1
s2: 11 7 5 3 2 1
s3: 11 7 5 3 2 1
s4: 1 2 3 5 7 11
```

9.4 Tree

A *tree* is a branching structure. It has a starting node called a *root* node. An n-ary tree can have up to n branches from each node to other nodes. A binary tree is a 2-ary tree. Every node in a tree is the root of a subtree. A tree is noncyclic, in other words there is only one path between any two nodes in a tree.

A binary tree is useful for classification by proposition. If $P(x, y)$ is a proposition regarding x and y, then, when a node represents x all items y, where $P(x, y)$ is false, should be accessible only via the left branch from the node and all items y, where $P(x, y)$ is true, should be accesible only via the right branch from the node. This can be used to sort elements. Binary trees can also be searched more quickly than linear structures such as a linked list.

In general an n-ary tree has a search time $O(log_n s)$ where s is the number of elements in the tree. For a linear structure such as the linked list the search time is $O(n)$. Diagrammatically a binary tree can be viewed as follows.

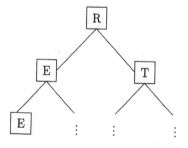

Figure 9.3: Diagrammatic Representation of a Binary Tree

We implement a binary tree as a class in C++. The class has methods for creating a new binary tree, destroying a binary tree, copying one binary tree to another (assignment operator), adding an item and removing an item from the tree (additem and removeitem), determining if an item is present in the tree (find) and iterating through the tree (first, last, next and previous).

```
// tree.h

#ifndef TREE_HEADER
#define TREE_HEADER

#include "list.h"
#include "stack.h"

using namespace std;

template <class T>
struct treenode
{
 T data;
 list<treenode<T>*> leftchildren;
 list<treenode<T>*> rightchildren;
};

template <class T>
class tree
{
 protected:
         treenode<T> *root;
         treenode<T> *current;
         stack<treenode<T>*> traverse;
         void additem(T,treenode<T>*);
         void copy_subtree(treenode<T>*,treenode<T>*);
         void delete_subtree(treenode<T>*);
         treenode<T> *find(treenode<T>*,T);
         unsigned int limit;
 public:
         tree(unsigned int);
         tree(const tree&);
         ~tree();
         tree &operator=(const tree&);
         void additem(T);
         int removeitem(void);
         int find(T);
         void first(void);
         void last(void);
         void next(void);
         void previous(void);
         T &data(void);
};

template <class T>
void tree<T>::additem(T t,treenode<T> *&tn)
{
```

```
 int i=0;
 list<treenode<T>*> *l;
 if(tn==NULL)
 {
  tn=new treenode<T>;
  tn->data=t;
  return;
 }
 if(t<tn->data) l=&(tn->leftchildren);
 else l=&(tn->rightchildren);

 if(l->getsize()==0)
 {
  l->additem(new treenode<T>);
  (*l)[0]->data=t;
  return;
 }
 else
 {
  while((i<l->getsize())&&((*l)[i]->data<t)) i++;
  if((l->getsize()<limit-1)&&
    (tn->leftchildren.getsize()+tn->rightchildren.getsize()<limit))
  {
   l->insertitem(new treenode<T>,i);
   (*l)[i]->data=t;
  }
  else additem(t,(*l)[i]);
 }
}


template <class T>
void tree<T>::copy_subtree(treenode<T> *&t1,const treenode<T> *&t2)
{
 int i;
 if(t2==NULL) {t1=NULL; return;}
 t1=new treenode<T>;
 t1.data=t2.data;
 for(i=0;i<t2->leftchildren.getsize();i++)
 {
  t1->leftchildren.additem(NULL);
  copy_subtree(t1->leftchildren[i],t2->leftchildren[i]);
 }
 for(i=0;i<t2->rightchildren.getsize();i++)
 {
  t1->rightchildren.additem(NULL);
  copy_subtree(t1->rightchildren[i],t2->rightchildren[i]);
 }
}
```

```
template <class T>
void tree<T>::delete_subtree(treenode<T> *t1)
{
 int i;
 if(t1==NULL) return;
 for(i=0;i<t1->leftchildren.getsize();i++)
  delete_subtree(t1->leftchildren[i]);
 for(i=0;i<t1->rightchildren.getsize();i++)
  delete_subtree(t1->rightchildren[i]);
 delete t1;
}

template <class T>
treenode<T> *find(treenode<T>* tn,T t)
{
 int i;
 list<treenode<T>*> *l;
 treenode<T> *result=NULL;
 if(tn->data==t) return tn;
 if(t<tn->data) l=&(tn->leftchildren);
 else l=&(tn->rightchildren);
 for(i=0;(i<l->getsize())&&(result==NULL);i++) result=find((*l)[i],t);
 return result;
}

template <class T>
tree<T>::tree(unsigned int n) {root=current=NULL; limit=n;}

template <class T>
tree<T>::tree(const tree &t)
{
 int i;
 stack<treenode<T>*> s;
 treenode<T> *tp;
 root=current=NULL;
 limit=t.limit;
 copy_subtree(root,t.root);
 current=root;
}

template <class T>
tree<T>::~tree()
{
 delete_subtree(root);
}

template <class T>
```

```cpp
tree<T> &tree<T>::operator=(const tree<T> &t)
{
 if(&t==this) return;
 delete_subtree(root);
 copy_subtree(root,t.root);
 current=root;
}

template <class T>
void tree<T>::additem(T t)
{
 additem(t,root);
}

template <class T>
int tree<T>::removeitem();

template <class T>
int tree<T>::find(T t)
{
 treenode<T> *t;
 result=find(root,t);
 if(result!=NULL)
 {
  current=result;
  return 1;
 }
 return 0;
}

template <class T>
void tree<T>::first(void)
{
 current=root;
 while(current->leftchildren.getsize()>0)
 {
  current->leftchildren.first();
  current=current->leftchildren->data();
 }
}

template <class T>
void tree<T>::last(void)
{
 current=root;
 while(current->rightchildren.getsize()>0)
 {
  current->rightchildren.last();
```

```
  current=current->rightchildren->data();
 }
}

template <class T>
void tree<T>::next(void);

template <class T>
void tree<T>::previous(void);

template <class T>
T &tree<T>::data(void)
{
 return current->data;
}
#endif
```

Now the ADT is used in an example program to illustrate the available operations.

```cpp
// treeeg.cpp

#include <iostream>
#include "tree.h"

using namespace std;

void main(void)
{
 int i;
 Tree<int> t;

 t.insert(4);t.insert(1);t.insert(2);t.insert(7);t.insert(5);
 Tree<int> t2(t);
 Tree<int> t3;

 t3=t;
 if(t2==t) cout << "t2==t" << endl;
 if(t3==t) cout << "t3==t" << endl;
 for(i=0;i<t.size();i++)
  cout << "t[" << i << "] = " << t[i] << endl;
 cout<<endl;
}
```

The program output is:

```
t2==t
t3==t
t[0] = 1
t[1] = 2
t[2] = 4
t[3] = 5
t[4] = 7
```

Chapter 10

Error Detection and Correction

10.1 Introduction

Due to external influences and the imperfection of physical devices, errors can occur
in data representation and data transmission. This chapter examines some methods
of limiting the effect of errors in data representation and transmission. Error control
coding should protect digital data against errors which occur during transmission
over a noisy communication channel or during storage in an unreliable memory.
The last decade has been characterized not only by an exceptional increase in data
transmission and storage requirements, but also by rapid developments in micro-
electronics providing us with both a need for, and the possibility to, implement
sophisticated algorithms for error control.

The data representation examined here is strings of bits (binary strings, binary
sequences)
$$a_{n-1}a_{n-2}\ldots a_0 \in B^n$$
where $a_i \in \{0,1\}$ $i = 0, 1, \ldots, n-1$ and

$$B^n = \{0,1\} \times \{0,1\} \times \ldots \times \{0,1\} \; (n \text{ times})$$

as defined before. Therefore an error is a bit flip, i.e. we have $\overline{a_i}$ for some i.

We discuss single bit error detection in the form of parity checks, Hamming codes
for single bit error correction and finally the noiseless coding theorem which describes the limitations of coding systems and the requirements on codes to reduce the
probability of error. Another commonly used error detection scheme, the weighted
checksum, is also discussed.

10.2 Parity Function

In data transmission it is important to identify errors in the transmission. If the probability of error is low enough, or example if we know that the probability of error is $\frac{1}{n}$ then bit strings of length n or longer are unlikely to have more than one error. If this error can be detected the data can be transmitted again until it is transmitted without error. The *parity function* can be used for this purpose. The parity function can be used to detect an odd number of errors in a bit string.

The result of the parity function is a single bit stored in an extra bit a_n, the bit is stored or transmitted with the data. If an odd number of errors occur the result of the parity function over $a_{n-1}a_{n-2}\ldots a_0$ will not concur with a_n. The parity of the bit string must be calculated when the data is sent or stored, and when the data is received or retrieved. The bit reserved for the parity information can take the values 0 or 1. To ensure the meaning of the bit is consistent we introduce the following definitions.

Definition. The *even-parity* function of a bit string is given by:

$$P_{even}(a_{n-1}a_{n-2}\ldots a_0) := a_{n-1} \oplus a_{n-2} \oplus \ldots \oplus a_0.$$

Definition. The *odd-parity* function of a bit string is given by

$$P_{odd}(a_{n-1}a_{n-2}\ldots a_0) := \overline{P_{even}(a_{n-1}a_{n-2}\ldots a_0)}.$$

The odd-parity function sets a_n such that $a_n a_{n-1}\ldots a_0$ has an odd number of 1s. The even-parity function sets a_n such that $a_n a_{n-1}\ldots a_0$ has an even number of 1s. a_n is called the parity bit. Either parity function can be used, but consistency must be ensured so that results are meaningful.

Example. Consider the bit string 1101. $P_{odd}(1101) = 0$. The stored string is then 01101. Suppose an error occurs giving 01001 then $P_{odd}(1001) = 1$ and an error is detected. Suppose an error occurs in the parity bit giving 11101. $P_{odd}(1101) = 0$ and once again an error is detected. If two errors occur, for example 11001, then $P_{odd}(1001) = 1$ and the errors are not detected. ♣

10.3 Hamming Codes

The *Hamming code* [3, 67] is a well-known type of error correction algorithm used for detecting and correcting memory errors. The algorithm was developed by R.W. Hamming and is able to detect single-bit errors and correct them. The algorithm is also able to detect double-bit errors and nibble-bit errors but is not able to correct them. First we have to introduce the Hamming distance.

Definition. The *Hamming distance* d_H of two bit strings $a_{n-1}a_{n-2}\ldots a_0$ and $b_{n-1}b_{n-2}\ldots b_0$ of the same length n is the number of positions that differ, formally

$$d_H(a_{n-1}a_{n-2}\ldots a_0, b_{n-1}b_{n-2}\ldots b_0) := \sum_{i=0}^{n-1}(a_i - b_i)^2.$$

We can easily see that d_H is a metric on B^n. For all $a, b, c \in B^n$ we have

- $d_H(a, b) \geq 0$

- $d_H(a, b) = 0$ iff $a = b$

- $d_H(a, b) = d_H(b, a)$

- $d_H(a, c) \leq d_H(a, b) + d_H(b, c)$

The first three properties are easy to see. The last property follows from the fact that

$$(a - c)^2 = (a - b + b - c)^2 = (a - b)^2 + 2(a - b)(b - c) + (b - c)^2 \leq (a - b)^2 + (b - c)^2$$

for $a, b, c \in \{0, 1\}$.

Example. Let $A = 10111010$ and $B = 01110101$. The Hamming distance is 6.   ♣

The following C++ program calculates the Hamming distance.

```
// hdist.cpp

#include <iostream.h>

void main(void)
{
 unsigned long x = 186; //10111010b
 unsigned long y = 117; //01110101b
 int dH = 0;
```

```
for(int i=8*sizeof(unsigned long)-1; i >= 0 ;i--)
{
  // Add 1 to the Hamming distance if the bit in position
  // i differs for x and y. The AND (&) operator isolates
  // the bit and the XOR (^) operator performs the comparison.
  dH += (((1 << i) & x) ^ ((1 << i) & y)) > 0) ? 1:0;
}

  cout << "dH(" << x << "," << y << ") = " << dH << endl;
}
```

The Hamming distance can be used as a tool for error correction. For a set $C \subset B^n$, of allowable bit strings for data representation, we define the minimum distance

$$\delta(C) := \min_{a,b \in C}\{d_H(a,b)\}.$$

It is then possible to detect up to $\delta(C)$ errors in a bit string from C. The *minimum distance principle for error correction* is to select $c \in C$ for a bit string $x \in B^n$ such that $d_H(c,x)$ is a minimum.

Theorem. If the minimum distance principle for error correction is used and

$$\delta(C) \geq 2e + 1$$

then up to e errors in a bit string from C can be corrected.

Proof. Let a_e be the bit string $a \in C$ with up to e errors. Let $b \in C$ and $b \neq a$ then

$$
\begin{aligned}
d_H(a, a_e) + d_H(a_e, b) &\geq d_H(a,b) \\
e + d_H(a_e, b) &\geq \delta(C) \\
&\geq 2e + 1.
\end{aligned}
$$

Therefore $d_H(a_e, b) \geq e + 1 > e$.   ♠

For $\delta(C) = 3$ only one error can be corrected. C is called a *code* and the elements of C are called *code words*.

Theorem. An upper bound of the number s of code words of length n which can correct up to e errors if the minimum distance principle is used, is given by

$$s \leq \frac{2^n}{\sum_{i=0}^{e}\binom{n}{i}}.$$

Proof. Since the codewords can correct up to e errors we have $d_H(a,b) > e$ for any two codewords a and b. We consider the number of binary sequences of length n

which would be corrected to a specific codeword c. This is simply the number of binary sequences of length n derived from c with up to e errors

$$\sum_{i=0}^{e}\binom{n}{i}.$$

There are s codewords, and a maximum of 2^n possible binary sequences of length n which gives

$$s\sum_{i=0}^{e}\binom{n}{i}\leq 2^n.$$

Thus the bound for s follows.   ♠

For $e = 1$ we find

$$s \leq \frac{2^n}{\sum_{i=0}^{e}\binom{n}{i}} = \frac{2^n}{1+n}.$$

Suppose $n + 1$ is a power of 2, i.e. $n + 1 = 2^m$. The above bound reduces to

$$s \leq 2^{2^m - m - 1}.$$

A *Hamming code* is the best code that can detect and correct one error in the sense that it contains the most code words. Let H_r be an $r \times (2^r - 1)$ matrix with entries $h_{i,j} \in \{0,1\}$, no two columns the same and no zero columns.

Example. For $r = 2$ and $r = 3$ we have

$$H_2 = \begin{pmatrix} 0 & 1 & 1 \\ 1 & 0 & 1 \end{pmatrix}$$

$$H_3 = \begin{pmatrix} 0 & 0 & 0 & 1 & 1 & 1 & 1 \\ 0 & 1 & 1 & 0 & 0 & 1 & 1 \\ 1 & 0 & 1 & 0 & 1 & 0 & 1 \end{pmatrix}$$

♣

The Hamming code is now given by

$$C_{H_r} := \{\, \mathbf{c} \in B^n \mid H_r \mathbf{c} = \mathbf{0} \,\}$$

where we use column representation of bit strings

$$\mathbf{c} := \begin{pmatrix} c_0 \\ c_1 \\ \vdots \\ c_{n-1} \end{pmatrix} \quad \text{and} \quad \mathbf{0} := \begin{pmatrix} 0 \\ 0 \\ \vdots \\ 0 \end{pmatrix}.$$

The Hamming code C_{H_r} has $|C_{H_r}| = 2^{2^r - r - 1}$ code words. Since addition is modulo 2, we find

$$\begin{pmatrix} 1 & 1 \end{pmatrix} \begin{pmatrix} 1 \\ 1 \end{pmatrix} = 1 \oplus 1 = 0.$$

Example. Using the matrices given above, for H_2 and H_3 we find

$$C_{H_2} = \left\{ \begin{pmatrix} 0 \\ 0 \\ 0 \end{pmatrix}, \begin{pmatrix} 1 \\ 1 \\ 1 \end{pmatrix} \right\}$$

$$C_{H_3} = \left\{ \begin{pmatrix} 0 \\ 0 \\ 0 \\ 0 \\ 0 \\ 0 \\ 0 \end{pmatrix}, \begin{pmatrix} 1 \\ 0 \\ 1 \\ 1 \\ 0 \\ 1 \\ 0 \end{pmatrix}, \begin{pmatrix} 1 \\ 0 \\ 0 \\ 0 \\ 0 \\ 1 \\ 1 \end{pmatrix}, \begin{pmatrix} 0 \\ 1 \\ 0 \\ 0 \\ 1 \\ 0 \\ 1 \end{pmatrix}, \begin{pmatrix} 0 \\ 0 \\ 1 \\ 0 \\ 1 \\ 1 \\ 0 \end{pmatrix}, \begin{pmatrix} 0 \\ 0 \\ 1 \\ 1 \\ 0 \\ 0 \\ 1 \end{pmatrix}, \begin{pmatrix} 0 \\ 1 \\ 0 \\ 1 \\ 0 \\ 1 \\ 0 \end{pmatrix}, \begin{pmatrix} 1 \\ 1 \\ 1 \\ 1 \\ 1 \\ 1 \\ 1 \end{pmatrix}, \right.$$

$$\left. \begin{pmatrix} 1 \\ 0 \\ 0 \\ 1 \\ 1 \\ 0 \\ 0 \end{pmatrix}, \begin{pmatrix} 0 \\ 0 \\ 0 \\ 1 \\ 1 \\ 1 \\ 1 \end{pmatrix}, \begin{pmatrix} 1 \\ 1 \\ 1 \\ 0 \\ 0 \\ 0 \\ 0 \end{pmatrix}, \begin{pmatrix} 1 \\ 1 \\ 0 \\ 0 \\ 1 \\ 1 \\ 0 \end{pmatrix}, \begin{pmatrix} 1 \\ 1 \\ 0 \\ 1 \\ 0 \\ 0 \\ 1 \end{pmatrix}, \begin{pmatrix} 0 \\ 1 \\ 1 \\ 0 \\ 0 \\ 0 \\ 1 \end{pmatrix}, \begin{pmatrix} 0 \\ 1 \\ 1 \\ 1 \\ 1 \\ 0 \\ 0 \end{pmatrix}, \begin{pmatrix} 1 \\ 0 \\ 1 \\ 0 \\ 0 \\ 1 \\ 1 \end{pmatrix} \right\}$$

Suppose the bit string 1101010 is received. We test if it is a valid code.

$$H_3 \begin{pmatrix} 0 \\ 1 \\ 0 \\ 1 \\ 0 \\ 1 \\ 1 \end{pmatrix} = \begin{pmatrix} 1 \\ 1 \\ 1 \end{pmatrix}.$$

So it is not a valid code. Assuming at most one error the code word 1101010 must have been $0101010 \in C_{H_3}$. ♣

In the previous example the result of the test was nonzero. The last row determines the even parity of the bits in positions 1, 3, 5 and 7 where the first bit is numbered as 1. The second row determines the even parity of bits 2, 3, 6 and 7 and the first bits 4, 5, 6, 7. For the last row the first bit of all the positions listed is 1. For the second row the second bit in the positions listed is 1. For the first row the third bit in the positions listed is 1. Thus if a bit string fails the test, the resulting bit string can be used to determine the position of the error. This is possible because the columns of H_3 are numerically ascending.

Example. From the above example the test result was 111. The result indicates the error is in the last position, giving the desired code 0101010. ♣

The Hamming code C_{H_r} forms an *abelian group* with group operation

$$a_{2^r-1}a_{2^r-2}\ldots a_0 \oplus b_{2^r-1}b_{2^r-2}\ldots b_0 = (a_{2^r-1} \oplus b_{2^r-1})(a_{2^r-2} \oplus b_{2^r-2})\ldots(a_0 \oplus b_0)$$

where $a_{2^r-1}a_{2^r-2}\ldots a_0, b_{2^r-1}b_{2^r-2}\ldots b_0 \in C$.

For all $a, b, c \in C$

- $a \oplus 0 = a$

- $a \oplus a = 0$, therefore $-a = a$

- $a \oplus (b \oplus c) = (a \oplus b) \oplus c$ due to the associativity of $\oplus$

- $H_r(a \oplus b) = (H_r a) \oplus (H_r b) = 0 \oplus 0 = 0$

♠

10.4   Weighted Checksum

Central to the *weighted checksum* representation is the weight matrix W. It is a $t \times n$ matrix that generates t checksums from a column vector of length n. The weighted checksum representation of a column vector is found by appending these checksums to the end of the vector, making it a separable code. The number of checksums is typically much smaller than the data it is calculated on. So it relies on a probabilistic model to catch most, but not all, errors in the data.

Given a column vector $\mathbf{a} = (a_1 \; a_2 \; \ldots \; a_n)^T$, and a $t \times n$ weight matrix W, the column coded version of $\mathbf{a}$ is

$$\mathbf{a}_c = \begin{pmatrix} \mathbf{a} \\ W\mathbf{a} \end{pmatrix}.$$

Let

$$H = \begin{pmatrix} W & -I_t \end{pmatrix}.$$

An encoded vector $\mathbf{a}_c$ containing valid data is guaranteed to satisfy the equation $H\mathbf{a}_c = 0$, which is seen as follows

$$
\begin{aligned}
H\mathbf{a}_c &= \begin{pmatrix} W & -I_t \end{pmatrix} \begin{pmatrix} \mathbf{a} \\ W\mathbf{a} \end{pmatrix} \\[2mm]
&= (WI_n - I_tW)\,\mathbf{a} \\
&= 0
\end{aligned}
$$

Matrices can be encoded in a similar manner. Each data matrix A has a set of column, row and full weighted checksum matrices A_c, A_r, and A_f.

$$A_c = \begin{pmatrix} A \\ WA \end{pmatrix}$$

$$A_r = \begin{pmatrix} A & AW^T \end{pmatrix}$$

$$A_f = \begin{pmatrix} A & AW^T \\ WA & WAW^T \end{pmatrix}$$

Matrix addition, multiplication, LU decompositions, transpose, and multiplication by a scalar all preserve the weighted checksum property.

10.5 Noiseless Coding Theorem

In this section an overview of Shannon's *noiseless coding theorem* is given following Schumacher [144]. Suppose A is a source of messages $\{a_0, \ldots, a_n\}$ with probabilities $p(a_0), \ldots, p(a_n)$. The probabilities $p(a_0), \ldots, p(a_n)$ satisfy

$$p(a_i) \geq 0 \quad \text{and} \quad \sum_{i=0}^{n} p(a_i) = 1.$$

Definition. The *Shannon entropy* $E_S(A)$ of A is defined by

$$E_S(A) := -\sum_a p(a) \log_2 p(a).$$

The Shannon entropy is also called the missing information.

Example. Let the probabilities for the messages be

$$\left\{ \frac{1}{4}, \frac{1}{16}, \frac{1}{16}, \frac{1}{4}, \frac{1}{8}, \frac{1}{4} \right\}.$$

Thus $p(a_0) = \frac{1}{4}$, $p(a_1) = \frac{1}{16}$, $\ldots$, $p(a_5) = \frac{1}{4}$. Then

$$E_S(A) = -\frac{3}{4} \log_2 \frac{1}{4} - \frac{1}{8} \log_2 \frac{1}{8} - \frac{2}{16} \log_2 \frac{1}{16}$$

$$= 2.375$$

It takes 3 bits to specify which message was received. The value 2.375 can be interpreted as the average number of bits needed to communicate this information. This can be achieved by assigning shorter codes to those messages of higher probability and longer messages to those of lower probability.

First we introduce the weak law of large numbers.

The weak law of large numbers

Let $x_1, x_2, \ldots, x_N$ be N independent, identically distributed random variables, each with mean $\bar{x}$ and finite variance σ. Given $\delta, \epsilon > 0$ then there exists $N_0(\delta, \epsilon)$ such that for $N > N_0$

$$p\left[\left|\frac{1}{N}\sum_{i=1}^{N} x_i - \bar{x}\right| > \delta\right] < \epsilon.$$

Now suppose A produces a sequence of independent messages $\mathbf{a} = a_1 a_2 \ldots a_N$ with probability

$$p(\mathbf{a}) = p(a_1)p(a_2)\ldots p(a_N).$$

Define the random variable $\alpha := -\log_2 p(a)$ for a generated by A with $\bar{\alpha} = E_S(A)$. It follows that for $\delta, \epsilon > 0$ there exists $N_0(\delta, \epsilon)$ such that for $N > N_0$

$$p\left[\left|-\frac{1}{N}\log_2 p(\mathbf{a}) - E_S(A)\right| > \delta\right] < \epsilon$$

with $-\log_2 p(\mathbf{a}) = \sum_{i=1}^{n}\alpha_i$. We assume now $N > N_0$. Define

$$\Gamma := \left\{\mathbf{a} \mid \left|-\frac{1}{N}\log_2 p(\mathbf{a}) - E_S(A)\right| \leq \delta\right\}$$

So with probability greater than $1 - \epsilon$ a sequence $\mathbf{a}$ is in Γ and satisfies

$$-\delta \leq -\frac{1}{N}\log_2 p(\mathbf{a}) - E_S(A) \leq \delta$$

$$2^{-N(E_S(A)-\delta)} \geq p(\mathbf{a}) \geq 2^{-N(E_S(A)+\delta)}$$

Let $\gamma = |\Gamma|$ denote the number of elements in Γ. The bounds of γ are given by

$$1 \geq \sum_{\mathbf{a}\in\Gamma} p(\mathbf{a}) \geq \sum_{\mathbf{a}\in\Gamma} 2^{-N(E_S(A)+\delta)} = \gamma 2^{-N(E_S(A)+\delta)}$$

and

$$1 - \epsilon \leq \sum_{\mathbf{a}\in\Gamma} p(\mathbf{a}) \leq \sum_{\mathbf{a}\in\Gamma} 2^{-N(E_S(A)-\delta)} = \gamma 2^{-N(E_S(A)-\delta)}.$$

Thus we find

$$(1 - \epsilon)2^{N(E_S(A)-\delta)} \leq \gamma \leq 2^{N(E_S(A)+\delta)}.$$

Noiseless coding theorem

Let A be a message source and $\epsilon, \delta > 0$.

1. If $E_S(A) + \delta$ bits are available to encode messages from A then there exists $N_0(\delta, \epsilon)$ such that for all $N > N_0$ sequences of messages from A of length N can be coded into binary sequences with probability of error less than ϵ.

Using the above results we have

$$\gamma \leq 2^{N(E_S(A)+\delta)}$$

and each element of Γ can be encoded uniquely in a bit string of length $E_S(A) + \delta$. The other sequences are encoded as bit strings of length $E_S(A) + \delta$ but will not be correctly decoded. Since these sequences are not in Γ they have probability less than ϵ.

2. If $E_S(A) - \delta$ bits are available to encode messages from A then there exists $N_0(\delta, \epsilon)$ such that for all $N > N_0$ sequences of messages from A of length N are coded into binary sequences with probability of error greater than $1 - \epsilon$.

Let $\lambda, \theta > 0$, $\lambda < \delta$. Then $2^{N(E_S(A)-\delta)}$ sequences of messages from A can be encoded uniquely. The rest will not be correctly decoded. There exists N_0 such that for $N > N_0$,

$$p(\mathbf{a}) \leq 2^{-N(E_S(A)-\lambda)}$$

for $\mathbf{a} \in \Gamma$ and

$$2^{N(\lambda-\delta)} < \frac{\epsilon}{2}.$$

Let p_c denote the probability that the sequence is correctly decoded. Then

$$
\begin{aligned}
p_c &\leq 2^{N(E_S(A)-\delta)} 2^{-N(E_S(A)-\lambda)} \\
&< \theta + 2^{N(E_S(A)-\delta)} 2^{-N(E_S(A)-\lambda)} \\
&= \theta + 2^{N(\lambda-\delta)}
\end{aligned}
$$

So for $\theta = \frac{\epsilon}{2}$, $p_c < \epsilon$ and the probability of error is greater than $1 - \epsilon$.

♠

The *fidelity* F of a coding-decoding scheme is defined to be the probability that a message sequence is decoded correctly, in other words the probability of error is $1 - F$.

10.6   Example Programs

The following C++ program generates the Hamming code of a given length. The
function **increment** takes an array of type **char** as input where the entries are 0 or 1,
and does a binary increment on the entries. The function **genmatrix** generates the
matrix used to generate the Hamming codes, i.e. the matrix with column entries of
0 and 1 and no zero columns. The function **hammingcode** iterates through all binary
codes determining which codes satisfy the criteria of the Hamming code using the
generated matrix.

```cpp
// hamming.cpp

#include <iostream>

using namespace std;

void increment(char *c,int n)
{
 int i,added = 0;
 for(i=0;(i<n) && (!added);i++)
  if(c[i] == 1)
   c[i]=0;
  else
  {
   added = 1;
   c[i] = 1;
  }
}

void genmatrix(char **m,int x)
{
 int i,j;
 char *c = new char[x];
 for(i=0;i < x;i++)
  c[i] = 0;
 for(i=0;i < ((1<<x)-1);i++)
 {
  increment(c,x);
  for(j=0;j < x;j++)
   m[j][i] = c[j];
 }
 delete[] c;
}

void hammingcode(int x)
{
 int size = (1<<x)-1;
```

```
int number = 1<<size;
char *c = new char[size];
char **m = new char*[x];
int h,i,j,sum,iszero;

for(i=0;i < size;i++)
 c[i] = 0;
for(i=0;i < x;i++)
 m[i] = new char[size];
genmatrix(m,x);
for(h=0;h<number;h++)
{
 iszero = 1;
 for(i=0;i < x;i++)
 {
  sum = 0;
  for(j=0;j < size;j++) sum += m[i][j]*c[j];
  if(sum%2 == 1) iszero = 0;
 }
 if(iszero)
 {
  cout << "( ";
  for(i=0;i < size-1;i++)
   if(c[i]) cout << "1" << ", ";
   else cout << "0" << ", ";
  cout << char('0'+c[size-1]) << " )" << endl;
 }
 increment(c,size);
}
for(i=0;i < x;i++) delete[] m[i];
delete[] m;
delete[] c;
}

void main(void)
{
 cout << "Hamming codes of length 3:" << endl;
 hammingcode(2);
 cout << "Hamming codes of length 7:" << endl;
 hammingcode(3);
}
```

The program output is

```
Hamming codes of length 3:
( 0, 0, 0 )
( 1, 1, 1 )
Hamming codes of length 7:
( 0, 0, 0, 0, 0, 0, 0 )
( 1, 1, 1, 0, 0, 0, 0 )
( 1, 0, 0, 1, 1, 0, 0 )
( 0, 1, 1, 1, 1, 0, 0 )
( 0, 1, 0, 1, 0, 1, 0 )
( 1, 0, 1, 1, 0, 1, 0 )
( 1, 1, 0, 0, 1, 1, 0 )
( 0, 0, 1, 0, 1, 1, 0 )
( 1, 1, 0, 1, 0, 0, 1 )
( 0, 0, 1, 1, 0, 0, 1 )
( 0, 1, 0, 0, 1, 0, 1 )
( 1, 0, 1, 0, 1, 0, 1 )
( 1, 0, 0, 0, 0, 1, 1 )
( 0, 1, 1, 0, 0, 1, 1 )
( 0, 0, 0, 1, 1, 1, 1 )
( 1, 1, 1, 1, 1, 1, 1 )
```

which is the same as results calculated earlier in this chapter.

The following C++ program implements a weighted checksum. The function **encode** takes a matrix (2-dimensional array) and a vector (1-dimensional array) as arguments and calculates the vector with checksum information using matrix multiplication. The function **checksum** takes a matrix and a vector as arguments. It determines the matrix for the checksum test, and determines if matrix multiplication with the supplied vector gives the zero vector (the checksum test is satisfied).

```cpp
// checksum.cpp

#include <iostream.h>

void encode(int n,int t,int **w,int *a,int *ac)
{
 int i,j;

 for(i=0;i < n;i++)
  ac[i] = a[i];
 for(i=0;i < t;i++)
 {
  ac[n+i] = 0;
  for(j=0;j < n;j++)
   ac[n+i] += w[i][j]*a[j];
 }
}

int checksum(int n,int t,int **w,int *ac)
{
 int i,j,sum;

 for(i=0;i < t;i++)
 {
  sum = 0;
  for(j=0;j < n;j++)
   sum += w[i][j]*ac[j];
  sum -= ac[n+i];
  if(sum != 0)
   return 0;
 }
 return 1;
}

void main(void)
{
 int data[7] = {3,8,1,7,9,200,5};
 int datac[10];
 int **W = new int*[3];
 int i;
```

```
for(i=0;i<3;i++)
 W[i] = new int[7];

W[0][0] = 1; W[1][0] = 1; W[2][0] = 1;
W[0][1] = 0; W[1][1] = 1; W[2][1] = 1;
W[0][2] = 1; W[1][2] = 0; W[2][2] = 1;
W[0][3] = 0; W[1][3] = 0; W[2][3] = 0;
W[0][4] = 1; W[1][4] = 1; W[2][4] = 0;
W[0][5] = 0; W[1][5] = 1; W[2][5] = 0;
W[0][6] = 1; W[1][6] = 0; W[2][6] = 1;

encode(7,3,W,data,datac);
if(checksum(7,3,W,datac))
 cout << "Checksum satisfied." <<endl;
else
 cout << "Checksum failed." <<endl;

i = datac[4];
datac[4] = 0;
if(checksum(7,3,W,datac))
 cout << "Checksum satisfied." <<endl;
else
 cout << "Checksum failed." <<endl;
datac[4] = i;

i = datac[9];
datac[9] = 0;
if(checksum(7,3,W,datac))
 cout << "Checksum satisfied." <<endl;
else
 cout << "Checksum failed." <<endl;

for(i=0;i<3;i++)
 delete W[i];
delete W;
}
```

Java includes a class `java.util.zip.CRC32` which implements the CRC-32 cyclic redundancy check checksum algorithm. The method

```
void update(byte[] b)
```

in class CRC32 is used to update the CRC-32 calculation when the bytes in the `byte` array are added to the data used to calculate the checksum. The method

```
byte[] getBytes()
```

in class `String` is used to provide the data for the calculation. The method

```
void reset()
```

in clas CRC32 resets the calculation so that the CRC-32 checksum can be calculated with new data. The method

```
long getValue()
```

is used to get the value of the checksum for the given data. If the value is not the expected value then the checksum indicates an error.

```java
// Cksum.java

class Cksum
{
 public static void main(String[] args)
 {
   long csum;
   java.util.zip.CRC32 code;
   String data = "Checksum example";
   String output;

   code = new java.util.zip.CRC32();
   code.update(data.getBytes());
   csum = code.getValue();

   output = "\"" + data + "\"" + " has a CRC32 checksum of ";
   output += Long.toString(csum);
   System.out.println(output);

   code.reset();
   data = "Ch-cksum exmaple";
   code.update(data.getBytes());

   output = "\"" +data + "\"" + " has a CRC32 checksum of ";
   output += Long.toString(code.getValue());
   System.out.println(output);
```

```
  if(csum == code.getValue())
   System.out.println("Checksum satisfied.");
  else
   System.out.println("Checksum failed.");
 }
}
```

The program output is

```
"Checksum example" has a CRC32 checksum of 1413948801
"Ch-cksum exmaple" has a CRC32 checksum of 2843844351
Checksum failed.
```

Chapter 11
Cryptography

11.1 Introduction

Cryptology is the science which is concerned with methods of providing secure storage and transport of information. *Cryptography* can be defined as the area within cryptology which is concerned with techniques based on a secret key for concealing or enciphering data. Only someone who has access to the key is capable of deciphering the encrypted information. In principle this is impossible for anyone else to do. Cryptanalysis is the area within cryptology which is concerned with techniques for deciphering encrypted data without prior knowledge of which key has been used.

Suppose A (the transmitter, normally called Alice) wishes to send a message enciphered to B (the receiver, normally called Bob). Often the original text is simply denoted by M, and the encrypted message by C. A possible method is for A to use a secret key K for encrypting the message M to C, which can then be transmitted and decrypted by B (assuming B possesses the key K). We denote by $C = E_K(M)$ the message M encrypted using the key K, and $M = D_K(C)$ the message C decrypted using the key K. We assume that an attacker (normally called Eve) can easily read any communication between Alice and Bob. The communication method must attempt to send the message in a form which Eve cannot understand and possibly also include authentication of the transmitter and receiver.

11.2 Classical Cypher Systems

We can distinguish between two types of classical cypher systems, namely transposition systems and substitution systems. A transposition cipher is based on changing the sequence of the characters in the message. In other words the enciphered message is a permutation of the original message. A substitution cipher does not change the order of the components of a message but replaces the original components with new components. We give three example programs. In the first example we consider the transposition cipher where the positions of the symbols in a message are rearranged. In the second example example the cipher substitutes one symbol for another, in this case a cyclic substitution is used. The substitution only depends on the symbol being replaced. In the third example a more advanced substitution is performed using the symbol to be replaced and the symbol's position in the message.

Example. The function **transpose** takes a text message m, a permutation p and the size of the permutation l as arguments. If the length **len** of the string m is not a multiple of l the permutation cannot be applied to the last **len%l** bytes of the string, where % is the modulus operator. The function returns without enciphering the text in this case. To overcome this the string could be lengthened with (for example) spaces. To decipher the message the same algorithm can be applied with the inverse permutation.

```
// transpose.cpp

#include <iostream>
#include <string>

using namespace std;

int transpose(string &m,int *p,int l)
{
 int i,j,len;
 char *temp = new char[l];

 len = m.length();
 if(len%l) return 0;
 for(i=0;i < len;i++)
 {
  temp[i%l] = m[l*(i/l)+p[i%l]];
  if((i%l) == l-1)
   for(j=i-l+1;j < i+1;j++) m[j] = temp[j%l];
 }
 delete[] temp;
 return 1;
}

void main(void)
```

```
{
  string m = "A sample message";
  int p1[2] = {1,0}, p1i[2] = {1,0};
  int p2[4] = {3,1,0,2}, p2i[4] = {2,1,3,0};
  int p3[8] = {5,1,7,0,2,3,4,6}, p3i[8] = {3,1,4,5,6,0,7,2};
  cout << "m = " << m << endl;
  transpose(m,p1,2);
  cout << "Enciphering m using p1 = " << m << endl;
  transpose(m,p1i,2);
  cout << "Deciphered using p1i = " << m << endl;
  transpose(m,p2,4);
  cout << "Enciphering m using p2 = " << m << endl;
  transpose(m,p2i,4);
  cout << "Deciphered using p2i = " << m << endl;
  transpose(m,p3,8);
  cout << "Enciphering m using p3 = " << m << endl;
  transpose(m,p3i,8);
  cout << "Deciphered using p3i = " << m << endl;
}
```

The program output is

```
m = A sample message
Enciphering m using p1 =  Aaspmelm seaseg
Deciphered using p1i = A sample message
Enciphering m using p2 = a Asepmlsm eeasg
Deciphered using p2i = A sample message
Enciphering m using p3 = p eAsamlame essg
Deciphered using p3i = A sample message
```

A keyword may be provided with the message to derive the permutation. For example the permutation may be specified by arranging the letters of the first word in alphabetical order. For example if the reference word is "word" and is placed at the beginning of the message as "dowr" the permutation can be inferred to be p2 in the above example, and the rest of the message can be deciphered.

In this case the permutation serves as the key. There are $N!$ permutations of length N. The identity permutation is not of any use so the total number of useful keys are $N! - 1$.

A substitution cipher is based on replacing components of the message. For example, we can replace characters in a text message with other characters. A one-to-one into mapping serves this purpose. A simple substitution cipher is the Caesar substitution. A cyclic shift of the alphabet is used. In other words if 'A' is the first letter, 'B' the second letter and so on, then the n-th letter maps to the $((n+k) \bmod 26)$-th letter, where k is an integer which defines the substitution.

Example. In this example the function substitute takes the message m to encipher, and the number n by which to shift the alphabet. The substitution is only applied to the letters 'A'-'Z' and 'a'-'z'.

```cpp
// substitute.cpp

#include <iostream>
#include <string>

using namespace std;

void substitute(string &m,int n)
{
  int i,l;

  l = m.length();
  while(n < 0) n += 26;
  for(i=0;i < l;i++)
   if((m[i] >= 'A')&&(m[i] <= 'Z'))
    m[i] = (m[i]-'A'+n)%26+'A';
   else if((m[i] >= 'a')&&(m[i] <= 'z'))
    m[i] = (m[i]-'a'+n)%26+'a';
}

void main(void)
{
  string m = "A sample message";
  cout << "m = " << m << endl;
  substitute(m,1);
  cout << "Caesar cipher with n=1 = " << m << endl;
  substitute(m,-1);
  substitute(m,-1);
  cout << "Caesar cipher with n=-1 = " << m << endl;
  substitute(m,1);
  substitute(m,10);
  cout << "Caesar cipher with n=10 = " << m << endl;
  substitute(m,-10);
  cout << "m = " << m << endl;
}
```

The program output is

```
m = A sample message
Caesar cipher with n=1 = B tbnqmf nfttbhf
Caesar cipher with n=-1 = Z rzlokd ldrrzfd
Caesar cipher with n=10 = K ckwzvo wocckqo
m = A sample message
```

If each alphabet is viewed as a key then there are only 26 keys. The first alphabet is the one we already use, so 25 useful keys are left. If permutations of the alphabet are used instead of only shifts a total of 26! − 1 useful keys are available.

A more advanced substitution is obtained using a *Vigenère table*. The substitution rule changes with each position in the message. Each symbol is used as an index for the column of the table. A keyword is repeated below the message. The symbol in the keyword string is used as an index for the row of the table. The table has the standard alphabet as the first row and the previous row shifted left for each row following. In other words the first row is 'A' 'B' 'C' ..., the second row is 'B' 'C' 'D' ... and so on.

A word can be used for a key to identify for each symbol to encode which row of the Vigenère table to use. For example, the word "CIPHER" indicates that the third, ninth, sixteenth, eighth, fifth and eighteenth rows are to be used for enciphering. Thus the symbol at position i is encoded using the row identified by the symbol in the $i \bmod l$ position of the key word, where l is the number of symbols in the key word.

Example. We modify the previous program to use the Vigenère table and a keyword. The function **vigenere** takes three arguments. The argument **decipher** determines if the function enciphers or deciphers the message. The argument **m** is the message to be enciphered, and **k** is used as the index for the row in the Vigenère table.

```cpp
// vigenere.cpp

#include <iostream>
#include <string>

using namespace std;

void vigenere(string &m,string k,int decipher)
{
  int i,l,n;

  n = k.length();
  l = m.length();
  for(i=0;i < l;i++)
    if((m[i] >= 'A')&&(m[i] <= 'Z'))
      if(decipher)
        m[i] = (m[i]-'A'+26-(k[i%n]-'A'))%26+'A';
```

```
    else
      m[i] = (m[i]-'A'+k[i%n]-'A')%26+'A';
   else if((m[i]>='a')&&(m[i]<='z'))
     if(decipher)
       m[i] = (m[i]-'a'+26-(k[i%n]-'A'))%26+'a';
     else
       m[i] = (m[i]-'a'+k[i%n]-'A')%26+'a';
}

void main(void)
{
 string m = "A sample message";
 string k = "CIPHER";
 cout << "m = " << m << endl;
 vigenere(m,"CIPHER",0);
 cout << "Cipher with Vigenere table and keyword CIPHER = "
      << m << endl;
 vigenere(m,"CIPHER",1);
 cout << "m = " << m << endl;
}
```

The program output is

```
m = A sample message
Cipher with Vigenere table and keyword CIPHER = C hhqgnm tijuivl
m = A sample message
```

11.3 Public Key Cryptography

In a public key system two keys are used, one for enciphering and one for deciphering the message. A system which relies on a public key and a private key is called an asymmetrical cipher system. These systems rely on a function to be easy to calculate but the inverse function is difficult to calculate without extra information.

The RSA system is a well known public key system. It uses the fact that the product of two prime numbers is easy to calculate, but to factor the product into the two prime numbers is difficult. First two prime numbers p and q are generated, and the product $n = pq$ calculated. Then e is determined as follows.

$$3 < e < (p-1)(q-1), \quad gcd(e, (p-1)(q-1)) = 1.$$

Lastly d must be determined from

$$ed = 1 \mod (p-1)(q-1).$$

Suppose we have a message with non-negative integer value M. The ciphered message is represented by

$$C = M^e \mod n.$$

The message is deciphered as follows

$$M = C^d \mod n.$$

Definition. *Euler's totient function* $\varphi(n)$ is the number of positive integers smaller than n and relatively prime to n. For a prime number p we have $\varphi(p) = p - 1$. Thus for $\varphi(n)$ we find

$$\varphi(n) = \varphi(p)\varphi(q) = (p-1)(q-1)$$

where $n = pq$ as given above.

Theorem. For all $a, n \in \mathbf{N}$ with $0 < a < n$ and $gcd(a, n) = 1$

$$a^{\varphi(n)} = 1 \mod n.$$

The theorem is called *Euler's theorem.* For the proof we refer to [171]. The theorem is of interest because it can be used to prove that encipherment and decipherment using the RSA system are inverse operations. In other words if we have a message M enciphered

$$C = M^e \mod n$$

and deciphered according to

$$M' = C^d \mod n$$

with $ed = 1 \mod n$ then $M' = M$. Again we refer to [171].

The public key in this system is (e, n) and the private key d. The method can be improved to include verification of the sender and remove transport of the private key from the sender to the receiver. Suppose the sender has a public key (e_1, n_1) and a private key d_1. Similarly suppose the receiver has public key (e_2, n_2) and private key d_2. Let the message to be encoded be M. Thus an encoded message would be

$$C = E_{(e_2, n_2)}(E_{d_1}(M)).$$

In other words the sender encodes the message using a private key and then using the public key of the receiver. The receiver can decode the message using

$$M = D_{(e_1, n_1)}(D_{d_2}(C)).$$

The receiver decodes the message by first using a private key and then using the public key of the sender. Using this method only public keys are exchanged. Since the receiver can only decode the message using the sender's public key the message source can be verified.

The RSA system relies on the fact that two large prime numbers p and q can be found. It is generally quite slow to check if numbers are prime, since the obvious method is check for any factors. Define the *Jacobi symbol* as follows

$$J(1, p) := 1$$

$$J(a, p) := \begin{cases} (-1)^{(p^2-1)/8} J(a/2, p) & a \text{ even} \\ (-1)^{(a-1)(p-1)/4} J(p \bmod a, a) & a \text{ odd} \end{cases}$$

Suppose we wish to test if p is prime, select $a \in \{1, 2, \ldots, p-1\}$ and calculate $gcd(a, p)$. If

$$gcd(a, p) \neq 1$$

then p is not a prime number. Otherwise if

$$J(a, p) \neq a^{(p-1)/2} \bmod p$$

p is not prime. If p is prime then

$$gcd(a, p) = 1$$

and

$$J(a, p) = a^{(p-1)/2} \bmod p$$

for all $a \in \{1, 2, \ldots, p-1\}$. If p is not prime then the test will fail in more than 50% of the cases. Every time a is successfully tested the probability that p is a prime number increases.

First the prime numbers must be generated to implement the algorithm. To perform faster encryption a table of prime numbers is used. The prime numbers are generated with the following C++ program, and then can be read by a program which needs prime numbers. The program takes one parameter on the command line to indicate how many prime numbers to generate. The program output is a list of prime numbers which can be used in other C++ programs. The standard error output stream is used to output how many prime numbers have been found. The program output can be redirected in UNIX and Windows systems with the command

```
genprime 10000 > primes.dat
```

which generates 10000 prime numbers and places them in the file primes.dat. The header file list.h contains the implementation of the ADT list class developed earlier.

```cpp
// gprime.cpp

#include <iostream>
#include <ctype.h>
#include <stdlib.h>
#include <math.h>
#include "list.h"

using namespace std;

typedef unsigned long type;

int main(int argc,char *argv[])
{
  list<type> l;
  int i,j,count,success;
  type n(5),sn;

  if(argc == 1) return 1;
  count = atoi(argv[1]);
  l.additem(2); l.additem(3);
  cout << count << endl;

  for(i=0;i < count-1;n+=type(2))
  {
    success = 1;
    sn = (type)(sqrt(n)+1);
    for(j=0;success&&(j<l.getsize())&&(l[j]<sn);j++)
     if((n%l[j]) == type(0)) success = 0;
    if(success)
    {
      l.additem(n);
      cout << n << endl;
```

```
   cerr << i << "              \r";
   i++;
 }
}
for(;i < count;n+=type(2))
{
 success = 1;
 sn = (type)(sqrt(n)+1);
 for(j=0;success&&(j<l.getsize())&&(l[j]<sn);j++)
  if((n%l[j]) == type(0)) success = 0;
 if(success)
 {
  l.additem(n);
  cout << n << endl;
  cerr << i << "              \r";
  i++;
 }
}
cerr << endl;
return 0;
}
```

Similarly the program **gkeys.cpp** generates an array of key values using the prime
numbers generated in **gprime.cpp**. The RSA program can then simply use an index
to specify the key. The generation of prime numbers and keys takes a long time,
it is much faster to do the long calculations once and then just use precalculated
results in the algorithm.

```
// gkeys.cpp

#include <fstream>
#include <stdlib.h>
#include <time.h>

using namespace std;

typedef unsigned long type;

type primelist(int i)
{
 type data;
 int j;
 ifstream primes("primes.dat");

 primes >> j;
 for(j=0;(j<=i)&&!primes.eof()&&!primes.fail();j++)
  primes >> data;
 primes.close();
```

```
 return data;
}

type GCD(type a,type b)
{
 type r(1);

 while(r != type(0))
 {
  r = a%b;
  if(r != type(0)) { a = b; b = r; }
 }
 return b;
}

void main(int argc,char *argv[])
{
 int i,j,count,maxprime,maxkeys = 0;
 int total;
 ifstream primes("primes.dat");

 primes >> maxprime;
 total = int((double(maxprime)*maxprime-1)/2);
 primes.close();

 if(argc == 1) return 1;
 count = atoi(argv[1]);
 if(count > total) count = total;
 cout << count << endl;

 srand(time(NULL));
 for(i=0;maxkeys<=count&&i<maxprime;i++)
  for(j=i+1;maxkeys<count&&j<maxprime;j++)
  {
   type temp,temp2,p,q,n,e,d;

   p = primelist(i);
   q = primelist(j);
   n = p*q;
   temp = (p-type(1))*(q-type(1));
   d = e = type(0);

   for(p=type(4);p < temp;p++)
    if(GCD(p,temp) == type(1))
     {
      e = p;
      for(q=type(1);p != temp;q++)
       if(((q*temp+1)%e) == 0)
```

```
      { d = (q*temp+1)/e;  p = temp; }
    }
  if((e != type(0))&&(d != type(0)))
  {
   maxkeys++;
   cout << n << " ";
   cout << e << " ";
   cout << d << endl;
  }
  cerr << (total--) << " left to try, " << maxkeys
       << " generated                \r";
  cerr.flush();
  }
  if(maxkeys<count) cout << "Not enough keys generated.";
  cerr << endl;
}
```

In the following program it is important to use the class **Verylong** [169], which
provides a theoretically unbounded integer type, since for even small prime numbers
$< 2^{16}$ the calculations used in the RSA system can exceed the bounds of the data
type **unsigned long** depending on the underlying hardware platform. The program
performs the RSA encoding of a message using the previously generated keys. We
again use a recursive implementation for raising an integer to an integer power, this
time using **Verylong** and modulo arithmetic.

```
// rsa.cpp

#include <fstream>
#include <stdlib.h>
#include <time.h>
#include <assert.h>
#include "verylong.h"

using namespace std;

void keylist(int i,Verylong &n,Verylong &e,Verylong &d)
{
 int j;
 ifstream keys("keys.dat");

 keys >> j;
 for(j=0;(j<=i)&&!keys.eof()&&!keys.fail();j++)
  { keys >> n; keys >> e; keys >> d;}
 keys.close();
}


Verylong powermodn(Verylong a,Verylong n,Verylong mod)
```

```
{
 Verylong temp;

 if(n == Verylong(0)) return Verylong(1);
 a %= mod;
 temp = powermodn(a,n/Verylong(2),mod);
 temp = (temp*temp)%mod;
 if(n%Verylong(2) == Verylong(1)) return (a*temp)%mod;
 return temp;
}

void rsa(Verylong *m,Verylong e,Verylong n,int len)
{
 int i;
 for(i=0;i < len;i++)
  m[i] = (powermodn(m[i],e,n));
}

void vltoc(Verylong *l,char *c,int n)
{
 int i;
 for(i=0;i <n ;i++) c[i] = char(int(l[i]));
}

void ctovl(char *c,Verylong *l,int n)
{
 int i;
 for(i=0;i < n;i++) l[i] = Verylong((unsigned)c[i]);
}

void main(void)
{
 int i,len,maxkeys;
 Verylong e,d,n;
 char m[18];
 Verylong mt[17];
 ifstream keys("keys.dat");

 keys >> maxkeys; keys.close();

 srand(time(NULL));
 keylist(rand()%maxkeys,n,e,d);

 strcpy(m,"A sample message");
 len = strlen(m);
 ctovl(m,mt,len);

 cout << "Initial message : " << endl;
```

```
  for(i=0;i < len-1;i++) cout << mt[i] << ",";
  cout << mt[i] << endl;
  cout << m << endl << endl;

  rsa(mt,e,n,len);

  cout << "Encrypted message : " << endl;
  for(i=0;i<len-1;i++) cout << mt[i] << ",";
  cout << mt[i] << endl << endl;

  rsa(mt,d,n,len);
  vltoc(mt,m,len);

  cout << "Decrypted message : " << endl;
  for(i=0;i<len-1;i++) cout << mt[i] << ",";
  cout << mt[i] << endl;
  cout << m << endl;
}
```

The program output is

Initial message :
65,32,115,97,109,112,108,101,32,109,101,115,115,97,103,101
A sample message

Encrypted message :
696340,554727,635395,510042,702669,39492,737693,78176,554727,702669,
78176,635395,635395,510042,635068,78176

Decrypted message :
65,32,115,97,109,112,108,101,32,109,101,115,115,97,103,101
A sample message

Chapter 12
Finite State Machines

12.1 Introduction

Finite state machines [49, 67] provide a visual representation of algorithms. Algorithms are implemented on a machine with a finite number of states representing the state of the algorithm. This provides an abstract way of designing algorithms. The chapter will only cover *deterministic* machines (the actions of the machines are determined uniquely).

The reason for studying these machines is to determine what are necessary requirements to be able to perform arbitrary functions. Certain machines (as will be illustrated) cannot perform certain functions. Computer scientists are interested in the requirements for functions to be performed and what functions can be performed. Finite state machines can be used to understand these problems. Finite state machines are concerned with taking an input, changing between internal states, and generating an output (which may just be the machine's final state). This describes all computing devices. Thus in an abstract way it is possible to consider what is computable. Any machine required to solve arbitrary problems must be described in terms of a basic set of features and operations which determine what the machine can do. From the description, algorithms to solve problems can be constructed. Furthermore the basic operations must be reasonable in the sense that it must be known that the operations can be performed in a finite amount of time. The features of a machine can, for example, be memory and the ability to output.

In this chapter we discuss finite automata, finite automata with output and Turing machines. It will become evident that with each improvement the machines can compute more. We show some problems which are computable by Turing machines and not by finite automata. Turing machines are used as the basis for deciding what is computable and what is not.

12.2   Finite Automata

In this section we discuss a simple machine type and consider some computations
these machines can perform. The basic operations are transitions between states.
The features are states which serve as memory.

Definition. A *finite automaton* consists of

- A finite set S of states. One state is designated as the *start state*. Some states
  may be designated as *final states*.

- An *alphabet* Σ of possible input symbols.

- A finite set of *transitions* for each state and symbol in the alphabet. Transitions are ordered triples (a, b, c) where $a, b \in S$ and $c \in \Sigma$, and b is uniquely
  determined by a and c.

An input string of elements of Σ is provided to the finite automaton. The finite
automaton reads each symbol in the string and causes a transition between states.
(a, b, c) represents a transition from state a to state b when the symbol c is read. If
all symbols of the string have been read and the finite automaton is not in the final
state then the finite automaton is said to *fail* and the input string is not *accepted*.
The finite automaton also fails if no transition exists for a symbol read. If the finite
automaton has not failed and all symbols in the input string have been read the
finite automaton terminates successfully and the input string is accepted.

Visually the finite automaton can be represented with circles for the states and
directed edges between states for the transitions. This visual representation is called
a *transition diagram*. A "−" in a state denotes the start state. A "+" in a state
denotes a final state.

Finite automata can be used to define languages. The language consists of all input
words accepted by the finite automata. The automaton can only accept input words,
it cannot provide any output except for failing or accepting. The only memory the
finite automaton possesses is the state it is currently in and its transitions. This
is obviously a limitation. More effective computing machines such as push-down
automata (using a stack as memory) and Turing machines can increase the number
of computable functions.

Now we provide some examples to show some of the uses of finite automata.

Example. We can use a finite automaton to perform a parity check. Let

$$S := \{S_{odd}, S_{even}\} \qquad \text{and} \qquad \Sigma := \{0, 1\}.$$

The start state is S_{even} and the final state S_{odd}. The table for the transitions is given by Table 12.1. The transition diagram is shown in Figure 12.1.

State	Input Symbol	Next State
S_{odd}	0	S_{odd}
S_{odd}	1	S_{even}
S_{even}	0	S_{even}
S_{even}	1	S_{odd}

Table 12.1: Parity Check Finite Automaton – Transitions

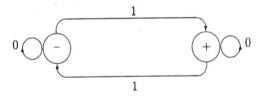

Figure 12.1: Parity Check Finite Automaton

The finite automaton only accepts bit strings which pass the odd parity test. If S_{even} were selected as the final state instead of S_{odd} the finite automaton would only accept bit strings which pass the even parity test. Note that it is not necessary to label the states in the transition diagram since this does not change the operation of the finite automata.   ♣

Example. Consider the finite automaton with $\Sigma := \{0,1\}$,

$$S := \{S_{Start}, S_{0,1}, S_{0,2}, S_{0,3}, S_{1,1}, S_{1,2}, S_{1,3}, S_{NA}\}$$

with the start state S_{Start} and final states $S_{0,3}$ and $S_{1,3}$, and transition Table 12.2. The transition diagram is given by Figure 12.2.

State	Input	NextState
S_{Start}	0	$S_{0,1}$
S_{Start}	1	$S_{1,1}$
$S_{0,1}$	0	$S_{0,2}$
$S_{0,1}$	1	S_{NA}
$S_{0,2}$	0	$S_{0,3}$
$S_{0,2}$	1	S_{NA}
$S_{0,3}$	0	S_{NA}
$S_{0,3}$	1	S_{NA}
$S_{1,1}$	0	S_{NA}
$S_{1,1}$	1	$S_{1,2}$
$S_{1,2}$	0	S_{NA}
$S_{1,2}$	1	$S_{1,3}$
$S_{1,3}$	0	S_{NA}
$S_{1,3}$	1	S_{NA}
S_{NA}	0	S_{NA}
S_{NA}	1	S_{NA}

Table 12.2: Hamming Code Finite Automaton – Transitions

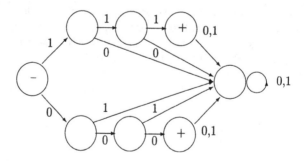

Figure 12.2: Hamming Code Finite Automaton

This finite automaton only accepts code words from the Hamming code C_{H_2}. ♣

12.3   Finite Automata with Output

Now we extend the abilities of the machine by letting it output symbols. This allows the machine to do something. The output may be used by other devices and more input may be generated so a machine that reacts to its environment can be constructed. Two extensions to finite automata that achieve this are Moore machines and Mealy machines.

Definition. A *Moore machine* consists of

- A finite set S of states. One state is designated as the *start state*.

- An *alphabet* Σ of possible input symbols.

- An *alphabet* Γ of possible output symbols.

- A finite set of *transitions* for each state and symbol in the alphabet Σ. Transitions are ordered triples (a, b, c) where $a, b \in S$ and $c \in \Sigma$, and b is uniquely determined by a and c.

- For each state the symbol from Γ to output when the state is entered.

The transition diagrams already introduced can be extended for Moore machines by writing the output symbol in the circle for each state. Unlike finite automata, a Moore machine does not accept or reject input strings, rather it processes them. If S is a state in a Moore machine then the notation $S-$ denotes the fact that S is a start state.

This machine has only the memory of which state it is in and its transitions. In this respect it is no more powerful than a finite automaton. But its relation to practical usage is stronger since now the machine is able to give us information about the input provided, beyond a simple accept or fail. The ability to output is also tied to memory. If a machine can read its own output at a later stage it may be able to compute more. These ideas are incorporated into the Turing machines.

These machines can be coupled so that the output of one machine can be used as input for another. A Moore machine exists for any pair of coupled Moore machines. The set of states for such a machine is the Cartesian product of the sets of states of each of the machines. Let $S_1 = \{s_{1,0}, s_{1,1}, \ldots\}$ and $S_2 = \{s_{2,0}, s_{2,1}, \ldots\}$, where $s_{1,0}$ and $s_{2,0}$ are the start states, be the states of the first and second Moore machines respectively. Let the output for $s_{i,j}$ be denoted by $o_{i,j}$. The machine with states $S_1 \times S_2$, start state $(s_{1,0}, s_{2,0})$, output $o_{2,j}$ for state $(s_{1,i}, s_{2,j})$, and transitions $(s_{1,i}, s_{2,j}) \to (s_{1,k}, s_{2,l})$ if the $s_{1,i} \to s_{1,k}$ is a transition for some input for the first machine and $s_{2,j} \to s_{2,l}$ is a transition for input $o_{1,k}$ in the second machine. Thus combining Moore machines provides no extra computing power to this class of machines.

Example. Table 12.3 describes a simple Moore machine which performs the NOT operation. Here

$$S := \{S_0, S_1, S_2\}, \quad \Sigma := \{0, 1\}, \quad \Gamma := \{0, 1\}.$$

State	Output	Input	Next State
S_0-	0	0	S_1
		1	S_2
S_1	1	0	S_1
		1	S_2
S_2	0	0	S_1
		1	S_2

Table 12.3: Moore Machine for the NOT Operation – Transitions

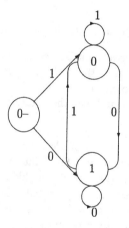

Figure 12.3: Moore Machine for the NOT Operation

♣

Example. This example shows how an n-bit incrementer which increments any n-bit number modulo 2^n ($2^n \equiv 0$ modulo 2^n) can be implemented with a Moore machine. The bits are fed in from low order to high order. For example the decimal number 11 with bit representation 1011 will be input as 1,1,0 and then 1. The transition table is given by Table 12.4. Here

$$S := \{S_0, S_1, S_2, S_3, S_4\}, \quad \Sigma := \{0, 1\}, \quad \Gamma := \{0, 1\}.$$

The transition diagram is given by Figure 12.4.

State	Output	Input	Next State
S_0-	0	0	S_1
		1	S_2
S_1	1	0	S_3
		1	S_4
S_2	0	0	S_1
		1	S_2
S_3	0	0	S_3
		1	S_4
S_4	1	0	S_3
		1	S_4

Table 12.4: n-bit Incrementer Moore Machine – Transitions

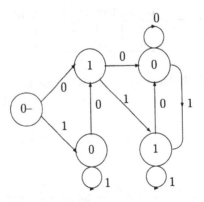

Figure 12.4: n-bit Incrementer Moore Machine

Definition. A *Mealy machine* consists of

- A finite set S of states. One state is designated as the *start state*.

- An *alphabet* Σ of possible input symbols.

- An *alphabet* Γ of possible output symbols.

- A finite set of *transitions* for each state and symbol in the alphabet Σ. Transitions are ordered triples (a, b, c) where $a, b \in S$ and $c \in \Sigma$, and b is uniquely determined by a and c.

- For each transition the symbol from Γ to output.

The transition diagrams already introduced can be extended for Mealy machines by writing the input and output symbols as an ordered pair (i, o) for each transition. Unlike finite automata a Mealy machine does not accept or reject input strings rather it processes them. Similarly to Moore machines, Mealy machines can be combined. Using a similar proof to the one for Moore machines, the combination of Mealy machines provides no extra computing power.

Example. Table 12.5 describes a simple Mealy machine which performs the NOT operation. Here

$$S := \{S_0\}, \quad \Sigma := \{0, 1\}, \quad \Gamma := \{0, 1\}.$$

The transition diagram is given in Figure 12.5.

State	Input	Output	Next State
S_0-	0	1	S_0
	1	0	S_0

Table 12.5: Mealy Machine for the NOT Operation – Transitions

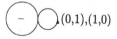

Figure 12.5: Mealy Machine for the NOT operation

♣

Example. This example shows how an n-bit incrementer which increments any n-bit number modulo 2^n ($2^n \equiv 0$ modulo 2^n) can be implemented with a Mealy machine. The bits are fed in from low order to high order. For example the number 11 with bit representation 1011 will be input as 1,1,0 and then 1. The transition table is given by Table 12.6. Here

$$S := \{S_0, S_1, S_2\}, \quad \Sigma := \{0, 1\}, \quad \Gamma = \{0, 1\}.$$

State	Input	Output	Next State
S_0-	0	1	S_1
	1	0	S_2
S_1	0	0	S_1
	1	1	S_1
S_2	0	1	S_1
	1	0	S_2

Table 12.6: n-bit Incrementer Mealy Machine – Transitions

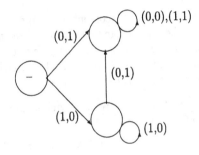

Figure 12.6: n-bit Incrementer Mealy Machine

♣

For every Moore machine there is an equivalent Mealy machine and conversely. For the proof we refer to [49]. This is simply a matter of showing how to gain the same output for Moore and Mealy machines with the same input.

12.4   Turing Machines

Turing machines are more powerful than the finite automata discussed in the previous section because they have memory.

Definition. A *Turing machine* consists of

- A finite set of states S one of which is designated the *start* state. Some states may be designated as *halt* states which cause the Turing machine to terminate execution.

- An *alphabet* Σ of possible input symbols.

- An *alphabet* Γ of possible output symbols.

- The *blank* symbol Δ.

- A *tape* or memory device which consists of adjacent *cells* labelled

$$cell[0],\ cell[1],\dots.$$

  Cells of the tape can contain a single symbol from $\Sigma \cup \Gamma \cup \{\Delta\}$. The input string is placed in the first cells of the tape, the rest of the cells are filled with Δ.

- A *tape head* that can read the contents of a tape cell, put a symbol from Γ or the Δ symbol in the tape cell and move one cell right or left. All these actions take place simultaneously. If the tape head tries to move left from $cell[0]$ the Turing machine is said to *crash*. If the head is at $cell[i]$ and moves left (right) then the head will be at $cell[i-1]$ ($cell[i+1]$).

- A finite set of *transitions* for states and symbols from $\Sigma \cup \Gamma \cup \{\Delta\}$. A transition is an ordered 5-tuple (a,b,c,d,e) with

$$a \in S, \quad b \in \Sigma \cup \Gamma \cup \{\Delta\}, \quad c \in S, \quad d \in \Gamma \cup \{\Delta\}$$

  and $e \in \{r,l\}$. Here a is the current state, b is the symbol read by the tape head, c is the next state, d is the symbol for the tape head to write in the current cell and $e = r$ ($e = l$) moves the tape head right (left). The elements c, d and e are uniquely determined by a and b. If an input symbol is read and no transition corresponds to the current state and symbol read the Turing machine is said to *crash*.

Input strings of symbols from Σ which cause the Turing machine to end on a halt state are said to be *accepted* by the Turing machine. Graphically states are represented with circles and transitions with directed edges between states labelled with a triple (a, b, c) where a is the symbol read from the tape, b is the symbol to write to the tape and c is the direction to move the tape head(l or r). A "$-$" in a state will represent a start state and a "$+$" in a state will represent a halt state. Obviously Turing machines can do at least as much as Mealy machines (and therefore also Moore machines) and finite automata.

Example. We can use a Turing machine to perform the parity check. Let

$$S := \{S_{odd}, S_{even}, S_{fin}\}$$

and

$$\Sigma := \{0,1\}, \quad \Gamma := \{0,1\}.$$

The start state is S_{even} and halt state S_{fin}. The table for the transitions is given by Table 12.7. The symbol r in the movement column instructs the tape head to move one cell right. The transition diagram is given by Figure 12.7.

State	Input Symbol	Output Symbol	Movement	Next State
S_{odd}	0	0	r	S_{odd}
S_{odd}	1	1	r	S_{even}
S_{even}	0	0	r	S_{even}
S_{even}	1	1	r	S_{odd}
S_{odd}	Δ	Δ	r	S_{fin}

Table 12.7: Parity Check Turing Machine – Transitions

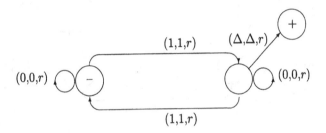

Figure 12.7: Parity Check Turing Machine

The Turing machine only accepts bit strings which pass the odd parity test. Note that it is not necessary to label the states in the transition diagram since this does not change the operation of the Turing machine.   ♣

Example. Now we use a Turing machine to calculate the parity bit for odd parity and place it in the cell of the tape immediately after the bit string used for input. Let

$$S := \{S_{odd}, S_{even}, S_{fin1}, S_{fin2}\}$$

and

$$\Sigma := \{0, 1\}, \quad \Gamma = \{0, 1\}.$$

The start state is S_{even} and halt states S_{fin1} and S_{fin2}. The table for the transitions is given by Table 12.8. The transition diagram is given by Figure 12.8.

State	Input Symbol	Output Symbol	Movement	Next State
S_{odd}	0	0	r	S_{odd}
S_{odd}	1	1	r	S_{even}
S_{odd}	Δ	0	r	S_{fin1}
S_{even}	0	0	r	S_{even}
S_{even}	1	1	r	S_{odd}
S_{even}	Δ	1	r	S_{fin2}

Table 12.8: Parity Calculation Turing Machine Transitions

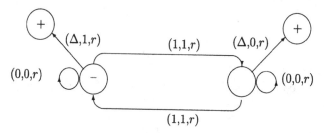

Figure 12.8: Parity Calculation Turing Machine

Example. Now we use a Turing machine to negate (NOT) a bit sequence (one's complement). The states are

$$S = \{S_{start}, S_{halt}\}.$$

S_{start} is the start state and S_{halt} is a halt state. The alphabets are

$$\Sigma := \{0,1\}, \quad \Gamma := \{0,1\}.$$

The transition table is given by Table 12.9. The transition diagram is given by Figure 12.9.

State	Input	Output	Movement	Next State
S_{start}	0	1	r	S_{start}
S_{start}	1	0	r	S_{start}
S_{start}	Δ	Δ	r	S_{halt}

Table 12.9: Turing Machine for the NOT Operation – Transitions

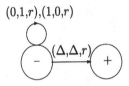

$(0,1,r),(1,0,r)$

(Δ,Δ,r)

Figure 12.9: Turing Machine for the NOT Operation

♣

Example. Now we consider a Turing machine which has no finite automaton equivalent. The Turing machine reverses a bit string. The states are

$$S := \{ S_{start}, S_{halt}, S_{0,1}, S_{0,2}, S_{0,3}, S_{1,1}, S_{1,2}, S_{1,3} \}$$

$\Sigma := \{0_I, 1_I\}$ and $\Gamma := \{0_O, 1_O, 0_I, 1_I\}$. S_{start} is the start state and S_{halt} is the halt state. The input and output alphabet are different so that the machine can differentiate between input and output symbols. The input and output will be interpreted as binary digits but using different alphabets means the machine can remember what it has already done. The transitions are given by Table 12.10. The transition diagram is given by Figure 12.10.   ♣

State	Input	Output	Movement	Next State
S_{start}	0_O	0_O	r	S_{halt}
	1_O	1_O	r	S_{halt}
	0_I	Δ	r	$S_{0,1}$
	1_I	Δ	r	$S_{1,1}$
$S_{0,1}$	0_I	0_I	r	$S_{0,1}$
	1_I	1_I	r	$S_{0,1}$
	0_O	0_O	l	$S_{0,2}$
	1_O	1_O	l	$S_{0,2}$
	Δ	Δ	l	$S_{0,2}$
$S_{0,2}$	0_I	0_O	l	$S_{0,3}$
	1_I	0_O	l	$S_{1,3}$
$S_{0,3}$	0_I	0_I	l	$S_{0,3}$
	1_I	1_I	l	$S_{0,3}$
	0_O	0_O	l	$S_{0,3}$
	1_O	1_O	l	$S_{0,3}$
	Δ	0_O	r	S_{start}
$S_{1,1}$	0_I	0_I	r	$S_{1,1}$
	1_I	1_I	r	$S_{1,1}$
	0_O	0_O	l	$S_{1,2}$
	1_O	1_O	l	$S_{1,2}$
	Δ	Δ	l	$S_{1,2}$
$S_{1,2}$	1_I	1_O	l	$S_{1,3}$
	0_I	1_O	l	$S_{0,3}$
$S_{1,3}$	0_I	0_I	l	$S_{1,3}$
	1_I	1_I	l	$S_{1,3}$
	0_O	0_O	l	$S_{1,3}$
	1_O	1_O	l	$S_{1,3}$
	Δ	1_O	r	S_{start}

Table 12.10: Bit Reversal Turing Machine Transitions

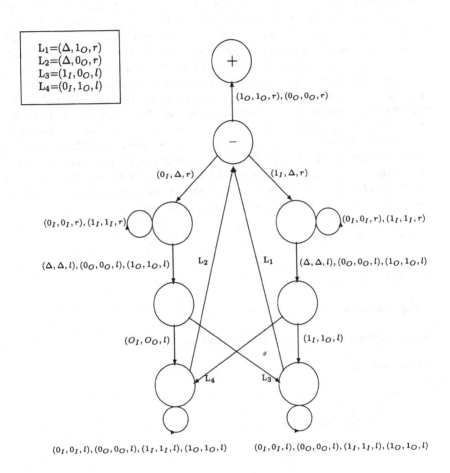

Figure 12.10: Bit Reversal Turing Machine

12.5 Example Programs

A general Turing machine is implemented in C++. The Turing machines for parity
check, parity calculation and bit string reversal are constructed and tested. The
program contains three classes, `Tapecell` which provides support for a dynamically growing tape, `Transition` which is used to implement transition tables and
`TuringMachine` which implements the Turing machine. The constructor for class
`Transition` takes as arguments an integer to identify the state, a symbol which
when read causes the transition, a symbol to output, a movement right (1) or left
(-1), the state to change to and a value to indicate if the state is a halt state (0 indicates the state is not a halt state). The `protected` methods of `TuringMachine` are
`tcrash` to print an error message when the machine crashes, `add` to extend the list
dynamically to accomodate new symbols on the tape, `lookup`, to find the transition
for the current symbol on the tape and current state of the machine and `ishalt` to
determine if the current state is a halt state. The constructor of `TuringMachine`
takes as arguments a transition table, an integer specifying how many transitions
the Turing machine has and an integer identifying the start state. The destructor
deallocates the list used for the tape. The method `run` takes as arguments a string
as input and an integer specifying the length of the input.

```cpp
// turing.cpp

#include <iostream>
#include <string>

using namespace std;

class TuringMachine; // forward declaration

class Tapecell
{
 protected:
        char symbol;
        Tapecell *next,*previous;

 friend class TuringMachine;
};

class Transition
{
 public:
    int state,nextstate;
    char input,output,movement,halt;
    Transition(int s = 0,char i = ' ',char o = ' ',
            char m = 1,int ns = 0,char h = 0)
            :state(s),input(i),output(o),
            nextstate(ns),movement(m),halt(h) { }
```

```
};

class TuringMachine
{
 protected:
   Tapecell *tape;
   Transition *table;
   int ccell,state,tentries,crash,sstate;
   void tcrash(char);
   void add(char);
   Transition *lookup(Tapecell *);
   int ishalt(int);

 public:
   TuringMachine(Transition *,int,int);
   ~TuringMachine();
   void run(const string &,int);
   static char left,right;
};

// constructor
TuringMachine::TuringMachine(Transition *ttable,int entr,int strt)
{
 int i;

 table = new Transition[entr];
 tape = (Tapecell *)NULL;
 for(i=0;i < entr;i++)
  table[i] = ttable[i];
 ccell = -1;
 sstate = strt;
 tentries = entr;
}

// destructor
TuringMachine::~TuringMachine()
{
 Tapecell *cell = tape;

 delete[] table;
 if(cell != (Tapecell *)NULL)
 while(cell->next != (Tapecell *)NULL)
 {
  cell = cell->next;
  if(cell->previous != (Tapecell *)NULL)
   delete cell->previous;
 }
 if(cell != (Tapecell *)NULL)
```

```
  delete cell;
}

void TuringMachine::run(const string &input,int len)
{
 int i,halt;
 Tapecell *cell = tape;
 Transition *trans;

 for(i=0;i < len;i++)
  add(input[i]);
 ccell = 0;
 crash = 0;
 state=sstate;
 if(cell == (Tapecell*)NULL)
 {
  add(' ');
  cell = tape;
 }
 halt = ishalt(state);
 while(!crash && !halt)
 {
  trans = lookup(cell);
  if(trans == (Transition *)NULL)
   tcrash(cell->symbol);
  else
  {
   if(!crash && !halt)
   {
    cell->symbol = trans->output;
    state = trans->nextstate;
    if(trans->movement < 0)
    {
     if(cell->previous == (Tapecell *)NULL)
      tcrash(cell->symbol);
     else
     {
      cell = cell->previous;
      ccell--;
     }
    }
    else if(trans->movement > 0)
    {
     if(cell->next == (Tapecell *)NULL)
      add(' ');
     cell = cell->next; ccell++;
    }
    else
```

```
     tcrash(cell->symbol);
   }
   halt = ishalt(state);
 }
}
if(!crash)
{
 cell = tape;
 cout << "Succesful completion, tape:" << endl;
 while(cell != (Tapecell *)NULL)
 {
  cout << cell->symbol;
  cell = cell->next;
 }
 cout << endl;
}
cell = tape;
if(cell != (Tapecell *)NULL)
while(cell->next != (Tapecell*)NULL)
{
 cell = cell->next;
 if(cell->previous != (Tapecell*)NULL)
  delete cell->previous;
}
if(cell != (Tapecell *)NULL)
 delete cell;
tape = (Tapecell *)NULL;
}

void TuringMachine::tcrash(char symbol)
{
 crash = 1;
 cout << "The Turing Machine crashed at state " << state
      << " and cell " << ccell
      << " with symbol \"" << symbol
      << "\"" << endl;
}

void TuringMachine::add(char symbol)
{
 if(tape == (Tapecell *)NULL)
 {
  tape = new Tapecell;
  tape->next = tape->previous = (Tapecell *)NULL;
  tape->symbol = symbol;
 }
 else
 {
```

```
  Tapecell *cell = tape;
  while(cell->next != (Tapecell *)NULL)
   cell = cell->next;
  cell->next = new Tapecell;
  cell->next->previous = cell;
  cell->next->next = (Tapecell *)NULL;
  cell->next->symbol = symbol;
 }
}


Transition *TuringMachine::lookup(Tapecell *cell)
{
 int i;
 for(i=0;i < tentries;i++)
  if((table[i].state == state)&&(table[i].input == cell->symbol))
   return &(table[i]);
 return (Transition*)NULL;
}


int TuringMachine::ishalt(int state)
{
 int i;
 for(i=0;i < tentries;i++)
  if((table[i].state == state)&&(table[i].halt == 1))
   return 1;
 return 0;
}


char TuringMachine::left = -1,TuringMachine::right = 1;

void main(void)
{
 // parity calculation Turing Machine Transitions
 Transition paritytable[8] = {
   Transition(1,'0','0',TuringMachine::right,1,0),
   Transition(1,'1','1',TuringMachine::right,0,0),
   Transition(1,' ','0',TuringMachine::right,2,0),
   Transition(0,'0','0',TuringMachine::right,0,0),
   Transition(0,'1','1',TuringMachine::right,1,0),
   Transition(0,' ','1',TuringMachine::right,3,0),
   Transition(2,' ',' ',TuringMachine::right,2,1),  //halt state
   Transition(3,' ',' ',TuringMachine::right,3,1)   //halt state
 };

 // string reverse Turing Machine Transitions
 Transition reversetable[29] = {
   Transition(0,'0','0',TuringMachine::right,50,0),
   Transition(0,'1','1',TuringMachine::right,50,0),
```

```
        Transition(0,'a',' ',TuringMachine::right,10,0),
        Transition(0,'b',' ',TuringMachine::right,11,0),
        Transition(10,'a','a',TuringMachine::right,10,0),
        Transition(10,'b','b',TuringMachine::right,10,0),
        Transition(10,'0','0',TuringMachine::left,20,0),
        Transition(10,'1','1',TuringMachine::left,20,0),
        Transition(10,' ',' ',TuringMachine::left,20,0),
        Transition(20,'a','0',TuringMachine::left,30,0),
        Transition(20,'b','0',TuringMachine::left,31,0),
        Transition(30,'a','a',TuringMachine::left,30,0),
        Transition(30,'b','b',TuringMachine::left,30,0),
        Transition(30,'0','0',TuringMachine::left,30,0),
        Transition(30,'1','1',TuringMachine::left,30,0),
        Transition(30,' ','0',TuringMachine::right,0,0),
        Transition(11,'a','a',TuringMachine::right,11,0),
        Transition(11,'b','b',TuringMachine::right,11,0),
        Transition(11,'0','0',TuringMachine::left,21,0),
        Transition(11,'1','1',TuringMachine::left,21,0),
        Transition(11,' ',' ',TuringMachine::left,21,0),
        Transition(21,'a','1',TuringMachine::left,30,0),
        Transition(21,'b','1',TuringMachine::left,31,0),
        Transition(31,'a','a',TuringMachine::left,31,0),
        Transition(31,'b','b',TuringMachine::left,31,0),
        Transition(31,'0','0',TuringMachine::left,31,0),
        Transition(31,'1','1',TuringMachine::left,31,0),
        Transition(31,' ','1',TuringMachine::right,0,0),
        Transition(50,' ',' ',TuringMachine::right,50,1)   //halt state
    };

    string paritycheck="01101001";
    string reversecheck="01101001";

    TuringMachine parity(paritytable,8,0);
    cout << "Parity calculation with input "
         << paritycheck << endl;
    parity.run(paritycheck,8);

    TuringMachine reverse(reversetable,29,0);
    cout << "Reverse input "
         << reversecheck << endl;
    reverse.run(reversecheck,8);

    paritycheck[6] = 'a';
    cout << "Crash parity calulation with input "
         << paritycheck << endl;
    parity.run(paritycheck,8);
}
```

The output of the program is

```
Parity calculation with input 01101001
Succesful completion, tape:
011010011
Reverse input 01101001
Succesful completion, tape:
10010110
Crash parity calulation with input 011010a1
The Turing Machine crashed at state 1 and cell 6 with symbol "a"
```

Chapter 13
Computability and Complexity

13.1 Introduction

Once we have the building blocks for a computing device, we can construct the device and give it tasks to perform. Some tasks are more difficult than others. Some tasks may even be impossible for the computing device to perform. This is the concept of *computability*. Since tasks can be represented as functions, we need to determine the computability of functions. The computable functions are obviously limited by the computing device, but if we choose a sufficiently general computing device it can serve as a measure for computability.

We also need a measure of the difficulty of tasks. This measure indicates how fast the task can be done. Some problems are inherently difficult such as prime number factorization as used in public key cryptography systems, and therefore take a long time to perform. This is referred to as the *complexity* of the problem. In general two measures of complexity are often used, the time complexity and space complexity. Time complexity describes the amount of time taken to do a task given the input. Space complexity refers to the amount of memory required to perform the task given the input. More precisely the measure of complexity is applied to algorithms, since some algorithms are more efficient than others.

Usually the complexity of an algorithm is described in terms of the size n of the input. The notation $f(n)$ is (of order) $O(g(n))$ is used to indicate that there exists $c \in \mathbf{R}$ with $c > 0$ and $N_0 \in \mathbf{N}$ such that for all $N > N_0$ $|f(N)| \leq c|g(N)|$. For example $(n+1)^2$ is $O(n)$ and $O(n^2)$.

The complexity of sequences of symbols has been analysed [106, 109]. Thus if an algorithm can be transformed into an appropriate sequence of symbols, the complexity of the sequence can be used as a measure of the complexity of the algorithm. An example is given in [161].

13.2 Computability

Computability is formulated with respect to given computing models. For example the Turing machine is a computing model. We could define computability in terms of what is computable with a Turing machine. A difficulty arises when we note that a Turing machine can compute more than a finite automaton. Other computing models exist, but if they are proven to be equivalent to the Turing machine model, the computable functions remain the same. The computing model must be reasonable in the sense that the components of the model must be achievable. We need to determine a reasonable computing model such that no other computing model can compute more functions.

13.2.1 Church's Thesis

Church's thesis states that the intuitively computable functions are exactly the partial recursive functions. Sometimes Church's thesis is called the Church-Turing thesis because it can be formulated as the intuitively computable functions are the functions which can be computed by Turing machines. To show that these two statements are equivalent requires that we show that every partial recursive function can be computed by a Turing machine and every Turing machine computes a partial recursive function. It is simple to see how to implement the successor function, at least it is simple to build a binary incrementer Turing machine (in the previous chapter we showed how to achieve this using Moore and Mealy machines). The projection operation is also not difficult to implement on a Turing machine. It can be achieved by reading from the least significant bit to the most significant bit and if the bit is 0 blank every second word (bit sequence) and if the bit is 1 blank every first word (bit sequence). We can introduce new symbols to indicate the end of words and the end of the words on the tape to simplify the implementation. The zero function is trivial to implement using a Turing machine. It is also necessary to show that primitive recursion and composition can be realised somehow with Turing machines. Composition should pose no problem, if new symbols are introduced again to make the task easier. The composition is a combination of the Turing machines implementing each of the functions in the composition, and a control structure. Primitive recursion can be implemented by writing $n, n - 1, \ldots, 0$ on the tape after the input number. The value for $n = 0$ is part of the Turing machine structure, independent of the contents of the tape. Once the function value is known for zero, the value at $n = 1$ can be calculated and so on, up to $n + 1$. So we expect that a Turing machine can compute all primitive recursive functions. A motivation for the thesis is that Turing machines can compute anything that we can. Given as much paper as needed we can compute certain functions using basic operations, for a Turing machine the paper is formally defined by the tape and the basic operations are formally defined by transitions. Any step in the computation is determined by the contents of the paper, a Turing machine operates uniquely according to the tape contents. Since we use the term "intuitively computable", the statement cannot be proven. A proof would require a definition of intuitive computability.

13.2.2 The Halting Problem

An interesting task for a computing model is simulation. If a computing model A can, in some way, simulate another computing model B then A is at least as powerful as B. There exists a Turing machine, called a *universal Turing machine* which can simulate any other Turing machine. As input, the table of transitions and the input of the simulated machine must be stored on the tape. Since we can number the states from 1 to n we can represent states by a bit string or simply a symbol duplicated i times for state i. We are completely free to choose the number of symbols for the representation of states and symbols. We also require a method of tracking which state the machine is in and which input symbol must be read next. The following universal Turing machine is due to Minsky [67, 117].

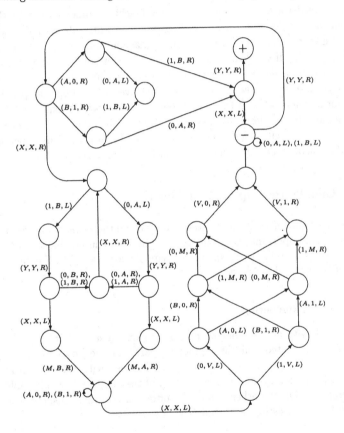

Figure 13.1: Universal Turing Machine

To simplify the machine, if there is no arc for a given state and input then the machine continues the last motion and replaces the symbol with itself (a transition to a state for this machine always has the same motion of the tape head). Also an arc with no label replaces the symbol on the tape with itself and moves the tape head left. We assume that the machine we wish to simulate uses only binary for input and output. For each state, a valid transition can be represented in a finite number of bits, i.e. a fixed number to represent the current and next state, and a single bit to represent the input, output and movement. The description here uses a tape which is infinite to the *left*, with the description of the Turing machine to be simulated starting at the rightmost position of the tape. The description consists of transitions represented in binary, where the end of a transition description is marked by the symbol X. The end of the table of transitions is marked by a Y. Additional symbols are used to mark the state of the machine. The start state is assumed to begin immediately under the tape head.

Now we consider some problems the Turing machine cannot solve. For the halting problem we consider if a Turing machine H exists which always halts, when given as input a representation of another Turing machine and its input, and will give an output indicating if the given Turing machine halts or not. A simple extension gives the machine H' which halts whenever the input machine does not halt, and never halts when the input machine does halt (achieved by a simple loop between two states for any symbol read from the tape). Furthermore we require that the input machine take its own description as input. If we use as input to the machine H', the machine H' itself with itself again as input, we obtain a machine which halts only when the machine does not halt. Thus such a Turing machine H' does not exist.

13.3 Gödel's Incompleteness Theorem

Gödel's Incompleteness Theorem states that not all theorems in number theory can be proved. An important part of the proof is the Gödel numbering given below. This can be used to describe any sequence of symbols, for example a theorem's proof, in terms of the natural numbers.

13.3.1 Gödel Numbering

We can work with an alphabet which contains only a single letter, e.g. the letter |. The words constructed from this alphabet (apart from the empty word) are: |, ||, |||, etc. These words can, in a trivial way, be identified with the natural numbers 0, 1, 2, ... Such an extreme standardization of the "material" is advisable for some considerations. On the other hand, it is often convenient to disperse the diversity of an alphabet consisting of several elements.

The use of an alphabet consisting of *one element* does not imply any essential limitation. We can associate the words W over an alphabet **A** consisting of N elements with natural numbers $G(W)$, in such a way that each natural number is associated with at most one word. Similar arguments apply to words of an alphabet consisting of *one element*. Such a representation of G is called a *Gödel numbering* [63] (also called *arithmetization*) and $G(W)$ is the *Gödel number* of the the word W with respect to G. The following are the requirements for an arithmetization of W:

1 If $W_1 \neq W_2$ then $G(W_1) \neq G(W_2)$.

2 There exists an algorithm such that for any given word W, the corresponding natural number $G(W)$ can be computed in a finite number of steps.

3 For any natural number n, it can be decided whether n is the Gödel number of a word W over **A** in a finite number of steps.

4 There exists an algorithm such that if n is the Gödel number of a word W over **A**, then this word W (which is unique by argument (1)) can be constructed in a finite number of steps.

Here is an example of a Gödel numbering. Consider the alphabet with the letters a, b, c. A word is constructed by any finite concatenation of these – that is, a placement of these letters side by side in a line. For example, *abcbba* is a word. We can then number the words as follows:

Given a word $x_1 x_2 \ldots x_n$ where each x_i is a, b or c, we assign to it the number

$$2^{d_0} \cdot 3^{d_1} \cdot \ldots \cdot p_n^{d_n}$$

where p_i is the ith prime number (and 2 is the 0th prime) and

$$d_i := \begin{cases} 1 & \text{if } x_i \text{ is } a \\ 2 & \text{if } x_i \text{ is } b \\ 3 & \text{if } x_i \text{ is } c \end{cases}$$

The empty word is given the number 0.

For example, the word *acbc* has number $2^1 \cdot 3^3 \cdot 5^2 \cdot 7^3 = 463050$, and *abc* has the number $2^1 \cdot 3^2 \cdot 5^3 = 2250$. The number 7350 represents *aabb* because $7350 = 2^1 \cdot 3^1 \cdot 5^2 \cdot 7^2$.

To show that this numbering satisfies the criteria given above, we use the *fundamental theorem of arithmetic*:

> *Any natural number ≥ 2 can be represented as a product of primes, and that product is, except for the order of the primes, unique.*

We may number all kinds of objects, not just alphabets. In general, the criteria for a numbering to be useful are:

1. No two objects have the same number.

2. Given any object, we can "effectively" find the number that corresponds to it.

3. Given any number, we can "effectively" determine if it is assigned to an object and, if so, to which object.

13.3.2 Gödel's Incompleteness Theorem

Now we give an overview of the incompleteness theorem [10]. We assume that number theory is consistent, i.e. a theorem and its logical negation cannot both be proved. If this were the case, the theory would not be interesting.

Since we have already considered a Gödel numbering we can associate numbers with theorems and proofs in number theory. Let the predicate $p(i,j)$ of two natural numbers be true if and only if i is the Gödel number associated with a formula $B(x)$ with one free variable x and j is the Gödel number associated with the proof of $B(i)$. Furthermore, if $p(i,j)$ is true then a proof can be constructed for these specific integers.

Now consider

$$\forall y \, \neg p(x, y)$$

with the Gödel number m. This states that there is no proof for the theorem x. Let

$$A := \forall y \, \neg p(m, y).$$

Thus A states that there is no proof of A. Suppose A can be proved and n is the Gödel number for the proof. Thus $p(m, n)$ is true and can be proved. But a proof exists for A which means we can prove $\forall y \, \neg p(m, y)$ which implies $\neg p(m, n)$, a contradiction. Suppose instead that $\neg A$ can be proved, i.e $\exists y \, p(m, y)$ can be proved. Thus there exists n such that $p(m, n)$ can be proved, and n is a proof of A which is a contradiction.

Thus if number theory is consistent there exists a theorem such as A which cannot be proved.

13.4 Complexity

13.4.1 Complexity of Bit Strings

Usually the complexity of an algorithm is expressed in terms of the size of the input. Many different definitions of complexity have been proposed in the literature. A few are algorithmic complexity (Kolmogorov-Chaitin) [41], the Lempel-Ziv complexity [109], the logical depth of Bennett [13], the effective measure of complexity of Grassberger [76], the complexity of a system based on its diversity [94], the thermodynamic depth [111], and a statistical measure of complexity [113].

We may describe the time complexity in terms of the total number of operations required for a certain input size, or we may choose some basic operation as the most expensive (such as multiplication or comparison) and use that to describe the complexity of an algorithm. We can represent any program as a bitstring, for example by calculating the Gödel number of the program and using the bit representation of this number. We can then use, as a measure of complexity, the compressibility of the bit string. Here we use the measure defined by Lempel and Ziv [109, 161].

Given a binary string $S = s_1 s_2, \ldots, s_n$ of finite length n, we denote by $S(i, j)$ the substring $s_i s_{i+1} \ldots s_j$ (or the empty word if $i > j$) of S and by $v(S)$ all substrings of S. If S_1 and S_2 are two strings $S_1 S_2$ denotes the concatenation (appending) of S_2 and S_1. The complexity in the sense of Lempel and Ziv of a finite string is evaluated from the point of view of a simple self-delimiting learning machine, which as it scans a given n digit string $S = s_1 s_2, \ldots, s_n$ from left to right, adds a new string to its memory every time it discovers a substring of consecutive digits not previously encountered. We begin with the complexity of the empty string as 0. Suppose we have already scanned the first r digits

$$R := s_1 s_2, \ldots, s_r \circ$$

where $\circ$ indicates that we know the complexity $c(R)$. We have to determine if the rest of the string $S(r + 1, n)$ can be produced by a simple copy operation. To do this we consider the substrings

$$Q_{r+i} := S(r + 1, r + i) \qquad 1 \leq i \leq n - r.$$

For $i < 1$ we use the empty string as Q_{r+i}. Initially we consider $i = 1$. The substring RQ_{r+i} can be produced by a simple copy operation if

$$Q_{r+i} \in v(RQ_{r+i-1}).$$

If this is the case and the substring begins at s_j with $j \leq r$, we can simply copy s_{j+k} to s_{r+k} for $k = 1, 2, \ldots i$, so we try $i + 1$. For $r + i = n$ we have the special case $c(RQ_{r+i}) = c(R) + 1$. If this is not the case, we have $c(RQ_{r+i}) = c(R) + 1$ and repeat the process using RQ_{r+i} as R and $i = 1$.

For example the bitstring consisting of only 0s or only 1s has complexity 2. Alternating 0s and 1s has complexity 3.

$$0 \circ 1 \circ 010101 \ldots.$$

The string 01101000011101001 has complexity 6.

$$0 \circ 1 \circ 10 \circ 100 \circ 001 \circ 1101001.$$

We give an implementation of this algorithm below.

```cpp
// complex.cpp

#include <iostream.h>
#include <string.h>

int substring(char *s,int r,int i)
{
 int j;

 for(j=0;j <= r;j++)
  if(strncmp(s+r+1,s+j,i) == 0) return 1;

 return 0;
}

int complexity(char *s)
{
 static char *laststring="";
 static int c,r;
 int n = strlen(s);
 int i;

 if(n == 0) return 0;
 if(r == n-1) return c;

 if(laststring!=s) { c = 1; r = 0; }
 laststring=s;

 for(i=1;i < n-r;i++)
  if(!substring(s,r,i))
  {
    c++;
    r+=i;
    return complexity(s);
  }

 return ++c;
}

void main(void)
{
 char *str1 = "0101010101";
 char *str2 = "10101010101010101010101";
 char *str3 = "01101000011101001";
 char *str4 = "1011001011";

 cout << str1 << " has complexity " << complexity(str1) << endl;
 cout << str2 << " has complexity " << complexity(str2) << endl;
```

```
  cout << str3 << " has complexity " << complexity(str3) << endl;
  cout << str4 << " has complexity " << complexity(str4) << endl;
}
```

The program output is

```
0101010101 has complexity 3
10101010101010101010 has complexity 3
01101000011101001 has complexity 6
1011001011 has complexity 5
```

13.4.2 NP-class of Problems

Definition. We can define the *time complexity* [114] $C_t(T, n)$ of a Turing machine T as the maximum number of transitions between states for an input of length n on the tape.

Definition. The *space complexity* [114] $C_s(T, n)$ of a Turing machine T as the maximum number of cells into which T writes.

Definition. A Turing machine T is called *polynomial* if there exists a polynomial $p(n)$ such that $C_t(T, n)$ is $O(p(n))$.

We consider problems on decidability, problems which require a 'yes' or 'no' answer.

Definition. The class of problems on decidability for which there exists a polynomial Turing machine is called the *P-class* of problems, denoted by the set P.

Definition. The *NP-class* of problems (denoted by the set NP) are those problems for which, when given a potential solution, there exists a polynomial Turing machine to determine if the solution is valid. The NP is for non-deterministic polynomial. Thus if we can find a potential solution, for example by construction using random numbers such that the probability of constructing an actual solution is sufficiently high, the validity of the solution can be efficiently checked.

Consider the problem of satisfiability. A logic formula consisting of n Boolean variables, requires (as a worst case) checking 2^n combinations of truth values before we know if the formula can be satisfied. The truth table method requires the evaluation of all 2^n combinations, for other techniques there always exists a formula for which the worst case holds. If we have an assignment of truth values to the Boolean variables, a polynomial operation can check if the assignment satisfies the formula.

Definition. A problem NPC on decidability is *NP-complete* if every problem in NP is polynomially reducable to NPC.

If A is NP-complete then we can reformulate any problem in NP to the form of A. Thus if we can solve A we can solve any other problem in NP, furthermore if a polynomial algorithm exists for A then a polynomial algorithm exists for every other problem in NP. An important question in complexity is if the classes P and NP are the same. This reduces to the question is $A \in P$ for A any NP-complete problem.

Cook's theorem [114, 183] states that the satisfiability problem is NP-complete. Since the satisfiability problem is in NP, there exists a polynomial Turing machine that can check the validity of a solution. The proof of the theorem consists of analysing the Turing machine and constructing a logical formula, in a polynomial number of operations, which describes the operation of the Turing machine. The formula introduces a polynomial number of Boolean variables. The formula is a conjunction of disjunctions which are the requirements on the Turing machine. For example, the Turing machine can only be in one state at a time. Thus if any problem A in NP is polynomially reducable to a satisfiability problem, A is also NP-complete.

Chapter 14

Neural Networks

14.1 Introduction

Artificial neural networks is an abstract simulation of a real nervous system that contains a collection of neuron nets communicating with each other via axon connections. Such a model bears a strong resemblance to axons and dendrites in a nervous system. The first fundamental modelling of neural nets was proposed in 1943 by McCulloch and Pitts in terms of a computational model of "nervous activity". The McCulloch-Pitts neuron is a binary device and each neuron has a fixed threshold logic. This model lead to the works of John von Neumann, Marvin Minsky, Frank Rosenlatt, and many others. Hebb postulated [85], that neurons were appropriately interconnected by self-organization and that "an existing pathway strengthens the connections between the neurons". He proposed that the connectivity of the brain is continually changing as an organism learns different functional tasks, and that cell assemblies are created by such changes. By embedding a vast number of simple neurons in an interactive nervous system, it is possible to provide computational power for very sophisticated information processing.

The neuron is the basic processor in neural networks. Each neuron has one ouput, which is generally related to the state of the neuron – its activation – and which may fan out to several other neurons. Each neuron receives several inputs over these connections, called synapses. The inputs are the activations of the incoming neurons multiplied by the weights of the synapses. The activation of the neuron is computed by applying a threshold function to this product. This threshold function is generally some form of nonlinear function.

The basic artificial neuron (Cichocki and Unbehauen [45], Fausett [65], Hassoun [82], Haykin [83], Rojas [139], Steeb [164]) can be modelled as a multi-input nonlinear device with weighted interconnections w_{ji}, also called *synaptic weights* or strengths. The cell body (soma) is represented by a nonlinear limiting or threshold function f. The simplest model of an artificial neuron sums the n weighted inputs and passes the result through a nonlinearity according to the equation

$$y_j = f\left(\sum_{i=1}^{n} w_{ji} x_i - \theta_j\right)$$

where f is a *threshold function*, also called an *activation function*, θ_j $(\theta_j \in \mathbf{R})$ is the external threshold, also called an offset or bias, w_{ji} are the synaptic weights or strengths, x_i are the inputs $(i = 1, 2, \ldots, n)$, n is the number of inputs and y_j represents the output. The activation function is also called the nonlinear transfer characteristic or the squashing function. The activation function f is a monotonically increasing function.

A threshold value θ_j may be introduced by employing an additional input x_0 equal to $+1$ and the corresponding weight w_{j0} equal to minus the threshold value. Thus we can write

$$y_j = f\left(\sum_{i=0}^{n} w_{ji}x_i\right)$$

where

$$w_{j0} = -\theta_j, \qquad x_0 = 1.$$

The basic artificial neuron is characterized by its nonlinearity and the threshold θ_j. The *McCulloch-Pitts model* of the neuron used only the binary (hard-limiting) function (step function or Heaviside function), i.e.

$$H(x) := \begin{cases} 1 & \text{if } x \geq 0 \\ 0 & \text{if } x < 0 \end{cases}.$$

In this model a weighted sum of all inputs is compared with a threshold θ_j. If this sum exceeds the threshold, the neuron output is set to 1, otherwise to 0. For bipolar representation we can use the sign function

$$\text{sign}(x) := \begin{cases} 1 & \text{if } x > 0 \\ 0 & \text{if } x = 0 \\ -1 & \text{if } x < 0 \end{cases}$$

The threshold (step) function may be replaced by a more general nonlinear function and consequently the output of the neuron y_j can either assume a value of a discrete set (e.g. $\{-1, 1\}$) or vary continuously (e.g. between -1 and 1 or generally between y_{min} and $y_{max} > y_{min}$). The activation level or the state of the neuron is measured by the output signal y_j, e.g. $y_j = 1$ if the neuron is firing (active) and $y_j = 0$ if the neuron is quiescent in the unipolar case and $y_j = -1$ for the bipolar case.

In the basic neural model the output signal is usually determined by a monotonically increasing sigmoid function of a weighted sum of the input signals. Such a sigmoid function can be described for example by

$$y_j = \tanh(\lambda u_j) \equiv \frac{1 - e^{-2\lambda u_j}}{1 + e^{-2\lambda u_j}}$$

for a symmetrical (bipolar) representation. For an unsymmetrical unipolar representation we have

$$y_j = \frac{1}{1 + e^{-\lambda u_j}}$$

where λ is a positive constant or variable which controls the steepness (slope) of the sigmoidal function. The quantity u_j is given by

$$u_j := \sum_{i=0}^{n} w_{ji} x_i.$$

The following program, thresh.cpp, gives an implementation of these threshold functions.

```cpp
// thresh.cpp

#include <iostream.h>
#include <math.h>

int H(double* w,double* x,int m)
{
   double sum = 0.0;
   for(int i=0; i<=m; i++)
   {
   sum += w[i]*x[i];
   }
   if(sum >= 0.0) return 1;
   else return 0;
}

int sign(double* w,double* x,int m)
{
   double sum = 0.0;
   for(int i=0; i<=m; i++)
   {
   sum += w[i]*x[i];
   }
   if(sum >= 0.0) return 1;
   else return -1;
}

double unipolar(double* w,double* x,int m)
{
   double lambda = 1.0;
   double sum = 0.0;
   for(int i=0; i<=m; i++)
   {
   sum += w[i]*x[i];
   }
   return 1.0/(1.0 + exp(-lambda*sum));
}

double bipolar(double* w,double* x,int m)
{
   double lambda = 1.0;
   double sum = 0.0;
   for(int i=0; i<=m; i++)
   {
   sum += w[i]*x[i];
   }
   return tanh(lambda*sum);
}
```

```
int main()
{
  int n = 5;              // length of input vector includes bias
  double theta = 0.5;   // threshold

  // allocation memory for weight vector w
  double* w = NULL;
  w = new double[n];

  w[0] = -theta;
  w[1] = 0.7; w[2] = -1.1; w[3] = 4.5; w[4] = 1.5;

  // allocation memory for input vector x
  double* x = NULL;
  x = new double[n];

  x[0] = 1.0;  // bias
  x[1] = 0.7; x[2] = 1.2; x[3] = 1.5; x[4] = -4.5;

  int r1 = H(w,x,n-1);
  cout << "r1 = " << r1 << endl;

  int r2 = sign(w,x,n-1);
  cout << "r2 = " << r2 << endl;

  double r3 = unipolar(w,x,n-1);
  cout << "r3 = " << r3 << endl;

  double r4 = bipolar(w,x,n-1);
  cout << "r4 = " << r4 << endl;

  delete [] w;
  delete [] x;

  return 0;
}
```

14.2 Hyperplanes

Hyperplanes are used to describe the function of a perceptron. They are used to classify points in space as being elements of one of two half spaces.

Definition. A *hyperplane* $H_{\mathbf{p},\alpha}$ is a subset of $\mathbf{R}^n$ defined by

$$\{\, \mathbf{x} \mid \mathbf{p}^T \mathbf{x} = \alpha,\ \mathbf{x} \in \mathbf{R}^n \,\}$$

with $\mathbf{p} \in \mathbf{R}^n$, $\alpha \in \mathbf{R}$ and $\mathbf{p}^T$ denotes the transpose of $\mathbf{p}$.

A hyperplane $H_{\mathbf{p},\alpha}$ defines two closed half spaces

$$\{\, \mathbf{x} \mid \mathbf{p}^T \mathbf{x} \geq \alpha,\ \mathbf{x} \in \mathbf{R}^n \,\}$$

$$\{\, \mathbf{x} \mid \mathbf{p}^T \mathbf{x} \leq \alpha,\ \mathbf{x} \in \mathbf{R}^n \,\}$$

and two open half spaces

$$H_{\mathbf{p},\alpha}^+ \ := \ \{\, \mathbf{x} \mid \mathbf{p}^T \mathbf{x} > \alpha,\ \mathbf{x} \in \mathbf{R}^n \,\}$$

$$H_{\mathbf{p},\alpha}^- \ := \ \{\, \mathbf{x} \mid \mathbf{p}^T \mathbf{x} < \alpha,\ \mathbf{x} \in \mathbf{R}^n \,\}$$

in $\mathbf{R}^n$.

Any point $\mathbf{x} \notin H_{\mathbf{p},\alpha}$ in $\mathbf{R}^n$ has the property that either $\mathbf{x} \in H_{\mathbf{p},\alpha}^+$ or $\mathbf{x} \in H_{\mathbf{p},\alpha}^-$.

These definitions can also be expressed in terms of a fixed point on the hyperplane. Suppose $\mathbf{a} \in \mathbf{R}^n$ is a point on the hyperplane $H_{\mathbf{p},\alpha}$. Any point $\mathbf{x}$ on the hyperplane must satisfy

$$\mathbf{p}^T \mathbf{x} - \mathbf{p}^T \mathbf{a} = \alpha - \alpha = 0.$$

Thus we obtain the definitions

$$\begin{aligned} H_{\mathbf{p},\mathbf{a}} &= \{\, \mathbf{x} \mid \mathbf{p}^T (\mathbf{x} - \mathbf{a}) = 0,\ \mathbf{x} \in \mathbf{R}^n \,\} \\ H_{\mathbf{p},\mathbf{a}}^+ &= \{\, \mathbf{x} \mid \mathbf{p}^T (\mathbf{x} - \mathbf{a}) > 0,\ \mathbf{x} \in \mathbf{R}^n \,\} \\ H_{\mathbf{p},\mathbf{a}}^- &= \{\, \mathbf{x} \mid \mathbf{p}^T (\mathbf{x} - \mathbf{a}) < 0,\ \mathbf{x} \in \mathbf{R}^n \,\}. \end{aligned}$$

Definition. Let $S_1, S_2 \subset \mathbf{R}^n$. If

$$S_1 \subseteq H_{\mathbf{p},\alpha}^+ \quad \text{and} \quad S_2 \subseteq H_{\mathbf{p},\alpha}^-$$

then S_1 and S_2 are said to be *properly separated* by $H_{\mathbf{p},\alpha}$.

Definition. Two sets of points A and B in the n-dimensional space $\mathbf{R}^n$ are called *linearly separable* if $n+1$ real numbers $w_0, w_1, \ldots, w_n$ exist, such that every point $(x_1, x_2, \ldots, x_n) \in A$ satisfies $\sum_{i=1}^n w_i x_i \geq w_0$ and every point $(x_1, x_2, \ldots, x_n) \in B$ satisfies $\sum_{i=1}^n w_i x_i < w_0$.

Definition. Two sets A and B of points in the n-dimensional space $\mathbf{R}^n$ are called *absolutely linearly separable* if $n+1$ real numbers $w_0, w_1, \ldots, w_n$ exist such that every point $(x_1, x_2, \ldots, x_n) \in A$ satisfies $\sum_{i=1}^n w_i x_i > w_0$ and every point $(x_1, x_2, \ldots, x_n) \in B$ satisfies $\sum_{i=1}^n w_i x_i < w_0$.

Definition. The open (closed) positive half space associated with the n-dimensional weight vector $\mathbf{w}$ is the set of all points $\mathbf{x} \in \mathbf{R}^n$ for which $\mathbf{w}^T \mathbf{x} > 0$ ($\mathbf{w}^T \mathbf{x} \geq 0$). The open (closed) negative half space associated with $\mathbf{w}$ is the set of all points $\mathbf{x} \in \mathbf{R}^n$ for which $\mathbf{w}^T \mathbf{x} < 0$ ($\mathbf{w}^T \mathbf{x} \leq 0$).

Example. Consider the plane in $\mathbf{R}^4$ described by

$$x_1 + x_2 - x_3 + 2x_4 = 4$$

with normal vector

$$\mathbf{p} = \begin{pmatrix} 1 \\ 1 \\ -1 \\ 2 \end{pmatrix}.$$

It is a hyperplane $H_{\mathbf{p},4}$. The point $(1,1,0,1)^T$ can be used to describe the two half spaces

$$H_{\mathbf{p},4}^+ = \{ (x_1, x_2, x_3, x_4)^T \in \mathbf{R}^4 \mid (x_1 - 1) + (x_2 - 1) - x_3 + 2(x_4 - 1) > 0 \}$$

$$H_{\mathbf{p},4}^- = \{ (x_1, x_2, x_3, x_4)^T \in \mathbf{R}^4 \mid (x_1 - 1) + (x_2 - 1) - x_3 + 2(x_4 - 1) < 0 \}.$$

To understand the separation better, we can examine the effect of the division on subspaces. The hyperplane divides the subspace corresponding to x_3 around the origin. The hyperplane divides the subspace corresponding to x_1 around 1. The same applies for the subspaces corresponding to x_2 and x_4. Thus we can classify the following points.

$$(s, t, u, v)^T \in H_{\mathbf{p},4}^+, \qquad s, t, v \geq 1, \ u \leq 0$$

$$(s, t, u, v)^T \in H_{\mathbf{p},4}^-, \qquad s, t, v \leq 1, \ u \geq 0$$

where at least one equality does not hold. Considering two- and three-dimensional subspaces leads to an even better description of the division of the vector space.

14.3 Perceptron

14.3.1 Introduction

The *perceptron* is the simplest form of a neural network used for the classification of special types of patterns said to be linearly separable (i.e. patterns that lie on opposite sides of a hyperplane). It consists of a single neuron with adjustable synaptic weights w_i and threshold θ.

Definition. A perceptron is a computing unit with threshold θ which, when receiving the n real inputs x_1, x_2, ..., x_n through edges with the associated weights w_1, w_2, ..., w_n, outputs 1 if the inequality

$$\sum_{i=1}^{n} w_i x_i \geq \theta$$

holds otherwise it outputs zero.

The origin of the inputs is not important irrespective of whether they come from other perceptrons or another class of computing units. The geometric interpretation of the processing performed by perceptrons is the same as with McCulloch-Pitts elements. A perceptron separates the input space into two half-spaces. For points belonging to one half-space the result of the computation is 0, for points belonging to the other it is 1.

We can also formulate this definition using the Heaviside step function

$$H(x) := \begin{cases} 1 & \text{for} \quad x \geq 0 \\ 0 & \text{for} \quad x < 0 \end{cases} .$$

Thus

$$H(\sum_{i=1}^{n} w_i x_i - \theta) = \begin{cases} 1 & \text{for} \quad (\sum_{i=1}^{n} w_i x_i - \theta) \geq 0 \\ 0 & \text{for} \quad (\sum_{i=1}^{n} w_i x_i - \theta) < 0 \end{cases}$$

With w_1, w_2, ..., w_n and θ given, the equation

$$\sum_{i=1}^{n} w_i x_i = \theta$$

defines a hyperplane which divides the Euclidean space $\mathbf{R}^n$ into two half spaces.

Example. The plane

$$x_1 + 2x_2 - 3x_3 = 4$$

divides $\mathbf{R}^3$ into two half spaces.

In many cases it is more convenient to deal with perceptrons of threshold zero only. This corresponds to linear separations which are forced to go through the origin of the input space. The threshold of the perceptron with a threshold has been converted into the weight $-\theta$ of an additional input channel connected to the constant 1. This extra weight connected to a constant is called the *bias* of the element. Thus the input vector $(x_1, x_2, \ldots, x_n)$ must be extended with an additional 1 and the resulting $(n + 1)$-dimensional vector

$$(1, x_1, x_2, \ldots, x_n)$$

is called the *extended input vector*, where

$$x_0 = 1$$

The extended weight vector associated with this perceptron is

$$(w_0, w_1, \ldots, w_n)$$

whereby $w_0 = -\theta$.

The threshold computation of a perceptron will be expressed using scalar products. The arithmetic test computed by the perceptron is thus

$$\mathbf{w}^T \mathbf{x} \geq \theta$$

if $\mathbf{w}$ and $\mathbf{x}$ are the weight and input vectors, and

$$\mathbf{w}^T \mathbf{x} \geq 0$$

if $\mathbf{w}$ and $\mathbf{x}$ are the extended weight and input vectors.

Example. If we are looking for the weights and threshold needed to implement the AND function with a perceptron, the input vectors and their associated outputs are

$$(0,0) \mapsto 0, \quad (0,1) \mapsto 0, \quad (1,0) \mapsto 0, \quad (1,1) \mapsto 1.$$

If a perceptron with threshold zero is used, the input vectors must be extended and the desired mappings are

$$(1,0,0) \mapsto 0, \quad (1,0,1) \mapsto 0, \quad (1,1,0) \mapsto 0, \quad (1,1,1) \mapsto 1.$$

A perceptron with three still unknown weights (w_0, w_1, w_2) can carry out this task.

Example. The *AND gate* can be simulated using the perceptron. The AND gate is given by

Input	Output
0 0	0
0 1	0
1 0	0
1 1	1

Thus the input patterns are

$$\mathbf{x}_0 = \begin{pmatrix} 0 \\ 0 \end{pmatrix}, \quad \mathbf{x}_1 = \begin{pmatrix} 0 \\ 1 \end{pmatrix}, \quad \mathbf{x}_2 = \begin{pmatrix} 1 \\ 0 \end{pmatrix}, \quad \mathbf{x}_3 = \begin{pmatrix} 1 \\ 1 \end{pmatrix}.$$

Let

$$\mathbf{w} = \begin{pmatrix} 1 \\ 1 \end{pmatrix}, \quad \theta = \frac{3}{2}.$$

Then

$$\mathbf{w}^T = (1, 1)$$

and the evaluation of $H(\mathbf{w}^T\mathbf{x}_j - \theta)$ for $j = 0, 1, 2, 3$ yields

$$H(\mathbf{w}^T\mathbf{x}_0 - \theta) = H(0 - \frac{3}{2}) = H(-\frac{3}{2}) = 0$$

$$H(\mathbf{w}^T\mathbf{x}_1 - \theta) = H(1 - \frac{3}{2}) = H(-\frac{1}{2}) = 0$$

$$H(\mathbf{w}^T\mathbf{x}_2 - \theta) = H(1 - \frac{3}{2}) = H(-\frac{1}{2}) = 0$$

$$H(\mathbf{w}^T\mathbf{x}_3 - \theta) = H(2 - \frac{3}{2}) = H(\frac{1}{2}) = 1.$$

Example. Consider the Boolean function

$$f(x_1, x_2, x_3) = (\overline{x_1} \cdot x_2) + (x_2 \cdot \overline{x_3}).$$

This Boolean function can be represented by

$$y = H(\mathbf{w}^T \mathbf{x} - \theta)$$

where $\mathbf{w}^T = (-1, 2, -1)$ and $\theta = \frac{1}{2}$ since

x_1	x_2	x_3	y
0	0	0	0
0	0	1	0
0	1	0	1
0	1	1	1
1	0	0	0
1	0	1	0
1	1	0	1
1	1	1	0

Table 14.1: Function Table for the Boolean Function $(\overline{x_1} \cdot x_2) + (x_2 \cdot \overline{x_3})$

Thus to find $\mathbf{w}$ and θ we have to solve the following inequalities

$$\begin{aligned}
0 &< \theta \\
w_1 &< \theta \\
w_2 &\geq \theta \\
w_3 &< \theta \\
w_1 + w_2 &\geq \theta \\
w_1 + w_3 &< \theta \\
w_2 + w_3 &\geq \theta \\
w_1 + w_2 + w_3 &< \theta
\end{aligned}$$

which admits the solution

$$\theta = \frac{1}{2}, \qquad w_1 = -1, \qquad w_2 = 2, \qquad w_3 = -1.$$

14.3.2   Boolean Functions

Which logical functions can be implemented with a single perceptron? A perceptron network is capable of computing any logical function since perceptrons are even more powerful than unweighted McCulloch-Pitts elements. If we reduce the network to a single element, which functions are still computable? Taking the boolean functions of two variables we can gain some insight into this problem.

Since we are considering logical functions of two variables, there are four possible combinations for the input. The outputs for the four inputs are four bits which uniquely distinguish each logical function. We use the number defined by these four bits as a subindex for the name of the functions. The function $(x_1, x_2) \mapsto 0$, for example, is denoted by f_0 (since 0 corresponds to the bit string 0000). The AND function is denoted by f_8 (since 8 corresponds to the bit string 1000), whereby the output bits are ordered according to the following ordering of the inputs: $(1,1)$, $(0,1)$, $(1,0)$, $(0,0)$.

The sixteen possible functions of two variables are thus

$$
\begin{aligned}
f_0(x_1, x_2) &= f_{0000}(x_1, x_2) = 0 \\
f_1(x_2, x_2) &= f_{0001}(x_1, x_2) = \overline{x_1 + x_2} \\
f_2(x_1, x_2) &= f_{0010}(x_1, x_2) = x_1 \cdot \overline{x_2} \\
f_3(x_1, x_2) &= f_{0011}(x_1, x_2) = \overline{x_2} \\
f_4(x_1, x_2) &= f_{0100}(x_1, x_2) = \overline{x_1} \cdot x_2 \\
f_5(x_1, x_2) &= f_{0101}(x_1, x_2) = \overline{x_1} \\
f_6(x_1, x_2) &= f_{0110}(x_1, x_2) = x_1 \oplus x_2 \\
f_7(x_1, x_2) &= f_{0111}(x_1, x_2) = \overline{x_1 \cdot x_2} \\
f_8(x_1, x_2) &= f_{1000}(x_1, x_2) = x_1 \cdot x_2 \\
f_9(x_1, x_2) &= f_{1001}(x_1, x_2) = \overline{x_1 \oplus x_2} \\
f_{10}(x_1, x_2) &= f_{1010}(x_1, x_2) = x_1 \\
f_{11}(x_1, x_2) &= f_{1011}(x_1, x_2) = x_1 + \overline{x_2} \\
f_{12}(x_1, x_2) &= f_{1100}(x_1, x_2) = x_2 \\
f_{13}(x_1, x_2) &= f_{1101}(x_1, x_2) = \overline{x_1} + x_2 \\
f_{14}(x_1, x_2) &= f_{1110}(x_1, x_2) = x_1 + x_2 \\
f_{15}(x_1, x_2) &= f_{1111}(x_1, x_2) = 1 \, .
\end{aligned}
$$

The function f_0 is the zero function whereas f_{14} is the inclusive OR-function. Perceptron-computable functions are those for which the points whose function value is 0 can be separated from the points whose function value is 1 using a line. For the AND function and OR function we can find such a separation.

Two of the functions cannot be computed in this way. They are the function XOR (exclusive OR) (function f_6) and the function XNOR f_9. No line can produce the necessary separation of the input space. This can also be shown analytically.

Let w_1 and w_2 be the weights of a perceptron with two inputs, and θ its threshold. If the perceptron computes the XOR function the following four inequalities must be fulfilled.

$$
\begin{aligned}
x_1 = 0 \quad x_2 = 0 \quad & w_1 x_1 + w_2 x_2 = 0 && \Rightarrow && 0 < \theta \\
x_1 = 1 \quad x_2 = 0 \quad & w_1 x_1 + w_2 x_2 = w_1 && \Rightarrow && w_1 \geq \theta \\
x_1 = 0 \quad x_2 = 1 \quad & w_1 x_1 + w_2 x_2 = w_2 && \Rightarrow && w_2 \geq \theta \\
x_1 = 1 \quad x_2 = 1 \quad & w_1 x_1 + w_2 x_2 = w_1 + w_2 && \Rightarrow && w_1 + w_2 < \theta.
\end{aligned}
$$

Since the threshold θ is positive, according to the first inequality, w_1 and w_2 are positive too, according to the second and third inequalities. Therefore the inequality $w_1 + w_2 < \theta$ cannot be true. This contradiction implies that no perceptron capable of computing the XOR function exists. An analogous proof holds for the function f_9.

A perceptron can only compute linearly separable functions, When $n = 2$, 14 out of the 16 possible Boolean functions are linearly separable. When $n = 3$, 104 out of 256 and when $n = 4$, 1882 out of 65536 possible functions are linearly separable. No formula for expressing the number of linearly separable functions as a function of n has yet been found.

Thus using

$$
y = H(\mathbf{w}^T \mathbf{x} - \theta)
$$

we cannot represent all Boolean functions. However we can realize the universal NAND-gate (or universal NOR-gate). Thus any boolean function can be realized using a network of linear threshold gates. For example the XOR gate can be constructed as in Figure 14.1.

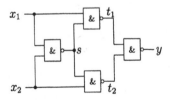

Figure 14.1: XOR Implementation Using NAND Operations

Now the NAND gate can be represented by

$$
w_1 = -\frac{1}{3}, \qquad w_2 = -\frac{1}{3}, \qquad \theta = -\frac{1}{2}.
$$

This is a solution to the set of inequaltities

$$0 > \theta, \qquad w_1 > \theta, \qquad w_2 > \theta, \qquad w_1 + w_2 < \theta.$$

Thus the XOR gate can be simulated by

$$
\begin{aligned}
y &= H(w_1 t_1 + w_2 t_2 - \theta) \\
&= H(w_1 H(w_1 x_1 + w_2 s - \theta) + w_2 H(w_1 s + w_2 x_2 - \theta) - \theta) \\
&= H(w_1 H(w_1 x_1 + w_2 H(w_1 x_1 + w_2 x_2 - \theta) - \theta) \\
&\quad + w_2 H(w_1 H(w_1 x_1 + w_2 x_2 - \theta) + w_2 x_2 - \theta) - \theta)
\end{aligned}
$$

The program below implements this equation.

```
// xor.cpp

#include <iostream>

using namespace std;

int H(double x) { return (x>=0); }
int NAND(int x1,int x2) { return H(-x1/3.0-x2/3.0+0.5); }

int XOR(int x1,int x2)
{
 int s=NAND(x1,x2);
 int t1=NAND(x1,s);
 int t2=NAND(x2,s);
 return NAND(t1,t2);
}

void main(void)
{
 cout << "XOR(0,0) = " << XOR(0,0) << endl;
 cout << "XOR(0,1) = " << XOR(0,1) << endl;
 cout << "XOR(1,0) = " << XOR(1,0) << endl;
 cout << "XOR(1,1) = " << XOR(1,1) << endl;
}
```

The program output is

```
XOR(0,0) = 0
XOR(0,1) = 1
XOR(1,0) = 1
XOR(1,1) = 0
```

14.3.3 Perceptron Learning

A learning algorithm is an adaptive method by which a network of computing units self-organize to implement the desired behavior. This is done in some learning algorithms by presenting some examples of the desired input-output mapping to the network. A correction step is executed iteratively until the network learns to produce the desired response.

Learning algorithms can be divided into *supervised* and *unsupervised* methods. Supervised learning denotes a method in which some input vectors are collected and presented to the network. The output computed by the network is observed and the deviation from the expected answer is measured. The weights are corrected according to the magnitude of the error in the way defined by the learning algorithm.

Unsupervised learning is used when, for a given input, the exact numerical output a network should produce is unknown. Assume, for example, that some points in two-dimensional space are to be classified into three clusters. We can use a classifier network with three output lines. Each of the three computing units at the output must specialize by firing only for inputs corresponding to elements of each cluster. If one unit fires, the others must keep silent. In this case we do not know a priori which unit is going to specialize on which cluster. Generally we do not even know how many well-defined clusters are present. The network must organize itself in order to be able to associate clusters with units.

Supervised learning is further divided into methods which use reinforcement or error correction. Reinforcement learning is used when after each presentation of an input-output example we only know whether the network produces the desired result or not. The weights are updated based on this information so that only the input vector can be used for weight correction. In learning with error correction, the magnitude of the error, together with the input vector, determines the magnitude of the corrections to the weights (corrective learning).

The perceptron learning algorithm is an example of supervised learning with reinforcement. Some variants use supervised learning with error correction.

The proof of convergence of the perceptron learning algorithm assumes that each perceptron performs the test $\mathbf{w}^T\mathbf{x} > 0$. So far we have been working with perceptrons which perform the test $\mathbf{w}^T\mathbf{x} \geq 0$. If a perceptron with threshold zero can linearly separate two finite sets of input vectors, then only a small adjustment to its weights is needed to obtain an absolute linear separation. This is a direct corollary of the following proposition.

Proposition. Two finite sets of points, A and B, in n-dimensional space which are linearly separable are also absolutely linearly separable.

A usual approach for starting the learning algorithm is to initialize the network weights randomly and to improve these initial parameters, looking at each step to see whether a better separation of the training set can be achieved. We identify points $(x_1, x_2, \ldots, x_n)$ in n-dimensional space with the vector $\mathbf{x}$ with the same coordinates.

Let P and N be two finite sets of points in $\mathbf{R}^n$ which we want to separate linearly. A weight vector is sought so that the points in P belong to its associated positive half-space and the points in N to the negative half-space. The error of a perceptron with weight vector $\mathbf{w}$ is the number of incorrectly classified points. The learning algorithm must minimize this error function $E(\mathbf{w})$. Now we introduce the perceptron learning algorithm. The training set consists of two sets, P and N, in n-dimensional extended input space. We look for a vector $\mathbf{w}$ capable of absolutely separating both sets, so that all vectors in P belong to the open positive half-space and all vectors in N to the open negative half-space of the linear separation.

Algorithm. Perceptron learning

start: The weight vector $\mathbf{w}(t = 0)$ is generated randomly
test: A vector $\mathbf{x} \in P \cup N$ is selected randomly,
             if $\mathbf{x} \in P$ and $\mathbf{w}(t)^T \mathbf{x} > 0$ goto *test*,
             if $\mathbf{x} \in P$ and $\mathbf{w}(t)^T \mathbf{x} \leq 0$ goto *add*,
             if $\mathbf{x} \in N$ and $\mathbf{w}(t)^T \mathbf{x} < 0$ goto *test*,
             if $\mathbf{x} \in N$ and $\mathbf{w}(t)^T \mathbf{x} \geq 0$ goto *subtract*,
add: set $\mathbf{w}(t + 1) = \mathbf{w}(t) + \mathbf{x}$ and $t := t + 1$, goto *test*
subtract: set $\mathbf{w}(t + 1) = \mathbf{w}(t) - \mathbf{x}$ and $t := t + 1$ goto *test*

This algorithm makes a correction to the weight vector whenever one of the selected vectors in P or N has not been classified correctly. The perceptron convergence theorem guarantees that if the two sets P and N are linearly separable the vector $\mathbf{w}$ is updated only a finite number of times. The routine can be stopped when all vectors are classified correctly.

Example. Consider the sets in the extended space

$$P = \{ (1, 2.0, 2.0), \ (1, 1.5, 1.5) \}$$

$$N = \{ (1, 0, 1), \ (1, 1, 0), \ (1, 0, 0) \}.$$

Thus in $\mathbf{R}^2$ we consider the two sets of points

$$\{ (2.0, 2.0), \ (1.5, 1.5) \}$$

and

$$\{ (0, 1), \ (1, 0), \ (0, 0) \}.$$

These two sets are separable by the line

$$x_1 + x_2 = \frac{3}{2}.$$

Thus $\mathbf{w}^T = (-\frac{3}{2}, 1, 1)$.

The following C++ program implements the algorithm.

```cpp
// classify.cpp

#include <iostream>
#include <stdlib.h>
#include <time.h>

using namespace std;

void classify(double **P,double **N,int p,int n,double *w,int d)
{
 int i,j,k,classified = 0;
 double *x,sum;

 srand(time(NULL));
 for(i=0;i<d;i++) w[i] = double(rand())/RAND_MAX;
 k = 0;
 while(!classified)
 {
  i = rand()%(p+n-1);
  if(i<p) x = P[i]; else x = N[i-p];
  for(j=0,sum=0;j < d;j++) sum += w[j]*x[j];
  if((i<p) && (sum<=0))
   for(j=0;j < d;j++) w[j] += x[j];
  if((i>=p) && (sum>=0))
   for(j=0;j < d;j++) w[j] -= x[j];
  k++;
  classified = 1;
  // check if the vectors are classified
  if((k%(2*p+2*n)) == 0)
  {
   for(i=0;(i < p)&&classified;i++)
   {
    sum = 0;
    for(j=0,sum=0;j < d;j++) sum += w[j]*P[i][j];
    if(sum <= 0) classified = 0;
   }
```

```
   for(i=0;(i<n)&&classified;i++)
   {
     sum = 0;
     for(j=0,sum=0;j < d;j++) sum += w[j]*N[i][j];
     if(sum >= 0) classified = 0;
   }
 }
 else classified = 0;
}
}

void main(void)
{
 double **P = new double*[2];

 P[0] = new double[3]; P[1] = new double[3];

 P[0][0] = 1.0; P[0][1] = 2.0; P[0][2] = 2.0;
 P[1][0] = 1.0; P[1][1] = 1.5; P[1][2] = 1.5;

 double **N = new double*[3];

 N[0] = new double[3]; N[1] = new double[3]; N[2] = new double[3];

 N[0][0] = 1.0; N[0][1] = 0.0; N[0][2] = 1.0;
 N[1][0] = 1.0; N[1][1] = 1.0; N[1][2] = 0.0;
 N[2][0] = 1.0; N[2][1] = 0.0; N[2][2] = 0.0;

 double *w = new double[3];

 classify(P,N,2,3,w,3);

 cout << "w = ( " << w[0] << " , " << w[1]
      << " , " << w[2] << " ) " << endl;

 delete[] P[0]; delete[] P[1];
 delete[] N[0]; delete[] N[1]; delete[] N[2];
 delete[] P; delete[] N;
 delete w;
}
```

The program output is

```
w = ( -1.59917 , 1.47261 , 1.2703 )
```

14.3.4 Quadratic Threshold Gates

Thus far we have considered linear threshold gates. We can consider using nonlinear threshold gates to simulate functions which cannot be simulated with linear threshold gates (for example the XOR operation). This can be accomplished by expanding the number of inputs to a linear threshold gate. For example, one can do this by feeding the products or AND of inputs as new inputs to the linear threshold gate. In this case, we require a fixed preprocessing layer of AND gates that artificially increases the dimensionality of the input space. We expect that the resulting Boolean function (which is now only partially specified) becomes a threshold function and hence realizable using a single linear threshold gate. The realization of a Boolean logic function by the preceeding process leads to a *quadratic threshold gate*. The general transfer characteristics for an n-input quadratic threshold gate are given by

$$y = \begin{cases} 1 & \sum_{i=1}^{n} w_i x_i + \sum_{i=1}^{n}\sum_{j=i}^{n} w_{ij} x_i x_j \geq \theta \\ 0 & \text{otherwise} \end{cases}$$

for $\mathbf{x} \in \mathbf{R}^n$ and

$$y = \begin{cases} 1 & \sum_{i=1}^{n} w_i x_i + \sum_{i=1}^{n}\sum_{j=i+1}^{n} w_{ij} x_i x_j \geq \theta \\ 0 & \text{otherwise} \end{cases}$$

for $\mathbf{x} \in \{0,1\}^n$. The only difference between the above two equations is the range of the index j of the second summation in the double-summation term. The bounds on the double summations eliminate $w_{ij} x_i x_j$ and $w_{ji} x_j x_i$ duplications. Quadratic threshold gates greatly increase the number of realizable Boolean functions when compared to linear threshold gates.

Example. Consider

$$y = \begin{cases} 1 & -x_1 - x_2 + 3x_1 x_2 \geq -\frac{1}{2} \\ 0 & \text{otherwise} \end{cases}$$

This quadratic threshold gate can be used to implement the XNOR operation. The function classifies points in $\mathbf{R}^2$ according to $g(x,y) \geq 0$ and $g(x,y) < 0$ where

$$g(x,y) = -x - y + 3xy + \frac{1}{2}.$$

The gate is illustrated in Figure 14.2.

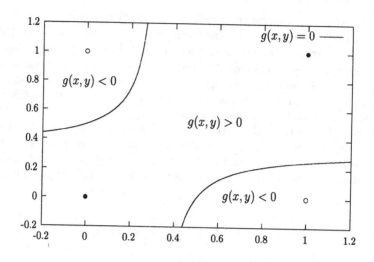

Figure 14.2: Quadratic Threshold Gate for XOR

The following program illustrates a quadratic threshold gate for the XNOR operation.

```cpp
// quadratic.cpp

#include <iostream.h>

double f(double *x,double *wv,double **wm,int n)
{
 double sum = 0.0;
 int i,j;

 for(i=0;i <= n;i++)
  sum += wv[i]*x[i];
 for(i=0;i <= n;i++)
  for(j=i;j <= n;j++)
   sum += wm[i][j]*x[i]*x[j];

 if(sum >= 0) return 1.0;
 return 0.0;
}

void main(void)
{
 int i;
```

```
int n = 2;
double T = 0.5;

double *x = NULL;
x = new double[n+1];
double *wv = NULL;
wv = new double[n+1];

double **wm = NULL;
wm = new double*[n+1];
for(i=0;i <= n;i++)
wm[i] = new double[n+1];

wv[0] = T; wv[1] = -1.0; wv[2] = -1.0;
wm[0][0] = 0.0; wm[0][1] = 0.0; wm[0][2] = 0.0;
wm[1][0] = 0.0; wm[1][1] = 0.0; wm[1][2] = 3.0;
wm[2][2] = 0.0;

x[0] = 1.0;
// case 1
x[1] = 0.0; x[2] = 0.0;
double r00 = f(x,wv,wm,n);
cout << "r00 = " << r00 << endl;
// case 2
x[1] = 0.0; x[2] = 1.0;
double r01 = f(x,wv,wm,n);
cout << "r01 = " << r01 << endl;
// case 3
x[1] = 1.0; x[2] = 0.0;
double r10 = f(x,wv,wm,n);
cout << "r10 = " << r10 << endl;
// case 4
x[1] = 1.0; x[2] = 1.0;
double r11 = f(x,wv,wm,n);
cout << "r11 = " << r11 << endl;

delete[] x;
delete[] wv;

for(i=0;i <= n;i++) delete[] wm[i];
delete[] wm;
}
```

14.3.5   One and Two Layered Networks

We now consider feed-forward networks structured in successive layers of computing units. The networks we consider must be defined in a more precise way in terms of their architecture. The atomic elements of any architecture are the computing units and their interconnections. Each computing unit collects the information from n input lines with an *integration function* $\Sigma : \mathbf{R}^n \to \mathbf{R}$. The total excitation computed in this way is then evaluated using an *activation function* $f : \mathbf{R} \to \mathbf{R}$. In perceptrons the integration function is the sum of the inputs. The activation, also called output function, compares the sum with a threshold. We can generalize f to produce all values between 0 and 1. In the case of Σ some functions other than addition can also be considered. In this case the networks can compute some difficult functions with fewer computing units.

Definition. A network architecture is a tuple (I, N, O, E) consisting of a set I of input sites, a set N of computing units, a set O of output sites and a set E of weighted directed edges. A directed edge is a tuple (u, v, w) whereby $u \in I \cup N, v \in N \cup O$ and $w \in \mathbf{R}$.

The input sites are entry points for information into the network and do not perform any computation. Results are transmitted to the output sites. The set N consists of all computing elements in the network. The edges between all computing units are weighted, as are the edges between input and output sites and computing units.

Layered architectures are those in which the set of computing units N is subdivided into ℓ subsets $N_1, N_2, \ldots, N_\ell$ in such a way that only connections from units in N_1 go to units in N_2, from units in N_2 to units in N_3, etc. The input sites are only connected to the units in the subset N_1, and the units in the subset N_ℓ are the only ones connected to the output sites. The units in N_ℓ are the output units of the network. The subsets N_i are called the *layers* of the network. The set of input sites is called the *input layer*, the set of output units is called the *output layer*. All other layers with no direct connections from or to the outside are called *hidden layers*. Usually the units in a layer are not connected to each other and the output sites are omitted from the graphical representation. A neural network with a layered architecture does not contain cycles. The input is processed and relayed from the layer to the other, until the final result has been computed.

In layered architectures normally all units from one layer are connected to all other units in the following layer. If there are m units in the first layer and n units in the second one, the total number of weights is mn. The total number of connections can be rather large.

14.3.6 Perceptron Learning Algorithm

Here we use $w_0 = -\theta$ for the first component of $\mathbf{w}$. For the input vectors $\mathbf{x}$ we use $x_0 = 1$ for the first component.

A simple perceptron learning algorithm is

1. Initialize the connection weight $\mathbf{w}$ to small random values.

2. Initialize acceptable error tolerance ϵ_0

3. Set $\epsilon_{\max} = 0$

4. For each of the input patterns $\{\,\mathbf{x}_j,\ j = 0, 1, \ldots, m-1\,\}$ do the following

   (a) Calculate the output y_j via

   $$y_j = H(\mathbf{w}^T \mathbf{x}_j)$$

   where H is the Heaviside function.

   (b) Calculate the difference between the output y_j and the desired output $\tilde{y}_j$ of the network

   $$d_j := \tilde{y}_j - y_j\,.$$

   (c) Calculate the changes in the connection strengths

   $$\Delta \mathbf{w}_j := \eta d_j \mathbf{x}_j$$

   where η is the learning rate.

   (d) Update the connection weight $\mathbf{w}$ according to

   $$\mathbf{w} \leftarrow \mathbf{w} + \Delta \mathbf{w}_j$$

   (e) Set

   $$\epsilon_{\max} \leftarrow \max(\epsilon_{\max}, \|d_j\|)$$

5. If $\epsilon_{\max} > \epsilon_0$ return to step 3.

Example. Consider the AND gate. Let

$$\mathbf{w}^T = (0.2, 0.1, 0.05), \quad \epsilon_0 = 0.01, \quad \eta = 0.5$$

with the input pattern

$$\mathbf{x}_0 = \begin{pmatrix} 1 \\ 0 \\ 0 \end{pmatrix}, \quad \mathbf{x}_1 = \begin{pmatrix} 1 \\ 1 \\ 0 \end{pmatrix}, \quad \mathbf{x}_2 = \begin{pmatrix} 1 \\ 0 \\ 1 \end{pmatrix}, \quad \mathbf{x}_3 = \begin{pmatrix} 1 \\ 1 \\ 1 \end{pmatrix}.$$

The desired output is

$$\tilde{y}_0 = 0, \quad \tilde{y}_1 = 0, \quad \tilde{y}_2 = 0, \quad \tilde{y}_3 = 1.$$

The calculations yield

1) (a) $y_0 = H(\mathbf{w}^T \mathbf{x}_0) = 1 \Rightarrow d_0 = \tilde{y}_0 - y_0 = -1 \Rightarrow \Delta\mathbf{w} = \eta d_0(1,0,0) = (-0.5, 0, 0)$

   $\Rightarrow \mathbf{w} = (-0.3, 0.1, 0.05)$

   (b) $y_1 = H(\mathbf{w}^T \mathbf{x}_1) = 0 \Rightarrow d_1 = \tilde{y}_1 - y_1 = 0 \Rightarrow \Delta\mathbf{w} = 0$

   (c) $y_2 = H(\mathbf{w}^T \mathbf{x}_2) = 0 \Rightarrow d_2 = \tilde{y}_2 - y_2 = 0 \Rightarrow \Delta\mathbf{w} = 0$

   (d) $y_3 = H(\mathbf{w}^T \mathbf{x}_3) = 0 \Rightarrow d_3 = \tilde{y}_3 - y_3 = 1 \Rightarrow \Delta\mathbf{w} = \eta d_3(1,1,1) = (0.5, 0.5, 0.5)$

   $\Rightarrow \mathbf{w} = (0.2, 0.6, 0.55), \quad \epsilon_{max} = 1$

2) (a) $y_0 = H(\mathbf{w}^T \mathbf{x}_0) = 1 \Rightarrow d_0 = -1 \Rightarrow \Delta\mathbf{w} = (-0.5, 0, 0)$

   $\Rightarrow \mathbf{w} = (-0.3, 0.6, 0.55)$

   (b) $y_1 = H(\mathbf{w}^T \mathbf{x}_1) = 1 \Rightarrow d_1 = -1 \Rightarrow \Delta\mathbf{w} = (-0.5, -0.5, 0)$

   $\Rightarrow \mathbf{w} = (-0.8, 0.1, 0.55)$

   (c) $y_2 = H(\mathbf{w}^T \mathbf{x}_2) = 0 \Rightarrow d_2 = 0 \Rightarrow \Delta\mathbf{w} = 0$

   (d) $y_3 = H(\mathbf{w}^T \mathbf{x}_3) = 0 \Rightarrow d_3 = 0 \Rightarrow \Delta\mathbf{w} = (0.5, 0.5.0.5)$

   $\Rightarrow \mathbf{w} = (-0.3, 0.6, 1.05), \quad \epsilon_{max} = 1$

3) (a) $y_0 = H(\mathbf{w}^T \mathbf{x}_0) = 0 \Rightarrow d_0 = 0 \Rightarrow \Delta\mathbf{w} = 0$

   (b) $y_1 = H(\mathbf{w}^T \mathbf{x}_1) = 1 \Rightarrow d_1 = -1 \Rightarrow \Delta\mathbf{w} = (-0.5, -0.5, 0)$

   $\Rightarrow \mathbf{w} = (-0.8, 0.1, 1.05)$

   (c) $y_2 = H(\mathbf{w}^T \mathbf{x}_2) = 1 \Rightarrow d_2 = -1 \Rightarrow \Delta\mathbf{w} = (-0.5, 0, -0.5)$

   $\Rightarrow \mathbf{w} = (-1.3, 0.1, 0.55)$

   (d) $y_3 = H(\mathbf{w}^T \mathbf{x}_3) = 0 \Rightarrow d_3 = 1 \Rightarrow \Delta\mathbf{w} = (0.5, 0.5, 0.5)$

   $\Rightarrow \mathbf{w} = (-0.8, 0.6, 1.05), \quad \epsilon_{max} = 1$

4) (a) $y_0 = H(\mathbf{w}^T \mathbf{x}_0) = 0 \Rightarrow d_0 = 0 \Rightarrow \Delta \mathbf{w} = 0$

   (b) $y_1 = H(\mathbf{w}^T \mathbf{x}_1) = 0 \Rightarrow d_1 = 0 \Rightarrow \Delta \mathbf{w} = 0$

   (c) $y_2 = H(\mathbf{w}^T \mathbf{x}_2) = 1 \Rightarrow d_2 = -1 \Rightarrow \Delta \mathbf{w} = (-0.5, 0.0, -0.5)$

   $\Rightarrow \mathbf{w} = (-1.3, 0.6, 0.55)$

   (d) $y_3 = H(\mathbf{w}^T \mathbf{x}_3) = 0 \Rightarrow d_3 = 1 \Rightarrow \Delta \mathbf{w} = (0.5, 0.5, 0.5)$

   $\Rightarrow \mathbf{w} = (-0.8, 1.1, 1.05), \quad \epsilon_{max} = 1$

5) (a) $y_0 = H(\mathbf{w}^T \mathbf{x}_0) = 0 \Rightarrow d_0 = 0 \Rightarrow \Delta \mathbf{w} = 0$

   (b) $y_1 = H(\mathbf{w}^T \mathbf{x}_1) = 1 \Rightarrow d_1 = -1 \Rightarrow \Delta \mathbf{w} = (-0.5, -0.5, 0.0)$

   $\Rightarrow \mathbf{w} = (-1.3, 0.6, 1.05)$

   (c) $y_2 = H(\mathbf{w}^T \mathbf{x}_2) = 0 \Rightarrow d_2 = 0 \Rightarrow \Delta \mathbf{w} = 0$

   (d) $y_3 = H(\mathbf{w}^T \mathbf{x}_3) = 1 \Rightarrow d_3 = 0, \quad \epsilon_{max} = 1$

6) (a) $y_0 = H(\mathbf{w}^T \mathbf{x}_0) = 0 \Rightarrow d_0 = 0 \Rightarrow \Delta \mathbf{w} = 0$

   (b) $y_1 = H(\mathbf{w}^T \mathbf{x}_1) = 0 \Rightarrow d_1 = 0 \Rightarrow \Delta \mathbf{w} = 0$

   (c) $y_2 = H(\mathbf{w}^T \mathbf{x}_2) = 0 \Rightarrow d_2 = 0 \Rightarrow \Delta \mathbf{w} = 0$

   (d) $y_3 = H(\mathbf{w}^T \mathbf{x}_3) = 1 \Rightarrow d_3 = 0 \Rightarrow \Delta \mathbf{w} = 0, \ \epsilon_{max} = 0$

Thus with

$$\mathbf{w}^T = (0.6, 1.05), \quad \theta = 1.3$$

we can simulate the AND gate.

In the extended space we have

$$\mathbf{w}^T = (w_0, w_1, w_2) = (-\theta, w_1, w_2), \qquad \mathbf{x}^T = (1, x_1, x_2).$$

In the program **percand.cpp** we use the notation of the extended space. Furthermore, the threshold is also initialized to a small random value at $t = 0$.

```cpp
// percand.cpp

#include <iostream.h>
#include <math.h>

double H(double z)
{
   if(z >= 0.0) return 1.0;
   else
   return 0.0;
}

double scalar(double* u, double* v,int n)
{
   double result = 0.0;
   for(int i=0; i<n; i++)
   result += u[i]*v[i];
   return result;
}

double distance(double* u, double* v,int n)
{
   double result = 0.0;
   for(int i=0; i<n; i++)
   result += fabs(u[i] - v[i]);
   return result;
}

void change(double** x,double* yt,double* w,double eta,int m,int n)
{
   double* d = NULL;   d = new double[m];

   for(int j=0; j<m; j++)
   {
   d[j] = yt[j] - H(scalar(w,x[j],n));
   for(int i=0; i<n; i++)
   {
   w[i] = w[i] + eta*d[j]*x[j][i];
   }
   }
   delete [] d;
}

int main()
```

```
{
    // number of input vectors (patterns) is m = 4
    // length of each input vector n = 3
    int m = 4;
    int n = 3;
    double** x = NULL;
    x = new double*[m];
    for(int k=0; k<m; k++)
    x[k] = new double[n];

    x[0][0] = 1.0; x[0][1] = 0.0; x[0][2] = 0.0;
    x[1][0] = 1.0; x[1][1] = 0.0; x[1][2] = 1.0;
    x[2][0] = 1.0; x[2][1] = 1.0; x[2][2] = 0.0;
    x[3][0] = 1.0; x[3][1] = 1.0; x[3][2] = 1.0;

    // desired output
    double* yt = NULL;
    yt = new double [m];
    yt[0] = 0.0; yt[1] = 0.0; yt[2] = 0.0; yt[3] = 1.0;

    // weight vector
    // w[0] = - theta (threshold)
    double* w = NULL;
    w = new double [n];
    // initialized to small random numbers
    w[0] = 0.01; w[1] = 0.005; w[2] = 0.006;

    // learning rate
    double eta = 0.5;

    double* wt = NULL;
    wt = new double[n];
    for(int i=0; i<n; i++)
    wt[i] = w[i];

    for(;;)
    {
    change(x,yt,w,eta,m,n);
    double dist = distance(w,wt,n);
    if(dist < 0.0001) break;
    for(i=0; i<n; i++)
    wt[i] = w[i];
    }

    // display the output of the weight vector
    for(i=0; i<n; i++)
    cout << "w[" << i << "] = " << w[i] << " ";
```

```
   delete [] w;
   delete [] wt;
   delete [] yt;

   for(i=0; i<m; i++)
   {
   delete [] x[i];
   }
   delete [] x;

   return 0;
}
```

The output is given by

```
w[0] = -1.49   w[1] = 1.005   w[2] = 0.506
```

Thus with

$$w_0 = -\theta = -1.49, \qquad w_1 = 1.005, \qquad w_2 = 0.506$$

we can simulate the AND gate.

14.3.7 The XOR Problem and Two-Layered Networks

The properties of a two-layered network can be discussed using the case of the XOR function as an example. A single perceptron cannot compute this function, but a two-layered network can. The network in Figure 4.5 is capable of doing this. The network consists of an input layer, a hidden layer and an output layer and three computing units. One of the units in the hidden layer computes the function $x_1 \wedge \neg x_2$, and the other the function $\neg x_1 \wedge x_2$. The third unit computes the OR function, so that the result of the complete network computation is

$$(x_1 \wedge \neg x_2) \vee (\neg x_1 \wedge x_2).$$

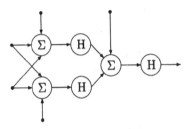

Figure 14.3: A Three-layered Network for the Computation of XOR

The calculations for the XOR gate are as follows. We work in the extended space. The input vectors are

$$\mathbf{x}_0 = \begin{pmatrix} 1 \\ 0 \\ 0 \end{pmatrix}, \quad \mathbf{x}_1 = \begin{pmatrix} 1 \\ 0 \\ 1 \end{pmatrix}, \quad \mathbf{x}_2 = \begin{pmatrix} 1 \\ 1 \\ 0 \end{pmatrix}, \quad \mathbf{x}_3 = \begin{pmatrix} 1 \\ 1 \\ 1 \end{pmatrix}.$$

1) input layer $\longrightarrow$ hidden layer. The weights are

$$w_{000} = -0.5, \quad w_{001} = 1.0, \quad w_{002} = -1.0$$

$$w_{010} = -0.5, \quad w_{011} = -1.0, \quad w_{012} = 1.0.$$

The weight has three indexes. The first index indicates the layer, in this case 0 for the input layer. The second index indicates to which node in the hidden layer it points where the number for the hidden node is incremented by 1 so that we can assign the index 0 to the bias in the hidden layer. The third index indicates the number of the neuron.

Consider the input vector $\mathbf{x}_0$

a) $H((w_{000}, w_{001}, w_{002}) \begin{pmatrix} 1 \\ 0 \\ 0 \end{pmatrix}) = H(-0.5) = 0 = z_0$

b) $H((w_{010}, w_{011}, w_{012}) \begin{pmatrix} 1 \\ 0 \\ 0 \end{pmatrix}) = H(-0.5) = 0 = z_1$

Consider the input vector $\mathbf{x}_1$

a) $H((w_{000}, w_{001}, w_{002}) \begin{pmatrix} 1 \\ 0 \\ 1 \end{pmatrix}) = H(-1.5) = 0 = z_0$

b) $H((w_{010}, w_{011}, w_{012}) \begin{pmatrix} 1 \\ 0 \\ 1 \end{pmatrix}) = H(+0.5) = 1 = z_1$

Consider the input vector $\mathbf{x}_2$

a) $H((w_{000}, w_{001}, w_{002}) \begin{pmatrix} 1 \\ 1 \\ 0 \end{pmatrix}) = H(+0.5) = 1 = z_0$

b) $H((w_{010}, w_{011}, w_{012}) \begin{pmatrix} 1 \\ 1 \\ 0 \end{pmatrix}) = H(-1.5) = 0 = z_1$

Consider the input vector $\mathbf{x}_3$

a) $H((w_{000}, w_{001}, w_{002}) \begin{pmatrix} 1 \\ 1 \\ 1 \end{pmatrix}) = H(-0.5) = 0 = z_0$

b) $H((w_{010}, w_{011}, w_{012}) \begin{pmatrix} 1 \\ 1 \\ 1 \end{pmatrix}) = H(-0.5) = 0 = z_1$

2) hidden layer $\longrightarrow$ output. The input pairs from the hidden layer are $(1, 0, 0)$, $(1, 0, 1)$, $(1, 1, 0)$ and $(1, 0, 0)$. Thus the first and the last patterns are the same. The weights are

$$w_{100} = -0.5, \qquad w_{101} = 1.0, \qquad w_{102} = 1.0.$$

Consider input pattern $(1, 0, 0)$ from hidden layer

a) $H((w_{100}, w_{101}, w_{102}) \begin{pmatrix} 1 \\ 0 \\ 0 \end{pmatrix}) = H(-0.5) = 0$

Consider input pattern $(1, 0, 1)$ from hidden layer

b) $H((w_{100}, w_{101}, w_{102}) \begin{pmatrix} 1 \\ 0 \\ 1 \end{pmatrix}) = H(+0.5) = 1$

Consider input pattern $(1, 1, 0)$ from hidden layer

c) $H((w_{100}, w_{101}, w_{102}) \begin{pmatrix} 1 \\ 1 \\ 0 \end{pmatrix}) = H(+0.5) = 1$

Consider input pattern $(1, 0, 0)$ from hidden layer (already considered above)

d) $H((w_{100}, w_{101}, w_{102}) \begin{pmatrix} 1 \\ 0 \\ 0 \end{pmatrix}) = H(-0.5) = 0$.

Thus we have simulated the XOR gate using a hidden layer.

```cpp
// XOR1.cpp

#include <iostream.h>

double H(double s)
{
   if(s >= 0.0) return 1.0;
   else
   return 0.0;
}


double map(double*** w,double* testpattern,int size2,int size3)
{
    int k;
    double* z = NULL;
    z = new double[size2];
    z[0] = 1.0;  z[1] = 0.0;  z[2] = 0.0;

    // input layer to hidden layer
    for(k=0; k<size3; k++)
    {
    z[1] += w[0][0][k]*testpattern[k];
    z[2] += w[0][1][k]*testpattern[k];
    }
    z[1] = H(z[1]);
    z[2] = H(z[2]);

    // hidden layer to output layer
    double y = 0.0;
    for(k=0; k<size3; k++)
    y += w[1][0][k]*z[k];

    delete [] z;

    y = H(y);
    return y;
}

int main()
{
   int size1, size2, size3;
   size1 = 2; size2 = 2; size3 = 3;
   int i, j, k;

   double*** w = NULL;
   w = new double** [size1];
   for(i=0; i<size1; i++)
   {
```

```
w[i] = new double* [size2];
for(j=0; j<size2; j++)
{
w[i][j] = new double [size3];
}
}
w[0][0][0] = -0.5; w[0][0][1] = 1.0; w[0][0][2] = -1.0;
w[0][1][0] = -0.5; w[0][1][1] = -1.0; w[0][1][2] = 1.0;

w[1][0][0] = -0.5; w[1][0][1] = 1.0; w[1][0][2] = 1.0;
w[1][1][0] = 0.0; w[1][1][1] = 0.0; w[1][1][2] = 0.0;

// input patterns
int p = 4; // number of input pattern
int n = 3; // length of each input pattern
double** x = NULL;
x = new double* [p];
for(int k=0; k<p; k++)
{
x[k] = new double [n];
}
x[0][0] = 1.0; x[0][1] = 0.0; x[0][2] = 0.0;
x[1][0] = 1.0; x[1][1] = 0.0; x[1][2] = 1.0;
x[2][0] = 1.0; x[2][1] = 1.0; x[2][2] = 0.0;
x[3][0] = 1.0; x[3][1] = 1.0; x[3][2] = 1.0;

double result = map(w,x[0],size2,size3);
cout << "result = " << result << endl;     // => 0

result = map(w,x[1],size2,size3);
cout << "result = " << result << endl;     // => 1

result = map(w,x[2],size2,size3);
cout << "result = " << result << endl;     // => 1

result = map(w,x[3],size2,size3);
cout << "result = " << result << endl;     // => 0

return 0;
}
```

14.4 Multilayer Perceptrons

14.4.1 Introduction

In a practical application of the back-propagation algorithm, learning results from the many presentations of a prescribed set of training examples to the multilayer perceptron. One complete presentation of the entire training set during the learning process is called an *epoch*. The learning process is maintained on an epoch-by-epoch basis until the synaptic weights and threshold levels of the network stabilize and the average squared error over the entire training set converges to some minimum value. Randomizing the order of presentation of training examples from one epoch to the next may improve the learning rate. This randomization tends to make the search in weight space stochastic over the learning cycles, thus avoiding the possibility of limit cycles in the evolution of the synaptic weight vectors. We follow in our notation closely Hassoun [82]. For a given training set, back-propagation learning may thus proceed in one of two basic ways.

Let

$$\{\, \mathbf{x}_k, \ \mathbf{d}_k \,\}$$

be the training data, where $k = 0, 1, \ldots, m - 1$. Here m is the number of training examples (patterns). The sets $\mathbf{x}_k$ ($k = 0, 1, \ldots, m-1$) are the input pattern and the sets $\mathbf{d}_k$ are the corresponding (desired) output pattern. One complete presentation of the entire training set during the learning process is called an epoch.

1. *Pattern Mode.* In the *pattern mode* of back-propagation learning, weight updating is performed after the presentation of each training example; this is the mode of operation for which the derivation of the back-propagation algorithm presented here applies. To be specific, consider an epoch consisting of m training examples (patterns) arranged in the order

$$\mathbf{x}_0, \mathbf{d}_0, \quad \mathbf{x}_1, \mathbf{d}_1, \quad \ldots, \quad \mathbf{x}_{m-1}, \mathbf{d}_{m-1} .$$

The first example $\mathbf{x}_0$, $\mathbf{d}_0$ in the epoch is presented to the network, and the sequence of forward and backward computations described below is performed, resulting in certain adjustments to the synaptic weights and threshold levels of the network. Then, the second example $\mathbf{x}(1)$, $\mathbf{d}(1)$ in the epoch is presented, and the sequence of forward and backward computations is repeated, resulting in further adjustments to the synaptic weights and threshold levels. This process is continued until the last training pattern $\mathbf{x}_{m-1}$, $\mathbf{d}_{m-1}$ is taken into account.

2. *Batch Mode.* In the *batch mode* of back-propagation learning, weight updating is performed after the presentation of all the training examples that constitute an epoch.

14.4.2 Cybenko's Theorem

Single-hidden-layer neural networks are universal approximators. A rigorous mathematical proof for the universality of feedforward layered neural networks employing continuous sigmoid type activation functions, as well as other more general activation units, was given by Cybenko [52]. Cybenko's proof is based on the Hahn-Banach theorem. The following is the statement of Cybenko's theorem.

Theorem. Let f be any continuous sigmoid-type function, for example

$$f(s) = 1/(1 + \exp(-\lambda s)), \qquad \lambda \geq 1.$$

Then, given any continuous real-valued function g on $[0,1]^n$ (or any other compact subset of $\mathbf{R}^n$) and $\epsilon > 0$, there exists vectors $\mathbf{w}_1, \mathbf{w}_2, \ldots, \mathbf{w}_N$, $\boldsymbol{\alpha}$, and $\boldsymbol{\theta}$ and a parameterized function

$$G(\cdot, \mathbf{w}, \boldsymbol{\alpha}, \boldsymbol{\theta}) : [0,1]^n \to \mathbf{R}$$

such that

$$|G(\mathbf{x}, \mathbf{w}, \boldsymbol{\alpha}, \boldsymbol{\theta}) - g(\mathbf{x})| < \epsilon \qquad \text{for all} \quad \mathbf{x} \in [0,1]^n$$

where

$$G(\mathbf{x}, \mathbf{w}, \boldsymbol{\alpha}, \boldsymbol{\theta}) = \sum_{j=1}^{N} \alpha_j f(\mathbf{w}_j^T \mathbf{x} + \theta_j)$$

and

$$\mathbf{w}_j \in \mathbf{R}^n, \quad \theta_j \in \mathbf{R}, \quad \mathbf{w} = (\mathbf{w}_1, \mathbf{w}_2, \ldots, \mathbf{w}_N)$$
$$\boldsymbol{\alpha} = (\alpha_1, \alpha_2, \ldots, \alpha_N), \quad \boldsymbol{\theta} = (\theta_1, \theta_2, \ldots, \theta_N).$$

For the proof we refer to the paper by Cybenko [52].

Thus a one hidden layer feedforward neural network is capable of approximating uniformly any continuous multivariate function to any desired degree of accuracy. This implies that any failure of a function mapping by a multilayer network must arise from inadequate choice of parameters, i.e., poor choices for $\mathbf{w}_1, \mathbf{w}_2, \ldots, \mathbf{w}_N$, $\boldsymbol{\alpha}$, and $\boldsymbol{\theta}$ or an insufficient number of hidden nodes.

Hornik et al. [92] employing the Stone-Weierstrass theorem and Funahashi [69] proved similar theorems stating that a one-hidden-layer feedforward neural network is capable of approximating uniformly any continuous multivariate function to any desired degree of accuracy.

14.4.3   Back-Propagation Algorithm

We consider one hidden layer. The notations we use follow closely Hassoun [82]. Thus we consider a two-layer feedforward architecture. This network receives a set of scalar signals

$$x_0, \ x_1, \ x_2, \ldots, x_{n-1}$$

where x_0 is a bias signal set to 1. This set of signals constitutes an input vector $\mathbf{x}_k \in \mathbf{R}^n$. The layer receiving this input signal is called the hidden layer. The hidden layer has J units. The output of the hidden layer is a J dimensional real-valued vector $\mathbf{z}_k = (z_0, z_1, \ldots, z_{J-1})$, where we set $z_0 = 1$ (bias signal). The vector $\mathbf{z}_k$ supplies the input for the output layer of L units. The output layer generates an L-dimensional vector $\mathbf{y}_k$ in response to the input vector $\mathbf{x}_k$ which, when the network is fully trained, should be identical (or very close) to the desired output vector $\mathbf{d}_k$ associated with $\mathbf{x}_k$.

The two activation functions f_h (input layer to hidden layer) and f_o (hidden layer to output layer) are assumed to be differentiable functions. We use the logistic functions

$$f_h(s) := \frac{1}{1 + \exp(-\lambda_h s)}, \qquad f_o(s) := \frac{1}{1 + \exp(-\lambda_o s)}$$

where $\lambda_h, \lambda_o \geq 1$. The *logistic function*

$$f(s) = \frac{1}{1 + \exp(-\lambda s)}$$

satisfies the nonlinear differential equation

$$\frac{df}{ds} = \lambda f(1 - f).$$

The components of the desired output vector $\mathbf{d}_k$ must be chosen within the range of f_o. We denote by w_{ji} the weight of the jth hidden unit associated with the input signal x_i. Thus the index i runs from 0 to $n-1$, where $x_0 = 1$ and j runs from 1 to $J-1$. We set $w_{0i} = 0$. Now we have m input/output pairs of vectors

$$\{\, \mathbf{x}_k, \ \mathbf{d}_k \,\}$$

where the index k runs from 0 to $m-1$. The aim of the algorithm is to adaptively adjust the $(J-1)n + LJ$ weights of the network such that the underlying function/mapping represented by the training set is approximated or learned. We can define an *error function* since the learning is supervised, i.e. the target outputs are available. We denote by w_{lj} the weight of the lth output unit associated with the input signal z_j from the hidden layer. We derive a supervised learning rule for adjusting the weights w_{ji} and w_{lj} such that the error function

$$E(\mathbf{w}) = \frac{1}{2} \sum_{l=0}^{L-1} (d_l - y_l)^2$$

is minimized (in a local sense) over the training set. Here $\mathbf{w}$ represents the set of all weights in the network.

Since the targets for the output units are given, we can use the delta rule directly for updating the w_{lj} weights. We define

$$\Delta w_{lj} := w_{lj}^{new} - w_{lj}^c.$$

Since

$$\Delta w_{lj} = -\eta_o \frac{\partial E}{\partial w_{lj}}$$

we find using the chain rule

$$\Delta w_{lj} = \eta_o (d_l - y_l) f_o'(net_l) z_j$$

where $l = 0, 1, \ldots, L-1$ and $j = 0, 1, \ldots, J-1$. Here

$$net_l := \sum_{j=0}^{J-1} w_{lj} z_j$$

is the weighted sum for the lth output unit, f_o' is the derivative of f_o with respect to net_l, and w_{lj}^{new} and w_{lj}^c are the updated (new) and current weight values, respectively. The z_j values are calculated by propagating the input vector $\mathbf{x}$ through the hidden layer according to

$$z_j = f_h \left(\sum_{i=0}^{n-1} w_{ji} x_i \right) = f_h(net_j)$$

where $j = 1, 2, \ldots, J - 1$ and $z_0 = 1$ (bias signal). For the hidden-layer weights w_{ji} we do not have a set of target values (desired outputs) for hidden units. However, we can derive the learning rule for hidden units by attempting to minimize the output-layer error. This amounts to propagating the output errors $(d_l - y_l)$ back through the output layer toward the hidden units in an attempt to estimate dynamic targets for these units. Thus a gradient descent is performed on the criterion function

$$E(\mathbf{w}) = \frac{1}{2} \sum_{l=0}^{L-1} (d_l - y_l)^2$$

where $\mathbf{w}$ represents the set of all weights in the network. The gradient is calculated with respect to the hidden weights

$$\Delta w_{ji} = -\eta_h \frac{\partial E}{\partial w_{ji}}, \qquad j = 1, 2, \ldots, J - 1, \qquad i = 0, 1, \ldots, n - 1$$

where the partial derivative is to be evaluated at the current weight values. We find

$$\frac{\partial E}{\partial w_{ji}} = \frac{\partial E}{\partial z_j} \frac{\partial z_j}{\partial net_j} \frac{\partial net_j}{\partial w_{ji}}$$

where

$$\frac{\partial net_j}{\partial w_{ji}} = x_i, \qquad \frac{\partial z_j}{\partial net_j} = f_h'(net_j).$$

We used the chain rule in this derivation. Since

$$\frac{\partial E}{\partial z_j} = - \sum_{l=0}^{L-1} (d_l - y_l) f_o'(net_l) w_{lj}$$

we obtain

$$\Delta w_{ji} = \eta_h \left(\sum_{l=0}^{L-1} (d_l - y_l) f_o'(net_l) w_{lj} \right) f_h'(net_j) x_i.$$

Now we can define an estimated target d_j for the jth hidden unit implicitly in terms of the backpropagated error signal as follows

$$d_j - z_j := \sum_{l=0}^{L-1}(d_l - y_l)f'_o(net_l)w_{lj}.$$

The complete approach for updating weights in a feedforward neural net utilizing these rules can be summarized as follows. We do a pattern-by-pattern updating of the weights.

1. *Initialization.* Initialize all weights to small random values and refer to them as current weights w^c_{lj} and w^c_{ji}.

2. *Learning rate.* Set the learning rates η_o and η_h to small positive values.

3. *Presentation of training example.* Select an input pattern $\mathbf{x}_k$ from the training set (preferably at random) propagate it through the network, thus generating hidden- and output-unit activities based on the current weight settings. Thus find z_j and y_l.

4. *Forward computation.* Use the desired target vector $\mathbf{d}_k$ associated with $\mathbf{x}_k$, and employ

$$\Delta w_{lj} = \eta_o(d_l - y_l)f'(net_l)z_j = \eta_o(d_l - y_l)\lambda_o f(net_l)(1 - f(net_l))z_j$$

to compute the output layer weight changes Δw_{lj}.

5. *Backward computation.* Use

$$\Delta w_{ji} = \eta_h \left(\sum_{l=0}^{L-1}(d_l - y_l)f'_o(net_l)w_{lj} \right) f'_h(net_j)x_i$$

or

$$\Delta w_{ji} = \eta_h \left(\sum_{l=0}^{L-1}(d_l - y_l)\lambda_o f_o(net_l)(1 - f_o(net_l))w_{lj} \right) \lambda_h f_h(net_j)(1 - f_h(net_j))x_i$$

to compute the hidden layer weight changes. The current weights are used in these computations. In general, enhanced error correction may be achieved if one employs the updated output-layer weights

$$w_{lj}^{new} = w_{lj}^c + \Delta w_{lj}.$$

However, this comes at the added cost of recomputing y_l and $f'(net_l)$.

6. *Update weights.* Update all weights according to

$$w_{ji}^{new} = w_{ji}^c + \Delta w_{ji}$$

and

$$w_{lj}^{new} = w_{lj}^c + \Delta w_{lj}$$

for the output and for the hidden layers, respectively.

7. *Test for convergence.* This is done by checking the output error function to see if its magnitude is below some given threshold. Iterate the computation by presenting new epochs of training examples to the network until the free parameters of the network stabilize their values. The order of presentation of training examples should be randomized from epoch to epoch. The learning rate parameter is typically adjusted (and usually decreased) as the number of training iterations increases.

An example of the back-propagation algorithm applied to the XOR problem is given in [164]. In the C++ program we apply the back-propagation algorithm to the parity function, where $m = 16$ is the number of input vectors each of length 5 (includes the bias input). The training set is given in Table 14.2. The number of hidden layer units is 5 which includes the bias input $z_0 = 1$. The neural network must calculate the parity bit such that the parity is even. By modifying m, n, J and L the program can easily be adapted to other problems. The arrays x[i] are the input values. The value x[i][0] is always 1 for the threshold. The arrays d[i] are the desired outputs for each input x[i]. In this case d[i] is the odd-parity bit calculated from x[i][1]-x[i][4]. In the program the value y[0], after each calculation, gives the neural network approximation of the parity calculation.

The following table gives the training set for the odd parity function over four bits. The equation is

$$P = \overline{A_3 \oplus A_2 \oplus A_1 \oplus A_0}$$

where P is the odd parity function and A_0, A_1, A_2 and A_3 are the inputs.

Inputs				Parity
0	0	0	0	1
0	0	0	1	0
0	0	1	0	0
0	0	1	1	1
0	1	0	0	0
0	1	0	1	1
0	1	1	0	1
0	1	1	1	0
1	0	0	0	0
1	0	0	1	1
1	0	1	0	1
1	0	1	1	0
1	1	0	0	1
1	1	0	1	0
1	1	1	0	0
1	1	1	1	1

Table 14.2: Training Set for Parity Function

```cpp
// backpr2.cpp
// back propagation

#include <iostream>
#include <math.h>       // for exp

using namespace std;

// activation function (input layer -> hidden layer)
double fh(double net)
{
   double lambdah = 10.0;
   return 1.0/(1.0 + exp(-lambdah*net));
}

// activation function (hidden layer -> output layer)
double fo(double net)
{
   double lambdao = 10.0;
   return 1.0/(1.0 + exp(-lambdao*net));
}

double scalar(double* a1, double* a2, int length)
{
   double result = 0.0;
```

```
    for(int i=0; i<length; i++)
    {
    result += a1[i]*a2[i];
    }
    return result;
}


int main()
{
    int k, i, j, l, t;  // summation index
    // k runs over all input pattern k = 0, 1, .. , m-1
    // l runs over all output units l = 0, 1, .. , L-1
    // j runs over all the hidden layer units j = 0, 1, .. , J-1
    // i runs over the length of the input vector i = 0, 1, .. , n-1

    // learning rates
    double etao = 0.05;
    double etah = 0.05;

    double lambdao = 10.0;
    double lambdah = 10.0;

    // memory allocations
    double** x = NULL;
    int m = 16;  // number of input vectors for Parity problem
    int n = 5;   // length of each input vector for Parity problem
    // input vectors
    x = new double* [m];
    for(k=0; k<m; k++) x[k] = new double [n];

    x[0][0] = 1.0; x[0][1] = 0.0; x[0][2] = 0.0; x[0][3] = 0.0;
    x[0][4] = 0.0;
    x[1][0] = 1.0; x[1][1] = 0.0; x[1][2] = 0.0; x[1][3] = 0.0;
    x[1][4] = 1.0;
    x[2][0] = 1.0; x[2][1] = 0.0; x[2][2] = 0.0; x[2][3] = 1.0;
    x[2][4] = 0.0;
    x[3][0] = 1.0; x[3][1] = 0.0; x[3][2] = 0.0; x[3][3] = 1.0;
    x[3][4] = 1.0;
    x[4][0] = 1.0; x[4][1] = 0.0; x[4][2] = 1.0; x[4][3] = 0.0;
    x[4][4] = 0.0;
    x[5][0] = 1.0; x[5][1] = 0.0; x[5][2] = 1.0; x[5][3] = 0.0;
    x[5][4] = 1.0;
    x[6][0] = 1.0; x[6][1] = 0.0; x[6][2] = 1.0; x[6][3] = 1.0;
    x[6][4] = 0.0;
    x[7][0] = 1.0; x[7][1] = 0.0; x[7][2] = 1.0; x[7][3] = 1.0;
    x[7][4] = 1.0;
    x[8][0] = 1.0; x[8][1] = 1.0; x[8][2] = 0.0; x[8][3] = 0.0;
    x[8][4] = 0.0;
```

```
x[9][0] = 1.0; x[9][1] = 1.0; x[9][2] = 0.0; x[9][3] = 0.0;
x[9][4] = 1.0;
x[10][0]= 1.0; x[10][1]= 1.0; x[10][2]= 0.0; x[10][3]= 1.0;
x[10][4]= 0.0;
x[11][0]= 1.0; x[11][1]= 1.0; x[11][2]= 0.0; x[11][3]= 1.0;
x[11][4]= 1.0;
x[12][0]= 1.0; x[12][1]= 1.0; x[12][2]= 1.0; x[12][3]= 0.0;
x[12][4]= 0.0;
x[13][0]= 1.0; x[13][1]= 1.0; x[13][2]= 1.0; x[13][3]= 0.0;
x[13][4]= 1.0;
x[14][0]= 1.0; x[14][1]= 1.0; x[14][2]= 1.0; x[14][3]= 1.0;
x[14][4]= 0.0;
x[15][0]= 1.0; x[15][1]= 1.0; x[15][2]= 1.0; x[15][3]= 1.0;
x[15][4]= 1.0;


// desired output vectors
// corresponding to set of input vectors x
double** d = NULL;
// number of outputs for Parity problem
int L = 1;
d = new double* [m];
for(k=0; k<m; k++) d[k] = new double [L];
d[0][0] = 1.0;  d[1][0] = 0.0;  d[2][0] = 0.0;  d[3][0] = 1.0;
d[4][0] = 0.0;  d[5][0] = 1.0;  d[6][0] = 1.0;  d[7][0] = 0.0;
d[8][0] = 0.0;  d[9][0] = 1.0;  d[10][0]= 1.0;  d[11][0]= 0.0;
d[12][0]= 1.0;  d[13][0]= 0.0;  d[14][0]= 0.0;  d[15][0]= 1.0;


// error function for each input vector
double* E = NULL;
E = new double [m];

double totalE = 0.0; // sum of E[k] k = 0, 1, .. , m


// weight matrix (input layer -> hidden layer);
// number of hidden layers includes 0
// current
int J = 5;
double** Wc = NULL;
Wc = new double* [J];
for(j=0; j<J; j++) Wc[j] = new double [n];

Wc[0][0] = 0.0;  Wc[0][1] = 0.0;  Wc[0][2] = 0.0;  Wc[0][3] = 0.1;
Wc[0][4] = -0.2;
Wc[1][0] = -0.2; Wc[1][1] = 0.5;  Wc[1][2] = -0.5; Wc[1][3] = 0.3;
Wc[1][4] = 0.1;
Wc[2][0] = -0.3; Wc[2][1] = -0.3; Wc[2][2] = 0.7;  Wc[2][3] = 0.1;
Wc[2][4] = -0.2;
Wc[3][0] = 0.2;  Wc[3][1] = 0.1;  Wc[3][2] = 0.5;  Wc[3][3] = -0.3;
```

```
  Wc[3][4] = -0.1;
  Wc[4][0] = -0.3; Wc[4][1] = -0.1; Wc[4][2] = 0.1;  Wc[4][3] = 0.3;
  Wc[4][4] = 0.2;

  // new
  double** Wnew = NULL;
  Wnew = new double* [J];
  for(j=0; j<J; j++) Wnew[j] = new double [n];

  // weight matrix (hidden layer -> output layer)
  // current
  double** Whc = NULL;
  Whc = new double* [L];
  for(l=0; l<L; l++) Whc[l] = new double [J];

  Whc[0][0] = -0.2; Whc[0][1] = 0.3; Whc[0][2] = 0.5;

  // new
  double** Whnew = NULL;   Whnew = new double* [L];
  for(l=0; l<L; l++) Whnew[l] = new double [J];

  // vector in hidden layer
  double* z = NULL;   z = new double [J];
  z[0] = 1.0;

  // vector output layer (output layer units)
  // for the Parity problem the output layer has only one element
  double* y = NULL;   y = new double [L];

  // increment matrix (input layer -> hidden layer)
  double** delW = NULL;   delW = new double* [J];
  for(j=0; j<J; j++) delW[j] = new double [n];

  // increment matrix (hidden layer -> output layer)
  double** delWh = NULL;   delWh = new double* [L];
  for(l=0; l<L; l++) delWh[l] = new double [J];

  // net vector (input layer -> hidden layer)
  double* netj = NULL;   netj = new double [J];
  netj[0] = 0.0;

  // net vector (hidden layer -> output layer)
  double* netl = NULL;   netl = new double [L];

  // training session
  int T = 10000; // number of iterations
  for(t=0; t<T; t++)
  {
```

```
// for loop over all input pattern
for(k=0; k<m; k++)
{

for(j=1; j<J; j++)
{
netj[j] = scalar(x[k],Wc[j],n);
z[j] = fh(netj[j]);
}

for(l=0; l<L; l++)
{
netl[l] = scalar(z,Whc[l],J);
y[l] = fo(netl[l]);
}

for(l=0; l<L; l++)
for(j=0; j<J; j++)
delWh[l][j] =
etao*(d[k][l]-y[l])*lambdao*fo(netl[l])*(1.0-fo(netl[l]))*z[j];

double* temp = NULL;
temp = new double [J];
for(j=0; j<J; j++)
temp[j] = 0.0;

for(j=0; j<J; j++)
for(l=0; l<L; l++)
temp[j] +=
(d[k][l]-y[l])*fo(netl[l])*(1.0-fo(netl[l]))*Whc[l][j];

for(j=0; j<J; j++)
for(i=0; i<n; i++)
delW[j][i] =
etah*temp[j]*lambdah*fh(netj[j])*(1.0-fh(netj[j]))*x[k][i];

for(i=0; i<n; i++)
delW[0][i] = 0.0;

// updating the weight matrices
for(j=0; j<J; j++)
for(i=0; i<n; i++)
Wnew[j][i] = Wc[j][i] + delW[j][i];

for(l=0; l<L; l++)
for(j=0; j<J; j++)
Whnew[l][j] = Whc[l][j] + delWh[l][j];
```

```
// setting new to current
for(j=0; j<J; j++)
for(i=0; i<n; i++)
Wc[j][i] = Wnew[j][i];

for(l=0; l<L; l++)
for(j=0; j<J; j++)
Whc[l][j] = Whnew[l][j];

E[k] = 0.0;
double sum = 0.0;
for(l=0; l<L; l++)
sum += (d[k][l] - y[l])*(d[k][l] - y[l]);

E[k] = sum/2.0;
totalE += E[k];
}  // end for loop over all input pattern
if(totalE < 0.0005) goto L;
else totalE = 0.0;
}  // end training session

L:
cout << "number of iterations = " << t << endl;

// output after training
for(j=0; j<J; j++)
for(i=0; i<n; i++)
cout << "Wc[" << j << "][" << i << "] = "
     << Wc[j][i] << endl;
cout << endl;

for(l=0; l<L; l++)
for(j=0; j<J; j++)
cout << "Whc[" << l << "][" << j << "] = "
     << Whc[l][j] << endl;

// testing the Parity function
// input (1,0,0,0,0)
for(j=1; j<J; j++)
{
netj[j] = scalar(x[0],Wc[j],n);
z[j] = fh(netj[j]);
}

for(l=0; l<L; l++)
{
netl[l] = scalar(z,Whc[l],J);
y[l] = fo(netl[l]);
```

```
cout << "y[" << 1 << "] = " << y[1] << endl;
}


// input (1,0,0,0,1)
for(j=1; j<J; j++)
{
netj[j] = scalar(x[1],Wc[j],n);
z[j] = fh(netj[j]);
}


for(l=0; l<L; l++)
{
netl[l] = scalar(z,Whc[l],J);
y[l] = fo(netl[l]);
cout << "y[" << 1 << "] = " << y[1] << endl;
}


// input (1,0,0,1,0)
for(j=1; j<J; j++)
{
netj[j] = scalar(x[2],Wc[j],n);
z[j] = fh(netj[j]);
}


for(l=0; l<L; l++)
{
netl[l] = scalar(z,Whc[l],J);
y[l] = fo(netl[l]);
cout << "y[" << 1 << "] = " << y[1] << endl;
}


// input (1,0,0,1,1)
for(j=1; j<J; j++)
{
netj[j] = scalar(x[3],Wc[j],n);
z[j] = fh(netj[j]);
}
for(l=0; l<L; l++)
{
netl[l] = scalar(z,Whc[l],J);
y[l] = fo(netl[l]);
cout << "y[" << 1 << "] = " << y[1] << endl;
}


// input (1,0,1,0,0)
for(j=1; j<J; j++)
{
netj[j] = scalar(x[4],Wc[j],n);
```

```cpp
    z[j] = fh(netj[j]);
    }
    for(l=0; l<L; l++)
    {
    netl[l] = scalar(z,Whc[l],J);
    y[l] = fo(netl[l]);
    cout << "y[" << l << "] = " << y[l] << endl;
    }


    // input (1,0,1,0,1)
    for(j=1; j<J; j++)
    {
    netj[j] = scalar(x[5],Wc[j],n);
    z[j] = fh(netj[j]);
    }
    for(l=0; l<L; l++)
    {
    netl[l] = scalar(z,Whc[l],J);
    y[l] = fo(netl[l]);
    cout << "y[" << l << "] = " << y[l] << endl;
    }


    // input (1,0,1,1,0)
    for(j=1; j<J; j++)
    {
    netj[j] = scalar(x[6],Wc[j],n);
    z[j] = fh(netj[j]);
    }
    for(l=0; l<L; l++)
    {
    netl[l] = scalar(z,Whc[l],J);
    y[l] = fo(netl[l]);
    cout << "y[" << l << "] = " << y[l] << endl;
    }


    // input (1,0,1,1,1)
    for(j=1; j<J; j++)
    {
    netj[j] = scalar(x[7],Wc[j],n);
    z[j] = fh(netj[j]);
    }
    for(l=0; l<L; l++)
    {
    netl[l] = scalar(z,Whc[l],J);
    y[l] = fo(netl[l]);
    cout << "y[" << l << "] = " << y[l] << endl;
    }
```

```
// input (1,1,0,0,0)
for(j=1; j<J; j++)
{
netj[j] = scalar(x[8],Wc[j],n);
z[j] = fh(netj[j]);
}
for(l=0; l<L; l++)
{
netl[l] = scalar(z,Whc[l],J);
y[l] = fo(netl[l]);
cout << "y[" << l << "] = " << y[l] << endl;
}

// input (1,1,0,0,1)
for(j=1; j<J; j++)
{
netj[j] = scalar(x[9],Wc[j],n);
z[j] = fh(netj[j]);
}
for(l=0; l<L; l++)
{
netl[l] = scalar(z,Whc[l],J);
y[l] = fo(netl[l]);
cout << "y[" << l << "] = " << y[l] << endl;
}

// input (1,1,0,1,0)
for(j=1; j<J; j++)
{
netj[j] = scalar(x[10],Wc[j],n);
z[j] = fh(netj[j]);
}
for(l=0; l<L; l++)
{
netl[l] = scalar(z,Whc[l],J);
y[l] = fo(netl[l]);
cout << "y[" << l << "] = " << y[l] << endl;
}

// input (1,1,0,1,1)
for(j=1; j<J; j++)
{
netj[j] = scalar(x[11],Wc[j],n);
z[j] = fh(netj[j]);
}
for(l=0; l<L; l++)
{
netl[l] = scalar(z,Whc[l],J);
```

```cpp
y[l] = fo(netl[l]);
cout << "y[" << l << "] = " << y[l] << endl;
}


// input (1,1,1,0,0)
for(j=1; j<J; j++)
{
netj[j] = scalar(x[12],Wc[j],n);
z[j] = fh(netj[j]);
}
for(l=0; l<L; l++)
{
netl[l] = scalar(z,Whc[l],J);
y[l] = fo(netl[l]);
cout << "y[" << l << "] = " << y[l] << endl;
}


// input (1,1,1,0,1)
for(j=1; j<J; j++)
{
netj[j] = scalar(x[13],Wc[j],n);
z[j] = fh(netj[j]);
}
for(l=0; l<L; l++)
{
netl[l] = scalar(z,Whc[l],J);
y[l] = fo(netl[l]);
cout << "y[" << l << "] = " << y[l] << endl;
}


// input (1,1,1,1,0)
for(j=1; j<J; j++)
{
netj[j] = scalar(x[14],Wc[j],n);
z[j] = fh(netj[j]);
}
for(l=0; l<L; l++)
{
netl[l] = scalar(z,Whc[l],J);
y[l] = fo(netl[l]);
cout << "y[" << l << "] = " << y[l] << endl;
}


// input (1,1,1,1,1)
for(j=1; j<J; j++)
{
netj[j] = scalar(x[15],Wc[j],n);
z[j] = fh(netj[j]);
```

```
    }
    for(l=0; l<L; l++)
    {
    netl[l] = scalar(z,Whc[l],J);
    y[l] = fo(netl[l]);
    cout << "y[" << l << "] = " << y[l] << endl;
    }

    return 0;
}
```

The output is

```
number of iterations = 10000
Wc[0][0] = 0
Wc[0][1] = 0
Wc[0][2] = 0
Wc[0][3] = 0.1
Wc[0][4] = -0.2
Wc[1][0] = -0.890614
Wc[1][1] = 0.199476
Wc[1][2] = -0.592286
Wc[1][3] = 0.605594
Wc[1][4] = 0.604114
Wc[2][0] = -0.379614
Wc[2][1] = -0.777377
Wc[2][2] = 0.777529
Wc[2][3] = 0.758172
Wc[2][4] = 0.760994
Wc[3][0] = 0.538437
Wc[3][1] = 0.372678
Wc[3][2] = 0.512117
Wc[3][3] = -0.656055
Wc[3][4] = -0.65043
Wc[4][0] = -0.0856427
Wc[4][1] = -0.165472
Wc[4][2] = 0.161642
Wc[4][3] = 0.151453
Wc[4][4] = 0.151421

Whc[0][0] = -2.05814
Whc[0][1] = 1.47181
Whc[0][2] = -2.45669
Whc[0][3] = 1.37033
Whc[0][4] = 3.96504
y[0] = 0.987144
```

```
y[0]  = 5.96064e-07
y[0]  = 5.32896e-07
y[0]  = 0.989954
y[0]  = 0.0183719
y[0]  = 0.986117
y[0]  = 0.98594
y[0]  = 0.0110786
y[0]  = 0.0200707
y[0]  = 0.998834
y[0]  = 0.998846
y[0]  = 0.00840843
y[0]  = 0.983464
y[0]  = 0.00589264
y[0]  = 0.00599696
y[0]  = 0.996012
```

The values y[0] approximate the parity function.

Chapter 15
Genetic Algorithms

15.1 Introduction

Evolutionary methods have gained considerable popularity as general-purpose robust optimization and search techniques. The failure of traditional optimization techniques in searching complex, uncharted and vast-payoff landscapes riddled with multimodality and complex constraints has generated interest in alternate approaches.

Genetic algorithms (Holland [89], Goldberg [72], Michalewicz [116], Steeb [164]) are self-adapting strategies for searching, based on the random exploration of the solution space coupled with a memory component which enables the algorithms to learn the optimal search path from experience. They are the most prominent, widely used representatives of evolutionary algorithms, a class of probabilistic search algorithms based on the model of organic evolution. The starting point of all evolutionary algorithms is the *population* (also called *farm*) of *individuals* (also called *animals*, *chromosomes*, *strings*). The individuals are composed of genes which may take on a number of values (in most cases 0 and 1) called alleles. The value of a gene is called its allelic value, and it ranges on a set that is usually restricted to $\{0, 1\}$. Thus these individuals are represented as binary strings of fixed length, for example

```
"10001011101"
```

Each individual can be uniquely represented as an unsigned integer. For example the bit string given above corresponds to the integer

$$1 \cdot 2^{10} + 0 \cdot 2^9 + 0 \cdot 2^8 + 0 \cdot 2^7 + 1 \cdot 2^6 + 0 \cdot 2^5 + 1 \cdot 2^4 + 1 \cdot 2^3 + 1 \cdot 2^2 + 0 \cdot 2^1 + 1 \cdot 2^0 = 1117.$$

If the binary string has length N, then 2^N binary strings can be formed. If we describe a DNA molecule the alphabet would be a set of 4 symbols, $\{A, C, G, T\}$ where A stands for Adenine, C stands for Cytosine, G stands for Guanine and T stands for Thymine. Strings of length N from this set allow for 4^N different individuals. We can also associate unsigned integers with these strings.

For example

"TCCGAT"

is associated with the integer

$$3 \cdot 4^5 + 1 \cdot 4^4 + 1 \cdot 4^3 + 2 \cdot 4^2 + 0 \cdot 4^1 + 3 \cdot 4^0 = 3427.$$

For the four colour problem we also use an alphabet of 4 symbols, $\{R, G, B, Y\}$ where R stands for red, G stands for green, B stands for blue and Y stands for yellow.

Each of the individuals represents a search point in the space of potential solutions to a given optimization problem. Then random operators model selection, reproduction, crossover and mutation. The optimization problem gives quality information (fitness function or short fitness) for the individuals and the selection process favours individuals of higher fitness to transfer their information (string) to the next generation. The fitness of each string is the corresponding function value. Genetic algorithms are specifically designed to treat problems involving large search spaces containing multiple local minima. The algorithms have been applied to a large number of optimization problems. Examples are solutions of ordinary differential equations, the smooth genetic algorithm, genetic algorithms in coding theory, Markov chain analysis, the DNA molecule.

In the fundamental approach to finding an optimal solution, a *fitness function* (also called *cost function*) is used to represent the quality of the solution. The objective function to be optimized can be viewed as a multidimensional surface where the height of a point on the surface gives the value of the function at that point. In case of a minimization problem, the wells represent high-quality solutions while the peaks represent low-quality solutions. In case of a maximization problem, the higher the point in the topography, the better the solution.

The search techniques can be classified into three basic categories.

(1) *Classical or calculus-based.* This uses a deterministic approach to find the best solution. This method requires the knowledge of the gradient or higher-order derivatives. The technique can be applied to well-behaved problems.

(2) *Enumerative.* With these methods, all possible solutions are generated and tested to find the optimal solution. This requires excessive computation in problems involving a large number of variables.

(3) *Random.* Guided random search methods are enumerative in nature; however, they use additional information to guide the search process. Simulated annealing and evolutionary algorithms are typical examples of this class of search methods.

15.2   The Sequential Genetic Algorithm

The genetic algorithm evolves a multiset of elements called a population of individuals or farm of animals. Each individual A_i $(i = 1, \cdots, n)$ of the population $\mathbf{A}$ represents a trial solution of the optimalization problem to be solved. Individuals are usually represented by strings of variables, each element of which is called a gene. The value of a gene is called its allelic value, and it ranges on a set that is usually restricted to $\{0, 1\}$.

The population of individuals is also called a farm of animals in the literature. Furthermore an individual or animal is also called a chromosome or string.

A genetic algorithm is capable of maximizing a given fitness function f computed on each individual of the population. If the problem is to minimize a given objective function, then it is required to map increasing objective function values into decreasing f values. This can be achieved by a monotonically decreasing function. The standard genetic algorithm is the following sequence:

Step 1. Randomly generate an initial population $\mathbf{A}(0) := (A_1(0), \cdots, A_n(0))$.

Step 2. Compute the fitness $f(A_i(t))$ of each individual $A_i(t)$ of the current population $\mathbf{A}(t)$.

Step 3. Generate an intermediate population $\mathbf{A}_r(t)$ by applying the reproduction operator.

Step 4. Generate $\mathbf{A}(t + 1)$ by applying some other operators to $\mathbf{A}_r(t)$.

Step 5: $t := t + 1$ if not *(end_test)* goto Step 2.

The most commonly used operators are the following:

1) Reproduction (selection). This operator produces a new population, $\mathbf{A}_r(t)$, extracting with repetition individuals from the old population, $\mathbf{A}(t)$. The extraction can be carried out in several ways. One of the most commonly used method is the roulette wheel selection, where the extraction probability $p_r(A_i(t))$ of each individual $A_i(t)$ is proportional to its fitness $f(A_i(t))$.

2) Crossover. This operator is applied in probability, where the crossover probability is a system parameter, p_c. To apply the standard crossover operator (several variations have been proposed) the individuals of the population are randomly paired. Each pair is then recombined, choosing one point in accordance with a uniformly distributed probability over the length of the individual strings (parents) and cutting them in two parts accordingly. The new individuals (offspring) are formed by the juxtaposition of the first part of one parent and the last part of the other parent.

3) Mutation. The standard mutation operator modifies each allele of each individual of the population in probability, where the mutation probability is a system parameter, p_m. Usually, the new allelic value is randomly chosen with uniform probability distribution.

4) Local search. The necessity of this operator for optimization problems is still under debate. Local search is usually a simple gradient-descent heuristic search that carries each solution to a local optimum. The idea behind this is that search in the space of local optima is much more effective than search in the whole solution space.

The purpose of *parent selection* (also called setting up the farm of animals) in a genetic algorithm is to give more reproductive chances, on the whole, to those population members that are the most fit. We use a binary string as a chromosome to represent real value of the variable x. The length of the binary string depends on the required precision. A population or farm could look like

```
"10101110011111110"
"00111101010100001"
. . . . . . . . . . . . . . . .
"11111110101010111" <- individual (chromosome, animal, string)
. . . . . . . . . . . . . . . .
"10101110010000110"
```

For the *crossover operation* the individuals of the population are randomly paired. Each pair is then recombined, choosing one point in accordance with a uniformly distributed probability over the length of the individual strings (parents) and cutting them in two parts, accordingly. The new individuals (offspring) are formed by the part of one part and the last part of the other. An example is

```
1011011000100101   parent
0010110110110111   parent
     |        |
1011010110110101   child
0010111000100111   child
```

The *mutation operator* modifies each allele (a bit in the bitstring) of each individual of the population in probability. The new allele value is randomly chosen with uniform probability distribution. An example is

```
1011011001011001   parent
     |
1011111001011001   child
```

The bit position is randomly selected. Whether the child is selected is decided by the fitness function.

We have to map the binary string into a real number x with a given interval $[a, b]$ $(a < b)$. The length of the binary string depends on the required precision. The total length of the interval is $b - a$. The binary string is denoted by

$$s_{N-1} s_{N-2} \ldots s_1 s_0$$

where s_0 is the least significant bit (LSB) and s_{N-1} is the most significant bit (MSB). In the first step we convert from base 2 to base 10

$$m = \sum_{i=0}^{N-1} s_i 2^i .$$

In the second step we calculate the corresponding real number on the interval $[a, b]$

$$x = a + m \frac{b - a}{2^N - 1} .$$

Obviously if the bit string is given by "000...00" we obtain $x = a$ and if the bitstring is given by "111...11" we obtain $x = b$.

We consider the two-dimensional case. The extension to higher dimensions is straightforward. Consider the two-dimensional domain

$$[a, b] \times [c, d]$$

which is a subset of $\mathbf{R}^2$. The coordinates are x_1 and x_2, i.e. $x_1 \in [a, b]$ and $x_2 \in [c, d]$. Given a bitstring

$$s_{N-1} s_{N-2} \ldots s_{N_1} s_{N_1-1} s_{N_1-2} \ldots s_1 s_0$$

of length

$$N = N_1 + N_2 .$$

The block

$$s_{N_1-1} s_{N_1-2} \ldots s_1 s_0$$

is identified with m_1, i.e.

$$m_1 = \sum_{i=0}^{N_1-1} s_i 2^i$$

and therefore

$$x_1 = a + m_1 \frac{b-a}{2^{N_1} - 1}.$$

The block

$$s_{N-1} s_{N-2} \dots s_{N_1}$$

is identified with the variable m_2, i.e.

$$m_2 = \sum_{i=N_1}^{N-1} s_i 2^{i-N_1}$$

and therefore

$$x_2 = c + m_2 \frac{d-c}{2^{N_2} - 1}.$$

where $N_2 = N - N_1$.

Example. In the one-dimensional case consider the binary string 10101101 of length 8 and the interval $[-1, 1]$. Therefore

$$m = 1 \cdot 2^0 + 1 \cdot 2^2 + 1 \cdot 2^3 + 1 \cdot 2^5 + 1 \cdot 2^7 = 173.$$

Thus

$$x = -1 + 173 \frac{2}{256 - 1} = 0.357.$$

Example. In the two-dimensional case consider the binary string 0000000000000000 with $N_1 = N_2 = 8$ and the domain $[-1, 1] \times [-1, 1]$. Then we find $m_1 = m_2 = 0$, $x_1 = -1$ and $x_2 = -1$.

Reversing a bit string can also be used as a technique to introduce variation in genetic algorithms. The operation is useful for implementing the Fourier transform. It is quite simple to reverse a bit sequence, for example the following C++ program implements the operation on integers. The size of the data type **unsigned int** is 4 bytes (32 bits). For each least significant bit of i is place in the least significant bit position of r. Then i is shifted right and r is shifted left. The process is repeated for each bit in i.

```
// reverse.cpp

#include <iostream.h>

unsigned int reverse(unsigned int i)
{
 int j;
 unsigned int r = 0;
 int len = sizeof(i)*8;

 for(j=0;j < len;j++)
 {
  r = r*2 + (i%2);
  i /= 2;
 }

 return r;
}

void main(void)
{
 cout << reverse(23) << endl;

 // The output is 3892314112
}
```

Since 23 is the bitstring

00000000 00000000 00000000 00010111

we obtain

11101000 00000000 00000000 00000000

which is 3892314112 in decimal.

15.3 Gray Code

The *Gray code* is an encoding of numbers so that adjacent numbers have a single digit differing by 1. It plays an important role in genetic algorithms. The *binary Gray code* can be used instead of the usual interpretation of binary values. The binary Gray code is an encoding of integers so that incrementing an integer value involves complementing exactly one bit in the bit string representation. For example the 3-bit binary Gray code is given in Table 15.1.

Decimal	Binary	Gray code
0	000	000
1	001	001
2	010	011
3	011	010
4	100	110
5	101	111
6	110	101
7	111	100

Table 15.1: 3 Bit Binary Gray Code

The advantage of the Gray code for genetic algorithms is that the mutation operator does not cause a large change in the numeric value of an animal in the population. Large changes are provided by additions of randomly initialized animals to the population at regular intervals. Thus mutation would provide a more local search.

The conversion from standard binary encoding to binary Gray code is achieved as follows. If we want to convert the binary sequence $b_{n-1}b_{n-2}\ldots b_0$ to its binary Gray code $g_{n-1}g_{n-2}\ldots g_0$, the binary Gray code is

$$b_{n-1} \oplus (b_{n-1} \oplus b_{n-2}) \oplus (b_{n-2} \oplus b_{n-3}) \oplus \ldots \oplus (b_1 \oplus b_0).$$

Thus $g_{n-1} = b_{n-1}$ and $g_i = b_{i+1} \oplus b_i$ for $0 < i \le n-1$. To use numerical values in calculations we need to apply the inverse Gray encoding. To convert the binary Gray code $g_{n-1}g_{n-2}\ldots g_0$ to the binary number $b_0b_1b_2\ldots b_{n-1}$ we use

$$b_i = g_{n-1} \oplus g_{n-2} \oplus \ldots \oplus g_i.$$

The following Java program gives an implementation. We apply the built in **BitSet** class in Java.

```
// Gray.java

import java.util.*;

public class Gray
{
 static int size;

 public static void main(String args[])
 {
  BitSet[] b=new BitSet[8];

  size=3;

  for(int i=0;i<8;i++)
  {
   b[i]=new BitSet(size);
   if((i&1)==1) b[i].set(0);
   if((i&2)==2) b[i].set(1);
   if((i&4)==4) b[i].set(2);
   System.out.println("binary to gray "+btos(b[i])+"  "
                              +btos(b[i]=graycode(b[i])));

  }
  for(int i=0;i<8;i++)
  {
   System.out.println("gray to binary "+btos(b[i])+"  "
                              +btos(inversegraycode(b[i])));

  }
 }

 private static String btos(BitSet b)
 {
  String s=new String();

  for(int i=0;i<size;i++)
  {
   if(b.get(i)) s="1"+s;
   else s="0"+s;
  }
  return s;
 }

 private static BitSet graycode(BitSet b)
 {
  BitSet g=new BitSet(size);
  BitSet gsr=new BitSet(size);

  //perform a right shift of g
```

```java
  for(int i=0;i<size;i++)
  {
   if(b.get(i))
   {
    g.set(i);
    if(i>0)
      gsr.set(i-1);
   }
  }
  g.xor(gsr);
  return g;
}

private static BitSet inversegraycode(BitSet b)
{
 BitSet ig=new BitSet(size);

 for(int i=0;i<size;i++)
 {
  int sum=0;
  for(int j=i;j<size;j++)
  {
   if(b.get(j)) sum++;
  }
  if((sum%2)==1)
   ig.set(i);
  else
   ig.clear(i);
 }
 return ig;
}
}
```

15.4 Schemata Theorem

A schema (Holland [89],Goldberg [72]) is a similarity template describing a subset of strings with similarities at certain string positions. We consider the binary alphabet $\{0, 1\}$. We introduce a schema by appending a special symbol to this alphabet. We add the * or don't care symbol which matches either 0 or 1 at a particular position. With this extended alphabet we can now create strings (schemata) over the ternary alphabet

{ 0, 1, * } .

A schema matches a particular string if at every location in the schema 1 matches a 1 in the string, a 0 matches a 0, and a * matches either. As an example, consider the strings and schemata of length 5. The schema

101

describes a subset with four members

01010, 01011, 11010, 11011

We consider a population of individuals (strings) A_j, $j = 1, 2, \ldots, n$ contained in the population $\mathbf{A}(t)$ at time (or generation) t $(t = 0, 1, 2, \ldots)$ where the boldface is used to denote a population. Besides notation to describe populations, strings, bit positions, and alleles, we need a convenient notation to describe the schemata contained in individual strings and populations. Let us consider a schema H taken from the three-letter alphabet

$$V := \{\, 0, 1, * \,\}.$$

For alphabets of cardinality k, there are $(k + 1)^l$ schemata, where l is the length of the string. Furthermore, recall that in a string population with n members there are at most $n \cdot 2^l$ schemata contained in a population because each string is itself a representative of 2^l schemata. These counting arguments give us some feel for the magnitude of information being processed by genetic algorithms.

All schemata are not created equal. Some are more specific than others. The schema 011*1** is a more definite statement about important similarity than the schema 0******. Furthermore, certain schemata span more of the total string length than others. The schema 1****1* spans a larger portion of the string than the schema 1*1****. To quantify these ideas, two schema properties are introduced: schema order and defining length.

Definition. The order of a schema H, denoted by $o(H)$, is the number of fixed positions (in a binary alphabet, the number of 1's and 0's) present in the template.

Example. The order of the schema 011*1** is 4, whereas the order of the schema 0****** is 1.

Definition. The defining length of a schema H, denoted by $\delta(H)$, is the distance between the first and last specific string position.

Example. The schema 011*1** has defining length $\delta = 4$ because the last specific position is 5 and the first specific position is 1. Thus $\delta(H) = 5 - 1 = 4$.

Schemata provide the basic means for analyzing the net effect of reproduction and genetic operators on building blocks contained within the population. Let us consider the individual and combined effects of reproduction, crossover, and mutation on schemata contained within a population of strings. Suppose at a given time step t there are $m(H, t)$ examples of a particular schema H contained within the population $\mathbf{A}(t)$. During reproduction, a string is copied according to its fitness, or more precisely a string A_i gets selected with probability

$$p_i = \frac{f_i}{\sum_{j=1}^{n} f_j}.$$

After picking a non-overlapping population of size n with replacement from the population $\mathbf{A}(t)$, we expect to have $m(H, t+1)$ representatives of the schema H in the population at time $t+1$ as given by

$$m(H, t+1) = \frac{m(H, t) n f(H)}{\sum_{j=1}^{n} f_j(t)}$$

where $f(H)$ is the average fitness of the strings representing schema H at time t. The average fitness of the entire population is defined as

$$\bar{f} := \frac{1}{n} \sum_{j=1}^{n} f_j.$$

Thus we can write the reproductive schema growth equation as follows

$$m(H, t+1) = m(H, t) \frac{f(H)}{\bar{f}(t)}.$$

Assuming that $f(H)/\bar{f}$ remains relatively constant for $t = 0, 1, \ldots$, the preceding equation is a linear difference equation $x(t + 1) = ax(t)$ with constant coefficient which has the solution $x(t) = a^t x(0)$. A particular schema grows as the ratio of the average fitness of the schema to the average fitness of the population. Schemata with fitness values above the population average will receive an increasing number of samples in the next generation, while schemata with fitness values below the population average will receive a decreasing number of samples. This behaviour is carried out with every schema H contained in a particular population $\mathbf{A}$ in parallel. In other words, all the schemata in a population grow or decay according to their schema averages under the operation of reproduction alone. Above-average schemata grow and below-average schemata die off. Suppose we assume that a particular schema H remains an amount $c\bar{f}$ above average with c a constant. Under this assumption we find

$$m(H, t+1) = m(H, t)\frac{(\bar{f} + c\bar{f})}{\bar{f}} = (1 + c)m(H, t).$$

Starting at $t = 0$ and assuming a stationary value of c, we obtain the equation

$$m(H, t) = m(H, 0)(1 + c)^t.$$

This is a geometric progression or the discrete analog of an exponential form. Reproduction allocates exponentially increasing (decreasing) numbers of trials to above-(below-) average schemata. The fundamental theorem of genetic algorithms is as follows (Goldberg [72]).

Theorem. By using the selection, crossover, and mutation of the standard genetic algorithm, then short, low-order, and above average schemata receive exponentially increasing trials in subsequent populations.

The short, low-order, and above average schemata are called building blocks. The fundamental theorem indicates that building blocks are expected to dominate the population. It is necessary to determine if the original goal of function optimization is promoted by this fact. The preceding theorem does not answer this question. Rather, the connection between the fundamental theorem and the observed optimizing properties of the genetic algorithm is provided by the following conjecture.

The Building Block Hypothesis. The globally optimal strings in Ω

$$f : \Omega \to \mathbf{R} \quad \text{with} \quad \Omega = \{0, 1\}^n$$

may be partitioned into substrings that are given by the bits of the fixed positions of building blocks.

15.5 Markov Chain Analysis

Vose [180] showed that the stochastic transition through genetic operations of crossover and mutation can be fully descibed by the transition matrix Q of size $N \times N$ where the matrix element $Q_{k,v}$ is the conditional probability that population v is generated from population k. The total number of different populations is denoted by N. Suppose populations consist of M individuals, each of length L over an alphabet of size α. We denote by $n_{k,j}$ the number of individuals in population k of type j where $0 \le j < \alpha^L$. We use the notation

$$0^{(n)} = \overbrace{00\ldots0}^{n \text{ times}}$$

The population can be represented by

$$0^{(n_{k,0})}10^{(n_{k,1})}1\ldots10^{(n_{k,\alpha^L-1})}$$

which is a $M + \alpha^L - 1$ bit representation. We use the '1' symbol to mark the end of the number of occurences of one individual and the beginning of the number of occurences of the next. Thus the number of different populations is

$$N = \binom{M + \alpha^L - 1}{\alpha^L - 1} = \binom{M + \alpha^L - 1}{M}.$$

When the new population consists only of individuals generated by selection, crossover and mutation the following equation for Q is obtained

$$Q_{k,v} = M! \prod_{j=0}^{\alpha^L-1} \frac{(p_{k,j})^{n_{v,j}}}{n_{v,j}!}$$

where $p_{k,j}$ is the probability that individual j occurs in population k, and $n_{v,j}$ is generated according to the multinomial distribution based on $p_{k,j}$. Furthermore Vose [180] derived

$$\mu^{LM} M! \prod_{j=0}^{\alpha^L-1} \frac{1}{n_{v,j}!} \le Q_{k,v} \le (1-\mu)^{LM} M! \prod_{j=0}^{\alpha^L-1} \frac{1}{n_{v,j}!}$$

for any population k. Here μ is the probability of mutation. Suzuki [168] analysed the modified elitist strategy for genetic algorithms. This strategy always selects the fittest individual i_k of a population k to be in the next generation (population), and the other $M - 1$ individuals are obtained by the operations of selection, crossover and mutaton operations. He obtained

$$Q_{k,v} = H(i_k - i_v)(M - 1)! \prod_{j=0}^{\alpha^L - 1} \frac{(p_{k,j})^{n_{v,j} - \delta_{j,i_k}}}{(n_{v,j} - \delta_{j,i_k})!}$$

where i_k is the fittest individual of population k, and

$$H(x) := \begin{cases} 1 & x \geq 0 \\ 0 & x < 0 \end{cases}$$

and δ_{j,i_k} denotes the Kronecker symbol, i.e.

$$\delta_{j,i_k} = \begin{cases} 1 & j = i_k \\ 0 & \text{otherwise} \end{cases}.$$

The matrix Q consists of submatrices $Q(i)$ of size $N(i) \times N(i)$ along the diagonal and zero above these matrices. For the size $N(i)$ we have

$$N(i) = \binom{M - 1 + \alpha^L - i}{M - 1}$$

where $Q(i)$ denotes the submatrix associated with the ith fittest individual of the i_k. The eigenvalues of each submatrix $Q(i)$ are eigenvalues of Q. Furthermore, the eigenvalues have magnitude not more than one. Denote by q_k^n the probability that the nth generation (population) is population k, and by K the set of all populations which include the fittest individual. To demonstrate the convergence of the genetic algorithm using the modified elitist strategy Suzuki [168] showed that there exists a constant C such that

$$\sum_{k \in K} q_k^n \geq 1 - C|\lambda_*|^n$$

where λ_* is the eigenvalue with greatest magnitude. Thus, with enough iterations, the probablity that a population includes the fittest individual is close to unity.

15.6   Bit Set Classes in C++ and Java

In genetic algorithms bitwise operations play the central role. In this section we describe these operations. The basic bit operations setbit, clearbit, swapbit and testbit can be implemented in C++ as follows. The bit position b runs from 0 to 31 starting counting from right to left in the bit string. In C, C++, and Java the bitwise operators are:

```
&  bitwise AND
|  bitwise OR    (inclusive OR)
^  bitwise XOR   (exclusive OR)
~  NOT operator (one's complement)
>> right-shift operator
<< left-shift operator
```

The operation setbit sets a bit at a given position b (i.e the bit at the position b is set to 1).

```
unsigned long b = 3;
unsigned long x = 15;
x |= (1 << b);          // shortcut for x = x | (1 << b);
```

The operation clearbit clears a bit at a given position b (i.e. the bit at the position b is set to 0).

```
unsigned long b = 3;
unsigned long x = 15;
x &= ~(1 << b);         // short cut for x = x & ~(1 << b);
```

The operation swapbit swaps the bit at the position b, i.e. if the bit is 0 it is set to 1 and if the bit is 1 it is set to 0.

```
unsigned long b = 3;
unsigned long x = 15;
x ^= (1 << b);          // short cut for x = x ^ (1 << b);
```

The operation testbit returns 1 or 0 depending on whether the bit at the position b is set or not.

```
unsigned long b = 3;
unsigned long x = 15;
unsigned long result = ((x & (1 << b)) != 0);
```

The operations `setbit`, `clearbit`, `swapbit` and `testbit` are written as functions. This leads to the following program.

```cpp
// mysetbit.cpp

#include <iostream.h>

inline void setbit(unsigned long& x, unsigned long b)
{
   x |= (1 << b);
}

inline void clearbit(unsigned long& x, unsigned long b)
{
   x &= ~(1 << b);
}

inline void swapbit(unsigned long& x, unsigned long b)
{
   x ^= (1 << b);
}

inline unsigned long testbit(unsigned long x, unsigned long b)
{
   return ((x & (1 << b)) != 0);
}

int main()
{
    unsigned long b = 3;
    unsigned long x = 10;   // binary 1010
    setbit(x,b);
    cout << "x = " << x << endl; // 10  => binary 1010

    clearbit(x,b);
    cout << "x = " << x << endl; // 2 => binary 10

    swapbit(x,b);
    cout << "x = " << x << endl; //      binary

    unsigned long r = testbit(x,b);
    cout << "r = " << r << endl; // 0

    unsigned long y = 17;   // 17 => binary 10001
    setbit(y,b);
    cout << "y = " << y << endl; // 25 => binary 11001
    clearbit(y,b);
    cout << "y = " << y << endl; // 17 => binary 10001
```

```
    unsigned long s = testbit(y,b);
    cout << "s = " << s << endl; // 0

    unsigned long z = 8;   // binary 8 => 1000
    unsigned long t = testbit(z,b);
    cout << "t = " << t << endl; // 1

    return 0;
}
```

Java has a BitSet class which includes the following methods (member functions):

void and(BitSet set)   performs a logical AND

void andNot(BitSet set)   clears all of the bits in this BitSet
                          whose corresponding bit is set in the
                          specified BitSet

void clear(int bitIndex) the bit with index bitIndex in this BitSet
                          is changed to the clear (false) state

boolean get(int bitIndex) returns the value of the bit with the
                          specified index

void or(Bitset set)   performs a logical OR of this bit set with the
                      bit set argument

void xor(BitSet set)   performs a logical XOR of this bit set with
                       the bit set argument

The constructors are

BitSet()   creates a new bit set

BitSet(int nbits)   creates a bit set whose initial size is the
                    specified number of bits

The BitSet class will be used in the program for the four colour problem.

In C++ we can use the standard template library's bitset class. The methods are

```
Constructors
bitset<N> s                  construct bitset for N bits
bitset<N> s(aBitSet)         copy constructor
bitset<N> s(ulong)           create bitset representing an
                             unsigned long value


Bit level operations
s.flip()                     flip all bits
s.flip(i)                    flip position i
s.reset(0)                   set all bits to false
s.reset(i)                   set bit position i to false
s.set()                      set all bits to true
s.set(i)                     set bit position i to true
s.test(i)                    test if bit position i is true


Operations on entire collection
s.any()                      return true if any bit is true
s.none()                     return true if all bits are false
s.count()                    return number of true bits


Assignment
=
s1&=s2                       bitwise AND and assign
s1|=s2                       bitwise inclusive OR and assign
s1^=s2                       bitwise exclusive OR and assign
s1<<=n                       shift left n and assign
s1>>=n                       shift right n and assign


Combination with other bitsets
s1 & s2                      bitwise AND
s1 | s2                      bitwise inclusive OR
s1 ^ s2                      bitwise exclusive OR
s == s2                      return true if two sets are the same


Other operations
~s                           bitwise complement of s
s << n                       shift set left by n
s >> n                       shift set right by n
s.to_string()                return string representation of set
```

The following small program shows an application of the bitset class.

```cpp
// bitset1.cpp

#include <iostream>
#include <bitset>
#include <string>
using namespace std;

int main()
{
    const unsigned long n = 32;
    bitset<n> s;
    cout << s.set() << endl;         // set all bits to 1

    cout << s.flip(12) << endl;      // flip at position 12

    bitset<n> t;
    cout << t.reset() << endl;       // set all bits to false

    t.set(23);
    t.set(27);

    bitset<n> u;
    u = s & t;
    cout << "u = " << u << endl;

    bitset<n> v;
    v = s | t;
    cout << "v = " << v << endl;

    bitset<n> w;
    w = s ^ t;
    cout << "w = " << w << endl;

    bitset<n> z;
    z = w ^ w;
    cout << "z = " << z << endl;

    cout << "z.to_string() = " << z.to_string();

    return 0;
}
```

15.7 A Bit Vector Class

```
// Bitvect.h
// Bit Vector Class

#include <string.h>

#ifndef __BITVECTOR
#define __BITVECTOR

const unsigned char _BV_BIT[8] = { 1,2,4,8,16,32,64,128 };

class BitVector
{
  protected:
    unsigned char *bitvec;
    int len;
  public:
    BitVector();
    BitVector(int nbits);
    BitVector(const BitVector& b);  // copy constructor
    ~BitVector();
    void SetBit(int bit,int val=1);
    int GetBit(int bit) const;
    void ToggleBit(int bit);
    BitVector operator&(const BitVector&) const;
    BitVector& operator &= (const BitVector&);
    BitVector operator | (const BitVector&) const;
    BitVector& operator |= (const BitVector&);
    BitVector operator ^ (const BitVector&) const;
    BitVector& operator ^= (const BitVector&);
    friend BitVector operator ~ (const BitVector&);
    BitVector& operator = (const BitVector&);
    int operator[](int bit) const;
    void SetLength(int nbits);
};

BitVector::BitVector()
{
  len = 0;
  bitvec = NULL;
}

BitVector::BitVector(int nbits)
{
  len = nbits/8+((nbits%8)?1:0);
  bitvec = new unsigned char[len];
}
```

```cpp
BitVector::BitVector(const BitVector &b)
{
  len = b.len;
  bitvec = new unsigned char[len];
  memcpy(bitvec,b.bitvec,len);
}


BitVector::~BitVector()
{
  if(bitvec != NULL) delete[] bitvec;
}

void BitVector::SetBit(int bit,int val)
{
  if(bit < 8*len)
  {
  if(val) bitvec[bit/8] |= _BV_BIT[bit%8];
  else bitvec[bit/8] &= ~_BV_BIT[bit%8];
  }
}

int BitVector::GetBit(int bit) const
{
  if(bit < 8*len) return ((bitvec[bit/8]&_BV_BIT[bit%8])?1:0);
  return -1;
}


void BitVector::ToggleBit(int bit)
{
  if(bit<8*len) bitvec[bit/8] ^= _BV_BIT[bit%8];
}


BitVector BitVector::operator & (const BitVector &b) const
{
  int i;
  int mlen = (len > b.len)?len:b.len;
  BitVector ret(mlen*8);
  for(i=0;i<mlen;i++)
  ret.bitvec[i] = bitvec[i]&b.bitvec[i];
  return ret;
}


BitVector& BitVector::operator &= (const BitVector &b)
{
  int i;
  int mlen = (len>b.len)?len:b.len;
  for(i=0;i<mlen;i++)
```

```
  bitvec[i] &= b.bitvec[i];
  return *this;
}

BitVector BitVector::operator | (const BitVector &b) const
{
  int i;
  int mlen = (len>b.len)?len:b.len;
  BitVector ret(mlen*8);
  for(i=0;i<mlen;i++)
  ret.bitvec[i] = bitvec[i]|b.bitvec[i];
  return ret;
}

BitVector& BitVector::operator |= (const BitVector &b)
{
  int i;
  int mlen = (len>b.len)?len:b.len;
  for(i=0;i<mlen;i++)
  bitvec[i] |= b.bitvec[i];
  return *this;
}

BitVector BitVector::operator ^ (const BitVector &b) const
{
  int i, mlen = (len>b.len)?len:b.len;
  BitVector ret(mlen*8);
  for(i=0;i<mlen;i++)
  ret.bitvec[i] = bitvec[i]^b.bitvec[i];
  return ret;
}

BitVector& BitVector::operator ^= (const BitVector &b)
{
  int i;
  int mlen = (len>b.len)?len:b.len;
  for(i=0;i<mlen;i++)
  bitvec[i] ^= b.bitvec[i];
  return *this;
}

BitVector operator ~ (const BitVector &b)
{
  int i;
  BitVector ret(b.len*8);
  for(i=0;i<b.len;i++)
  ret.bitvec[i] = ~b.bitvec[i];
  return ret;
```

```
}

BitVector& BitVector::operator = (const BitVector& b)
{
  if(bitvec == b.bitvec) return *this;
  if(bitvec != NULL) delete[] bitvec;
  len = b.len;
  bitvec = new unsigned char[len];
  memcpy(bitvec,b.bitvec,len);
  return *this;
}

int BitVector::operator[](int bit) const
{
  return GetBit(bit);
}

void BitVector::SetLength(int nbits)
{
  if(bitvec != NULL) delete[] bitvec;
  len = nbits/8 + ((nbits%8)?1:0);
  bitvec = new unsigned char[len];
}

#endif
```

15.8   Maximum of One-Dimensional Maps

As an example we consider the following fitness functions

$$f(x) = \cos(x)$$

and

$$g(x) = \cos(x) - \sin(2x)$$

in the interval $[0 : 2\pi]$. In this interval the function f has two global maxima at the value 0 and 2π. The function g has three maxima. The global maximum is at 5.64891 and the two local maxima are at 0 and 2.13862.

A simple C++ program would include the following functions

```
// fitness function of individual
double f(double)

// fitness function value of individual
double f_value(double (*func)(double),int* arr,int& N,
               double a, double b)

// x_value
double x_value(int* arr,int& N,double a,double b)

// setup of farm
void setup(int** farm, int M, int N)

// crossing two individuals
void crossings(int** farm, int M, int N)

// mutate an individual
void mutate(int** farm, int M, int N)
```

Here N is the length of the binary string and M is the size of the population, which is kept constant at each time step. For the given problem we select $N = 10$ and $M = 12$. The binary string "$s_{N-1}s_{N-2}...s_0$" is mapped into the integer number m and then into the real number x in the interval $[0 : 2\pi]$ as described above.

The farm is set up using a random number generator. In our implementation the crossing function selects the two fittest strings from the two parents and the two children. The parents are selected by a random number generator. With a population of 12 strings in the farm we find after 100 iterations both the maxima at 0 and 2π for the function f. A typical result is that five strings are related to the maximum at $x = 0$ and seven strings are related to the maximum at $x = 2\pi$. For the fitness function g we find the global maximum and the second highest maximum after 100 iterations.

```cpp
// genetic.cpp
// A simple genetic algorithm
// finding the global maximum of
// the function f in the interval [a,b].

#include <iostream.h>
#include <stdlib.h>
#include <time.h>       // for srand(), rand()
#include <math.h>       // for cos(), sin(), pow

// fitness function where maximum to be found
double f(double x)
{
   return cos(x) - sin(2*x);
}

// fitness function value for individual
double f_value(double (*func)(double),int* arr,int& N,
        double a,double b)
{
   double res;
   double m = 0.0;
   for(int j=0; j<N; j++)
   {
   double k = j;
   m += arr[N-j-1]*pow(2.0,k);
   }
   double x = a + m*(b-a)/(pow(2.0,N)-1.0);
   res = func(x);
   return res;
}

// x_value at global maximum
double x_value(int* arr,int& N,double a,double b)
{
   double m = 0.0;
   for(int j=0; j<N; j++)
   {
   double k = j;
   m += arr[N-j-1]*pow(2.0,k);
   }
   double x = a + m*(b-a)/(pow(2.0,N)-1.0);
   return x;
}

// setup the population (farm)
void setup(int** farm, int M, int N)
{
```

```
    time_t t;
    srand((unsigned) time(&t));
    for(int j=0; j<M; j++)
    {
    for(int k=0; k<N; k++)
    {
    farm[j][k] = rand()%2;
    }
    }
}

// cross two individuals
void crossings(int** farm,int& M,int& N,double& a,double& b)
{
    int K = 2;
    int** temp = NULL;
    temp = new int* [K];
    for(int i=0; i<K; i++)
    {
    temp[i] = new int[N];
    }

    double res[4];
    int r1 = rand()%M;
    int r2 = rand()%M;
    // random returns a value between
    // 0 and one less than its parameter
    while(r2 == r1) r2 = rand()%M;

    res[0] = f_value(f,farm[r1],N,a,b);
    res[1] = f_value(f,farm[r2],N,a,b);

    for(int j=0; j<N; j++)
    {
    temp[0][j] = farm[r1][j];
    temp[1][j] = farm[r2][j];
    }

    int r3 = rand()%(N-2) + 1;

    for(j=r3; j<N; j++)
    {
    temp[0][j] = farm[r2][j];
    temp[1][j] = farm[r1][j];
    }

    res[2] = f_value(f,temp[0],N,a,b);
    res[3] = f_value(f,temp[1],N,a,b);
```

```
    if(res[2] > res[0])
    {
    for(j=0; j<N; j++)
    farm[r1][j] = temp[0][j];
    res[0] = res[2];
    }

    if(res[3] > res[1])
    {
    for(j=0; j<N; j++)
    farm[r2][j] = temp[1][j];
    res[1] = res[3];
    }
    for(j=0; j<K; j++)
    delete [] temp[j];
    delete [] temp;
}


// mutate an individual
void mutate(int** farm,int& M,int& N,double& a,double& b)
{
    double res[2];
    int r4 = rand()%N;
    int r1 = rand()%M;
    res[0] = f_value(f,farm[r1],N,a,b);
    int v1 = farm[r1][r4];
    if(v1 == 0) farm[r1][r4] = 1;
    if(v1 == 1) farm[r1][r4] = 0;
    double a1 = f_value(f,farm[r1],N,a,b);
    if (a1 < res[0]) farm[r1][r4] = v1;

    int r5 = rand()%N;
    int r2 = rand()%M;
    res[1] = f_value(f,farm[r2],N,a,b);
    int v2 = farm[r2][r5];
    if(v2 == 0) farm[r2][r5] = 1;
    if(v2 == 1) farm[r2][r5] = 0;
    double a2 = f_value(f,farm[r2],N,a,b);
    if(a2 < res[1]) farm[r2][r5] = v2;
}


void main()
{
    int M = 12;     // population (farm) has 12 individuals (animals)
    int N = 10;     // length of binary string

    int** farm = NULL;        // allocate memory for population
```

```
farm = new int* [M];
for(int i=0; i<M; i++)
{
farm[i] = new int[N];
}

setup(farm, M, N);

double a = -1.0;  double b = 1.0;  // interval [a,b]

for(int k=0; k<1000; k++)
{
crossings(farm,M,N,a,b);
mutate(farm,M,N,a,b);
}  // end for loop

for(int j=0; j<N; j++)
{
cout << "farm[1][" << j << "] = " << farm[1][j] << endl;
}
cout << endl;

for(j=0; j<M; j++)
cout << "fitness f_value[" << j << "] = "
     << f_value(f,farm[j],N,a,b)
     << "  " << "x_value[" << j << "] = "
     << x_value(farm[j],N,a,b) << endl;

for(j=0; j<M; j++)
delete [] farm[j];
delete [] farm;
}
```

In the program given above we store a bit as `int`. This wastes a lot of memory space. A more optimal use of memory is to use a string, for example "1000111101". Then we use 1 byte for 1 or 0. An even more optimal use is to manipulate the bits themselves. In the following we use the class `BitVector` described above to manipulate the bits. The `BitVector` class is included in the header file `bitVect.h`.

```cpp
// findmax.cpp

#include <iostream.h>
#include <math.h>
#include <stdlib.h>
#include <time.h>
#include "bitvect.h"

double f(double x) { return cos(x)-sin(2*x); }

double f_value(double (*func)(double),const BitVector &arr,
               int &N,double a,double b)
{
  double res, m = 0.0;
  for(int j=0;j<N;j++)
  {
  double k = j;
  m += arr[N-j-1]*pow(2.0,k);
  }
  double x = a+m*(b-a)/(pow(2.0,N)-1.0);
  res = func(x);
  return res;
}

double x_value(const BitVector &arr,int &N,double a,double b)
{
  double m = 0.0;
  for(int j=0;j<N;j++)
  {
  double k = j;
  m += arr[N-j-1]*pow(2.0,k);
  }
  double x = a + m*(b-a)/(pow(2.0,N)-1.0);
  return x;
}

void setup(BitVector *farm,int M,int N)
{
  srand((unsigned)time(NULL));
  for(int j=0;j<M;j++)
  for(int k=0;k<N;k++)
  farm[j].SetBit(k,rand()%2);
```

```
}

void crossings(BitVector *farm,int &M,int &N,double &a,double &b)
{
  int K = 2, j;
  BitVector *temp = new BitVector[K];
  for(int i=0;i<K;i++) temp[i].SetLength(N);
  double res[4];
  int r1 = rand()%M;
  int r2 = rand()%M;
  while(r2 == r1) r2 = rand()%M;
  res[0] = f_value(f,farm[r1],N,a,b);
  res[1] = f_value(f,farm[r2],N,a,b);
  for(j=0;j<N;j++)
  {
  temp[0].SetBit(j,farm[r1][j]);
  temp[1].SetBit(j,farm[r2][j]);
  }
  int r3 = rand()%(N-2)+1;
  for(j=r3;j<N;j++)
  {
  temp[0].SetBit(j,farm[r2][j]);
  temp[1].SetBit(j,farm[r1][j]);
  }
  res[2] = f_value(f,temp[0],N,a,b);
  res[3] = f_value(f,temp[0],N,a,b);
  if(res[2]>res[0])
  {
  farm[r1] = temp[0];
  res[0] = res[2];
  }
  if(res[3] > res[1])
  {
  farm[r2] = temp[1];
  res[1] = res[3];
  }
  delete[] temp;
}

void mutate(BitVector *farm,int &M,int &N,double &a,double &b)
{
    double res[2];
    int r4 = rand()%N;
    int r1 = rand()%M;
    res[0] = f_value(f,farm[r1],N,a,b);
    int v1 = farm[r1][r4];
    farm[r1].ToggleBit(r4);
    double a1 = f_value(f,farm[r1],N,a,b);
```

```
    if(a1 < res[0]) farm[r1].ToggleBit(r4);
    int r5 = rand()%N;
    int r2 = rand()%M;
    res[1] = f_value(f,farm[r2],N,a,b);
    int v2 = farm[r2][r5];
    farm[r2].ToggleBit(r5);
    double a2 = f_value(f,farm[r2],N,a,b);
    if(a2 < res[1]) farm[r2].ToggleBit(r5);
}

void main(void)
{
  int M = 12;
  int N = 10;
  int i, j, k;

  BitVector* farm = new BitVector[M];

  for(i=0;i<M;i++) farm[i].SetLength(N);
  setup(farm,M,N);

  double a = 0.0, b = 6.28318;

  for(k=0;k<1000;k++)
  {
  crossings(farm,M,N,a,b);
  mutate(farm,M,N,a,b);
  }
  for(j=0;j<N;j++)
  cout << "farm[1]["<<j<<"]=" << farm[1][j] << endl;
  cout<<endl;
  for(j=0;j<M;j++)
  cout << "fitness f_value["<<j<<"]="
       << f_value(f,farm[j],N,a,b)
       <<"  x_value["<<j<<"]=" << x_value(farm[j],N,a,b) << endl;
  delete[] farm;
}
```

A typical output is

```
farm[1][0]=1
farm[1][1]=1
farm[1][2]=1
farm[1][3]=0
farm[1][4]=1
farm[1][5]=0
farm[1][6]=0
farm[1][7]=0
```

```
farm[1][8]=0
farm[1][9]=0

fitness f_value[0]=1.75411   x_value[0]=5.6997
fitness f_value[1]=1.75411   x_value[1]=5.6997
fitness f_value[2]=1.75411   x_value[2]=5.6997
fitness f_value[3]=1.75411   x_value[3]=5.6997
fitness f_value[4]=1.75411   x_value[4]=5.6997
fitness f_value[5]=1.75411   x_value[5]=5.6997
fitness f_value[6]=1   x_value[6]=0
fitness f_value[7]=0.59771   x_value[7]=0.196541
fitness f_value[8]=1.75411   x_value[8]=5.6997
fitness f_value[9]=1   x_value[9]=0
fitness f_value[10]=1.75411   x_value[10]=5.6997
fitness f_value[11]=1.75411   x_value[11]=5.6997
```

15.9   Maximum of Two-Dimensional Maps

Here we consider the problem how to find the maximum of a two-dimensional bounded function $f : [a, b] \times [c, d] \to \mathbf{R}$, where $a, b, c, d \in \mathbf{R}$, $a < b$ and $c < d$. We follow in our presentation closely Michalewicz [116]. Michalewicz also gives a detailed example.

We use the following notation. N is the length of the chromosome (binary string). The chromosome includes both the contributions from the x variable and y variable. The size of N depends on the required precision. M denotes the size of the farm (population) which is kept constant at each time step. First we have to decide about the precision. We assume further that the required precision is four decimal places for each variable. First we find the domain of the variable x, i.e. $b-a$. The precision requirement implies that the range $[a, b]$ should be divided into at least $(b-a) \cdot 10000$ equal size ranges. Thus we have to find an integer number N_1 such that

$$2^{N_1-1} < (b - a) \cdot 10000 \leq 2^{N_1} .$$

The domain of variable y has length $d - c$. The same precision requirement implies that we have to find an integer N_2 such that

$$2^{N_2-1} < (d - c) \cdot 10000 \leq 2^{N_2} .$$

The total length of a chromosome (solution vector) is then $N = N_1 + N_2$. The first N_1 bits code x and the remaining N_2 bits code y.

Next we generate the farm. To optimize the function f using a genetic algorithm, we create a population of `size` $= M$ chromosomes. All N bits in all chromosomes are initialized randomly using a random number generator.

Let us denote the chromosomes by $v_0, v_1, \ldots, v_{M-1}$. During the evaluation phase we decode each chromosome and calculate the fitness function values $f(x, y)$ from (x, y) values just decoded.

Now the system constructs a roulette wheel for the selection process. First we calculate the total fitness F of the population

$$F := \sum_{i=0}^{M-1} f(v_i) .$$

Next we calculate the probability of a selection p_i and the cumulative probability q_i for each chromosome v_i

$$p_i := \frac{f(v_i)}{F}, \qquad q_i := \sum_{k=0}^{i} p_k, \qquad i = 0, 1, \dots, M-1.$$

Obviously, $q_{M-1} = 1$. Now we spin the roulette wheel M times. First we generate a (random) sequence of M numbers for the range $[0..1]$. Each time we select a single chromosome for a new population as follows. Let r_0 be the first random number. Then $q_k < r_0 < q_{k+1}$ for a certain k. We selected chromosome $k+1$ for the new population. We do the same selection process for all the other $M-1$ random numbers. This leads to a new farm of chromosomes. Some of the chromosomes can now occur twice.

We now apply the recombination operator, crossover, to the individuals in the new population. For the probability of crossover we choose $p_c = 0.25$. We proceed in the following way: for each chromosome in the (new) population we generate a random number r from the range $[0..1]$. Thus we generate again a sequence of M random numbers in the interval $[0, 1]$. If $r < 0.25$, we select a given chromosome for crossover. If the number of selected chromosomes is even, so we can pair them. If the number of selected chromosomes were odd, we would either add one extra chromosome or remove one selected chromosome. Now we mate selected chromosomes randomly. For each of these two pairs, we generate a random integer number **pos** from the range $[0..N-2]$. The number **pos** indicates the position of the crossing point. We do now the same process for the second pair of chromosomes and so on. This leads to a new farm of chromosomes.

The next operator, mutation, is performed on a bit-by-bit basis. The probability of mutation $p_m = 0.01$, so we expect that (on average) 1% of bits would undergo mutation. There are $N \times M$ bits in the whole population; we expect (on average) $0.01 \cdot N \cdot M$ mutations per generation. Every bit has an equal chance to be mutated, so, for every bit in the population, we generate a random number r from the range $[0..1]$. If $r < 0.01$, we mutate the bit. This means that we have to generate $N \cdot M$ random numbers. Then we translate the bit position into chromosome number and the bit number within the chromosome. Then we swap the bit. This leads to a new population of the same size M.

Thus we have completed one iteration (i.e., one generation) of the while loop in the genetic procedure. Next we find the fitness function for the new population and the total fitness of the new population, which should be higher compared to the old population. The fitness value of the fittest chromosome of the new population should also be higher than the fitness value of the fittest chromosome in the old population. Now we are ready to run the selection process again and apply the genetic operators, evaluate the next generation and so on. A stopping condition could be that the total fitness does not change anymore.

```cpp
// twodim.cpp

#include <iostream.h>
#include <math.h>
#include <stdlib.h>
#include <time.h>

// function to optimize
double f(double x, double y)
{
   return exp(-(x-1.0)*(x-1.0)*y*y/2.0);
   return x*y;
}


// determines the chromosone length required
// to obtain the desired precision
int cLength(int precision, double rangeStart, double rangeEnd)
{
   int length = 0;
   double total = (rangeEnd - rangeStart)*pow(10.0,precision);
   while(total > pow(2.0,length)) length++;
   return length;
}

void setup(int** farm,int size,int length)
{
   int i, j;
   time_t t;
   srand((unsigned) time(&t));

   for(i=0; i<size; i++)
   for(j=0; j<length; j++)
   farm[i][j] = rand()%2;
}


void printFarm(int** farm,int length,int size)
{
   int i, j;
   for(i=0; i<size; i++)
   {
   cout << "\n";
   for(j=0; j<length; j++)
   {
   cout << farm[i][j];
   }
   }
}
```

```
double xValue(int* chromosome,int xLength,double* domain)
{
   int i;
   double m = 0.0;
   for(i=0; i<xLength; i++)
   {
   m += chromosome[xLength-i-1]*pow(2.0,i);
   }
   double x =
   domain[0] + m*(domain[1]-domain[0])/(pow(2.0,xLength)-1.0);
   return x;
}

double yValue(int* chromosome,int yLength,int length,double* domain)
{
   int i;
   double m = 0.0;
   for(i=0; i<yLength; i++)
   {
   m += chromosome[length-i-1]*pow(2.0,i);
   }
   double y =
   domain[2] + m*(domain[3]-domain[2])/(pow(2.0,yLength)-1);
   return y;
}

double fitnessValue(double (*f)(double,double), int* chromosome,
       int length, double* domain,int xLength,int yLength)
{
   double x = xValue(chromosome, xLength, domain);
   double y = yValue(chromosome, yLength, length, domain);
   double result = f(x,y);
   return result;
}

// A new farm is set up by using a roulette wheel
// parent selection process
void roulette(int** farm,int length,int size,double* domain,
              int xLength,int yLength)
{
   int i, j;
   // fitness matrix contains the fitness of each
   // individual chromosome on farm
   double* fitnessVector = NULL;
   fitnessVector = new double[size];

   for(i=0; i<size; i++)
   {
```

```
    fitnessVector[i] =
    fitnessValue(f,farm[i],length,domain,xLength,yLength);
    }


    // fitness vector contains the fitness of
    // each individual chromosome of the farm
    double totalFitness = 0.0;
    for(i=0; i<size; i++)
    {
    totalFitness += fitnessVector[i];
    }


    // calculate probability vector
    double* probabilityVector = NULL;
    probabilityVector = new double[size];
    for(i=0; i<size; i++)
    {
    probabilityVector[i] = fitnessVector[i]/totalFitness;
    }


    // calculate cumulative probability vector
    double cumulativeProb = 0.0;
    double* cum_prob_Vector = NULL;
    cum_prob_Vector = new double [size];

    for(i=0; i<size; i++)
    {
    cumulativeProb += probabilityVector[i];
    cum_prob_Vector[i] = cumulativeProb;
    }


    // setup random vector
    double* randomVector = NULL;
    randomVector = new double [size];
    time_t t;
    srand((unsigned) time(&t));

    for(i=0; i<size; i++)
    randomVector[i] = rand()/double(RAND_MAX);

    // create new population
    int count;
    int** newFarm = NULL;
    newFarm = new int* [size];
    for(i=0; i<size; i++)
    newFarm[i] = new int [length];

    for(i=0; i<size; i++)
```

```
{
count = 0;
while(randomVector[i] > cum_prob_Vector[count]) count++;
for(j=0; j<length; j++)
{
newFarm[i][j] = farm[count][j];
}
}

    for(i=0; i<size; i++)
    for(j=0; j<length; j++)
    farm[i][j] = newFarm[i][j];

    delete [] fitnessVector;
    delete [] probabilityVector;
    delete [] cum_prob_Vector;
    delete [] randomVector;

    for(i=0; i<size; i++)
    delete [] newFarm[i];
    delete [] newFarm;
} // end function roulette

void crossing(int** farm,int size,int length)
{
    int i, j, k, m;
    int count = 0;
    int* chosen = NULL;
    chosen = new int [size];

    double* randomVector = NULL;
    randomVector = new double [size];

    time_t t;
    srand((unsigned) time(&t));

    for(i=0; i<size; i++)
    randomVector[i] = rand()/double(RAND_MAX);

    // fill chosen with indexes of all random values < 0.25
    for(i=0; i<size; i++)
    {
    if(randomVector[i] < 0.25)
    {
    chosen[count] = i;
    count++;
    }
    }
```

```
// if chosen contains an odd number of chromosomes
// one more chromosome is to be selected
if((count%2 != 0) || (count == 1))
{
int index = 0;
while(randomVector[index] < 0.25) index++;
count++;
chosen[count-1] = index;
}


// cross chromosomes with index given in chosen
int** temp = NULL;
temp = new int* [2];
for(i=0; i<2; i++)
{
temp[i] = new int[length];
}


for(i=0; i<count; i=i+2)
{
for(j=0; j<length; j++)
{
temp[0][j] = farm[chosen[i]][j];
temp[1][j] = farm[chosen[i+1]][j];
}
int position = rand()%length;

for(k=position; k<length; k++)
{
temp[0][k] = farm[chosen[i+1]][k];
temp[1][k] = farm[chosen[i]][k];
}

for(m=0; m<length; m++)
{
farm[chosen[i]][m] = temp[0][m];
farm[chosen[i+1]][m] = temp[1][m];
}
}

delete [] chosen;
delete [] randomVector;

for(i=0; i<2; i++)
delete [] temp[i];
delete [] temp;
} // end function crossing
```

```
void mutate(int** farm,int size,int length)
{
   int i;
   int totalbits = size*length;

   double* randomVector = NULL;
   randomVector = new double [totalbits];

   time_t t;
   srand((unsigned) time(&t));

   for(i=0; i<totalbits; i++)
   randomVector[i] = rand()/double(RAND_MAX);

   int a, b;
   for(i=0; i<totalbits; i++)
   {
   if(randomVector[i] < 0.01)
   {
   if(i >= length)
   {
   a = i/length;  b = i%length;
   }
   else
   {
   a = 0;  b = i;
   }
   if(farm[a][b] == 0)
   farm[a][b] = 1;
   else
   farm[a][b] = 0;
   }
   }

   delete [] randomVector;
}

void printFinalResult(int** farm,int length,int size,double* domain,
                      int xLength,int yLength,int iterations)
{
   int i;

   double* fitnessVector = NULL;
   fitnessVector = new double [size];

   for(i=0; i<size; i++)
   fitnessVector[i] =
```

```
    fitnessValue(f,farm[i],length,domain,xLength,yLength);


    // search for chromosome with maximum fitness
    double x, y;
    int pos = 0;
    double max = fitnessVector[0];

    for(i=1; i<size; i++)
    {
    if(fitnessVector[i] > max)
    {
    max = fitnessVector[i];
    pos = i;
    }
    }


    x = xValue(farm[pos], xLength, domain);
    y = yValue(farm[pos], yLength, length, domain);

    // displaying the result
    cout << "\n\n After " << iterations
         << " iterations the fitnesses are: \n";
    for(i=0; i<size; i++)
    {
    cout << "\n fitness of chromosome "
         << i << ": " << fitnessVector[i];
    }

    cout << "\n\n The maximum fitness: f(" << x << "," << y << ") = "
         << max;

    delete [] fitnessVector;
}

int main()
{
    int size = 32;         // population size
    int precision = 6;  // precision
    int iterations = 10000;

    double domain[4];  // variables specifying domain
    double x1, x2, y1, y2;
    x1 = -2.0; x2 = 2.0;
    y1 = -2.0; y2 = 2.0;

    domain[0] = x1; domain[1] = x2;
    domain[2] = y1; domain[3] = y2;
```

```
int xLength = cLength(precision,domain[0],domain[1]);
cout << "\n\n the xLength is: " << xLength;

int yLength = cLength(precision,domain[2],domain[3]);
cout << "\n the yLength is: " << yLength;

// total length
int length = xLength + yLength;
cout << "\n the chromosone length is: " << length;

int i;

// allocate memory for farm
int** farm = NULL;
farm = new int* [size];
for(i=0; i<size; i++) { farm[i] = new int[length]; }

setup(farm,size,length);

cout << "\n\n The inital farm: \n";
printFarm(farm,length,size);
cout << endl;

// iteration loop
int t;

for(t=0; t<iterations; t++)
{
roulette(farm,length,size,domain,xLength,yLength);
crossing(farm,size,length);
roulette(farm,length,size,domain,xLength,yLength);
mutate(farm,size,length);
}

printFinalResult(farm,length,size,domain,xLength,yLength,iterations);

for(i=0; i<size; i++)
{ delete [] farm[i]; }
delete [] farm;

return 0;
}
```

15.10 The Four Colour Problem

A map is called n-colourable [42] if each region of the map can be assigned a colour from n different colours such that no two adjacent regions have the same colour. The four colour conjecture is that every map is 4-colourable. In 1890 Heawood proved that every map is 5-colourable. In 1976 Appel and Haken proved the four colour conjecture with extensive use of computer calculations.

We can describe the m regions of a map using a $m \times m$ adjacency matrix A where $A_{ij} = 1$ if region i is adjacent to region j and $A_{ij} = 0$ otherwise. We set $A_{ii} = 0$. For the fitness function we can determine the number of adjacent regions which have the same colour. The lower the number, the fitter the individual.

The program below solves the four colour problem for the map in Figure 15.1(a).

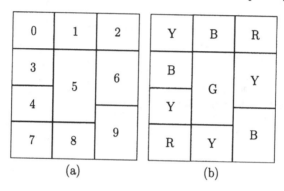

Figure 15.1: A Map for the Four Colour Problem

Individuals are represented as strings of characters, where each character represents the colour for the region corresponding to the characters position in the string. The member population is the number of individuals in the population, and mu is the probability that an individual is mutated. The method fitness evaluates the fitness of a string using the adjacency matrix to determine when adjacent regions have the same colour. If the fitness is equal to 0 we have found a solution. The adjacency matrix can be modified to solve for any map. The method mutate determines for each individual in the population whether the individual is mutated, and mutates a component of the individual by randomly changing the colour. The method crossing performs the crossing operation as discussed previously. The genetic algorithm is implemented in the method GA. The arguments are an adjacency matrix, a string specifying which colours to use and the number of regions on the map. It returns a string specifying a solution to the problem. One such solution is YBRBYGYRYB, where R stands for red, G for green, B for blue and Y for yellow. This corresponds to the colouring in Figure 15.1(b).

```
// Colour.java

public class Colour
{
 static int population=1000;
 static double mu=0.01;

 public static void main(String[] args)
 {
  int[][] adjM={{0,1,0,1,0,0,0,0,0,0},
                {1,0,1,0,0,1,0,0,0,0},
                {0,1,0,0,0,0,1,0,0,0},
                {1,0,0,0,1,1,0,0,0,0},
                {0,0,0,1,0,1,0,1,0,0},
                {0,1,0,1,1,0,1,0,1,1},
                {0,0,1,0,0,1,0,0,0,1},
                {0,0,0,0,1,0,0,0,1,0},
                {0,0,0,0,0,1,0,1,0,1},
                {0,0,0,0,0,1,1,0,1,0}};

  System.out.println(GA(adjM,"RGBY",10));
 }

 static int fitness(int[][] adjM,String s,int N)
 {
  int count = 0;
  for(int i=0;i < N-1;i++)
  {
   for(int j=i+1;j < N;j++)
   {
    if((s.charAt(i) == s.charAt(j)) && (adjM[i][j] == 1))
      count++;
   }
  }
  return count;
 }

 static void mutate(String[] p,String colors)
 {
  int j;
  for(int i=0;i<p.length;i++)
  {
   if(Math.random()<mu)
   {
    int pos=(int)(Math.random()*(p[i].length()-1));
    int mut=(int)(Math.random()*(colors.length()-2));
    char[] ca1=p[i].toCharArray();
    char[] ca2=colors.toCharArray();
```

```
    for(j=0;ca1[pos]!=ca2[j];j++) {};
    ca1[pos]=ca2[(j+mut)%colors.length()];
    p[i]=new String(ca1);
   }
  }
}


static void crossing(String[] p,int[][] adjM)
{
 int p1=(int)(Math.random()*(p.length-1));
 int p2=p1;
 int c1=(int)(Math.random()*(p[0].length()-1));
 int c2=c1;

 while(p2==p1) p2=(int)(Math.random()*(p.length-1));
 while(c2==c1) c2=(int)(Math.random()*(p[0].length()-1));
 if(c2<c1) {int temp=c2; c2=c1; c1=temp;}

 String[] temp=new String[4];
 temp[0]=p[p1];temp[1]=p[p2];
 temp[2]=p[p1].substring(0,c1)+p[p2].substring(c1+1,c2)
         +p[p1].substring(c2+1,p[p1].length()-1);
 temp[3]=p[p2].substring(0,c1)+p[p1].substring(c1+1,c2)
         +p[p2].substring(c2+1,p[p2].length()-1);

 int i,f;
 for(i=0,f=0;i<4;i++)
 {
  if(fitness(adjM,temp[i],temp[i].length())
     >fitness(adjM,temp[f],temp[f].length()))
   f=i;
 }
 {String tmp=temp[f]; temp[f]=temp[0]; temp[0]=tmp;}
 for(i=1,f=1;i<4;i++)
 {
  if(fitness(adjM,temp[i],temp[i].length())
     >fitness(adjM,temp[f],temp[f].length()))
   f=i;
 }
 {String tmp=temp[f]; temp[f]=temp[1]; temp[1]=tmp;}
 p[p1]=temp[2]; p[p2]=temp[3];
}


static String GA(int[][] adjM,String colors,int N)
{
 int maxfitness,mfi=0;
 String[] p = new String[population];
 char[] temp = new char[N];
```

```
for(int i=0;i < population;i++)
{
 for(int j=0;j < N;j++)
 {
  temp[j] = colors.charAt((int)((Math.random()*colors.length())));
 }
 p[i]=new String(temp);
}
maxfitness=fitness(adjM,p[0],p[0].length());
while(maxfitness!=0)
{
 mutate(p,colors);
 crossing(p,adjM);
 for(int i=0;i<p.length;i++)
 {
  if(fitness(adjM,p[i],p[i].length())<maxfitness)
  {
   maxfitness=fitness(adjM,p[i],p[i].length());
   mfi=i;
  }
 }
}
 return p[mfi];
}
}
```

15.11   Problems with Constraints

15.11.1   Introduction

Thus far, we have only discussed genetic algorithms for searching unconstrained objective functions. Many practical problems contain one or more constraints that must also be satisfied. A typical example is the traveling salesman problem, where all cities must be visited exactly once. In a more mathematical formulation, the traveling salesman problem is stated as follows. For a given $n \times n$ distance matrix $C = (c_{ij})$, find a cyclic permutation π of the set $\{1, 2, \ldots, n\}$ that minimizes the function

$$c(\pi) = \sum_{i=1}^{n} c_{i\pi(i)} \,.$$

The value $c(\pi)$ is usually referred to as the length (or cost or weight) of the permutation π. The traveling salesman problem is one of the standard problems in combinatorial optimization and has many important applications like routing or production scheduling with job-dependent set-up times. Another example is the knapsack problem, where the weight which can be carried is the constraint. The norm of an $n \times n$ matrix over the real numbers $\mathbf{R}$ is given by

$$\|A\| := \sup_{\|\mathbf{x}\|=1} \|A\mathbf{x}\| \,.$$

This is a problem with the constraint

$$\|\mathbf{x}\| = 1$$

i.e. the length of the vector $\mathbf{x} \in \mathbf{R}^n$ must be 1. This problem can be solved with the *Lagrange multiplier method*. The Lagrange multiplier method is as follows. Let M be a manifold and f be a real valued function of class $C^{(1)}$ on some open set containing M. We consider the problem of finding the extrema of the function $f|M$. This is called a problem of constrained extrema. Assume that f has a constrained extremum at $\mathbf{x}^* = (x_1^*, x_2^*, \ldots, x_n^*)$. Let $g_1(\mathbf{x}) = 0, \ldots, g_m(\mathbf{x}) = 0$ be the constraints (manifolds). Then there exist real numbers $\lambda_1, \ldots, \lambda_m$ such that $\mathbf{x}^*$ is a critical point of the function

$$F(\mathbf{x}) := f(\mathbf{x}) + \lambda_1 g_1(\mathbf{x}) + \ldots + \lambda_m g_m(\mathbf{x}) \,.$$

The numbers $\lambda_1, \ldots, \lambda_m$ are called Lagrange multipliers. For the problem to find the norm of an $n \times n$ matrix one considers the functions

$$F(\mathbf{x}) := \|A\mathbf{x}\| + \lambda \|\mathbf{x}\|$$

where λ is the Lagrange multiplier.

The most difficult problem in genetic algorithms is the inclusion of constraints. Constraints are usually classified as equality or inequality relations. Equality constraints may be included into the system. It would appear that inequality constraints pose no particular problem. A genetic algorithm generates a sequence of parameters to be tested using the system model, objective function, and the constraints. We simply run the model, evaluate the fitness function, and check to see if any constraints are violated. If not, the parameter set is assigned the fitness value corresponding to the objective function evaluation. If constraints are violated, the solution is infeasible and thus does not have a fitness. This procedure is fine except that many practical problems are highly constrained; finding a feasible point is almost as difficult as finding the best. As a result, we usually want to get some information out of infeasible solutions, perhaps by degrading their fitness ranking in relation to the degree of constraint violation. This is what is done in a *penalty method*. In a penalty method, a constrained problem in optimization is transformed to an unconstrained problem by associating a cost or penalty with all constraint violations. This cost is included in the objective function evaluation.

Consider, for example, the original constrained problem in minimization form

minimize $g(\mathbf{x})$ subject to

$$h_i(\mathbf{x}) \geq 0, \qquad i = 1, 2, \ldots, n$$

where $\mathbf{x}$ is an m vector. We transform this to the unconstrained form

minimize

$$g(\mathbf{x}) + r \sum_{i=1}^{n} \Phi[h_i(\mathbf{x})]$$

where Φ is the penalty function and r is the penalty coefficient. Other approaches use decoders or repair algorithms.

A detailed discussion of problems with constraints is given by Michalewicz [116]. He proposes that appropriate data structures and specialized genetic operators should do the job of taking care of constraints. He then introduces an approach to handle problems with linear constraints (domain constraints, equalities, and inequalities). We consider here the knapsack problem and traveling salesman problem applying genetic algorithms.

15.11.2 Knapsack Problem

Formally, the *knapsack problem* can be stated as follows.

Problem. Given M, the capacity of the knapsack,

$$\{\, w_i \mid w_i > 0, \ i = 0, 1, \ldots, n-1 \,\}$$

the weights of the n objects, and

$$\{\, v_i \mid v_i > 0, \ i = 0, 1, \ldots, n-1 \,\}$$

their corresponding values,

$$\text{maximize} \quad \sum_{i=0}^{n-1} v_i x_i$$

$$\text{subject to} \quad \sum_{i=0}^{n-1} w_i x_i \leq M$$

$$\text{where} \quad x_i \in \{0, 1\}.$$

Here $x_i = 0$ means that item i should not be included in the knapsack, and $x_i = 1$ means that it should be included.

As an example for the knapsack problem we consider the following problem. A hiker planning a backpacking trip feels that he can comfortably carry at most 20 kilograms. After laying out all the items that he wants to take and discovering that their total weight exceeds 20 kilograms, he assigns to each item a "value" rating, as shown in the Table. Which items should he take to maximize the value of what he can carry without exceeding 20 kilograms ?

Table. An Instance of the Knapsack Problem

Item	Tent	Canteen (filled)	Change of clothes	Camp stoves	Sleeping bag	Dried food
Weight	11	7	5	4	3	3
Value	20	10	11	5	25	50

Item	First-aid kit	Mosquito repellent	Flashlight	Novel	Rain gear	Water purifier
Weight	3	2	2	2	2	1
Value	15	12	6	4	5	30

Although we do not know yet how to obtain the solution, the way to fill the knapsack to carry the most value is to take the sleeping bag, food, mosquito repellent, first-aid kit, flashlight, water purifier, and change of clothes, for a total value of 149 with a total weight of 19 kilograms. An interesting aspect of the solution is that it is not directly limited by the weight restriction. There are ways of filling the knapsack with exactly 20 kilograms, such as substituting the change of clothes for the camp stove and rain gear, but this decreases the total value.

The following program uses a genetic algorithm to solve the problem. We use the header file `bitvect.h` given above.

```cpp
// knapsack.cpp

#include <fstream.h>
#include <time.h>
#include <stdlib.h>
#include "bitvect.h"

struct item
{
  char name[50];
  double weight;
  double value;
};

void readitems(char *file,item *&list,int &n,double &max)
{
  int i;
  ifstream data(file);
  data >> n;
  list = new item[n];

  for(i=0;i<n;i++)
  {
  data >> list[i].name;
  data >> list[i].weight;
  data >> list[i].value;
  }
  data >> max;
}

void destroyitems(item *list)
{
  delete[] list;
}

double value(const BitVector &b,int n,double max,item *list)
```

```
{
  int i;
  double tweight = 0.0;
  double tvalue = 0.0;

  for(i=0;i<n;i++)
  {
  if(b.GetBit(i))
  {
  tweight += list[i].weight;
  tvalue += list[i].value;
  }
  if(tweight > max)
  { tvalue = -1.0; i = n; }
  }
  return tvalue;
}


void mutate(BitVector *farm,int m,int n,item *list,double max)
{
  const int tries = 1000;
  int animal = rand()%m;
  int i = 0, pos, pos2;
  BitVector* newanim = new BitVector(farm[animal]);
  pos2 = pos = rand()%n;
  newanim -> ToggleBit(pos);

  while(i<tries)
  {
  while(pos2 == pos) pos2 = rand()%n;
  newanim -> ToggleBit(pos2);
  if(value(*newanim,n,max,list) > 0) i=tries;
  else { newanim -> ToggleBit(pos2);i++;pos2=pos; }
  }
  if(value(*newanim,n,max,list)>value(farm[animal],n,max,list))
  farm[animal] = *newanim;

  delete newanim;
}


void crossing(BitVector *farm,int m,int n,item *list,double max)
{
  const int tries = 1000;
  int animal1 = rand()%m;
  int animal2 = rand()%m;
  int i, pos;

  while(animal2 == animal1) animal2 = rand()%m;
```

```
BitVector *newanim1 = new BitVector(farm[animal1]);
BitVector *newanim2 = new BitVector(farm[animal2]);
pos = rand()%n;

for(i=pos;i<n;i++)
{
newanim1 -> SetBit(i,farm[animal2][i]);
newanim2 -> SetBit(i,farm[animal1][i]);
}

if(value(*newanim1,n,max,list) > value(farm[animal1],n,max,list))
farm[animal1] = *newanim1;
if(value(*newanim2,n,max,list) > value(farm[animal2],n,max,list))
farm[animal2] = *newanim1;

delete newanim1;
delete newanim2;
}

void setupfarm(BitVector *farm,int m,int n,item *list,double max)
{
  const int tries = 2000;
  double temp;
  int i,j,k;
  srand(time(NULL));
  for(i=0;i<m;i++)
  {
  for(j=0;j<n;j++) farm[i].SetBit(j,0);
  temp = 0.0;
  k = 0;
  while((temp < max) && (k < tries))
  {
  j = rand()%n;
  if(!farm[i].GetBit(j)) temp+=list[j].weight;
  if(temp < max) farm[i].SetBit(j);
  k++;
  }
  }
}

void main()
{
  item* list = NULL;
  int n, m = 100,i,iterations = 500, besti = 0;
  double max, bestv = 0.0, bestw = 0.0, temp;
  BitVector *farm = NULL;
  readitems("knapsack.dat",list,n,max);
```

```
    farm = new BitVector[m];

    for(i=0;i<m;i++)
    farm[i].SetLength(n);

    setupfarm(farm,m,n,list,max);

    for(i=0;i<iterations;i++)
    {
    crossing(farm,m,n,list,max);
    mutate(farm,m,n,list,max);
    }

    for(i=0;i<m;i++)
    if((temp=value(farm[i],n,max,list)) > bestv)
    { bestv=temp; besti=i; }

    cout<<"Items to take :"<<endl<<endl;
    for(i=0;i<n;i++)
    {
    if(farm[besti].GetBit(i))
    {
    cout << list[i].name << "," << endl;
    bestw += list[i].weight;
    }
    }
    cout << endl;
    cout << "for a weight of " << bestw
         << "kg and value of " << bestv <<endl;
    delete[] farm;
    destroyitems(list);
}
```

The input file knapsack.dat is

```
12
tent                    11 20
canteen_(filled)         7 10
change_of_clothes        5 11
camp_stoves              4  5
sleeping_bag             3 25
dried_food               3 50
first-aid_kit            3 15
mosquito_repellent       2 12
flashlight               2  6
novel                    2  4
rain_gear                2  5
water_purifier           1 30
20
```

The output is

```
Items to take :

change_of_clothes,
sleeping_bag,
dried_food,
first-aid_kit,
mosquito_repellent,
flashlight,
water_purifier,

for a weight of 19kg and value of 149
```

We can extend the knapsack problem to one with m knapsacks. The capacity of knapsack j is denoted by M_j. The problem statement becomes

$$\text{Maximize} \quad \sum_{i=0}^{m-1}\sum_{j=0}^{n-1} v_j x_{j,i} \quad \text{subject to} \quad \sum_{i=0}^{n-1} w_i x_{i,j} \leq M_j \quad \text{and} \quad \sum_{i=0}^{m-1} x_{j,i} = 1$$

where $x_{i,j} \in \{0,1\}$. Here $x_{i,j} = 0$ means that item i should not be included in knapsack j, and $x_{i,j} = 1$ means that it should be included. The meanings of w_i and v_i are the same as for the single knapsack problem.

15.11.3   Traveling Salesman Problem

The traveling salesman problem is a combinatorial optimization problem. Many combinatorial optimization problems like the traveling salesman problem can be formulated as follows. Let

$$\pi = \{ i_1, i_2, \dots, i_n \}$$

be some permutation from the set

$$\{1, 2, \dots, n\}.$$

The number of permutations is $n!$. Let Ω be a space of feasible solutions (states) and $f(\pi)$ the optimality function (criterion). It is necessary to find π^* such that

$$\pi^* = \{ i_1^*, i_2^*, \dots, i_n^* \} = \arg\{ f(\pi) \to \min_{\pi \subset \Omega} \}.$$

The structure of Ω and $f(\pi)$ depends on the problems considered. A typical problem is the traveling salesman problem. The traveling salesman problem is deceivingly simple to state. Given the distances separating a certain number of towns the aim is to find the shortest tour that visits each town once and ends at the town it started from. As there are several engineering and scientific problems equivalent to a traveling salesman problem. The problem is of practical importance. The number of all possible tours is finite, therefore in principle the problem is solvable. However, the brute force strategy is not only impractical but completely useless even for a moderate number of towns n, because the number of possible tours grows factorially with n. The traveling salesman problem is the best-known example of the whole class of problems called NP-complete (or NP-hard), which makes the problem especially interesting theoretically. The NP-complete problems are transformable into each other, and the computation time required to solve any of them grows faster than any power of the size of the problem. There are strong arguments that a polynomial time algorithm may not exist at all. Therefore, the aim of the calculations is usually to find near-optimum solutions.

The following C++ program `permut.cpp` finds all permutations of the numbers $1, 2, \dots, n$. The array element `p[0]` takes the value 0 at the beginning of the program. The end of the evaluation is indicated by `p[0]` = 1.

```
// permut.cpp
// permutation of the numbers 1, 2, ... , n

#include <iostream.h>

int main()
{
   int i, j, k, t, tau;
   unsigned long n = 3;

   int* p = NULL;  p = new int[n+1];

   // starting permutation
   // identity 1, 2, ... , n -> 1, 2, ... , n
   for(i=0; i<=n; i++)
   {
   p[i] = i;
   cout << "p[" << i << "] = " << p[i] << "  ";
   }
   cout << endl;

   int test = 1;

   do
   {
   i = n -1;
   while(p[i] > p[i+1]) i = i - 1;
   if(i > 0) test = 1; else test = 0;
   j = n;
   while(p[j] <= p[i]) j = j - 1;

   t = p[i];  p[i] = p[j];  p[j] = t;
   i = i + 1;  j = n;
   while(i < j)
   {
   t = p[i];  p[i] = p[j];  p[j] = t;
   i = i + 1; j = j - 1;
   }
   // display result
   for(tau=0; tau<=n; tau++)
   cout << "p[" << tau << "] = " << p[tau] << "  ";
   cout << endl;
   } while(test == 1);
   return 0;
}
```

Goldberg and Lingle [73] suggested a crossover operator, the so-called partially mapped crossover. They believe it will lead to an efficient solution of the traveling salesman problem. A partially mapped crossover proceeds as follows. We number the cities from 0 to $N-1$. Let $N = 10$.

We explain with an example how the partially mapped operator works. Assume that the parents are

(1 2 3 4 5 6 7 8 9 10 11 12)   : a1

(7 3 6 11 4 12 5 2 10 9 1 8)   : a2

a1 and a2 are integer arrays. Positions count from 0 to $n-1$, where $n = 12$. We select two random numbers $r1$ and $r2$

$$0 \le r1 \le (n-1), \ \ 0 \le r2 \le (n-1), \ \ r1 \le r2$$

Let $r1 = 3$, $r2 = 6$. Truncate parents using $r1$ and $r2$.

(1 2 3 | 4 5 6 7 | 8 9 10 11 12)

(7 3 6 | 11 4 12 5 | 2 10 9 1 8)

We obtain the subarrays

s1 = (4 5 6 7) and s2 = (11 4 12 5).

Next we do the crossing

(1 2 3 | 11 4 12 5 | 8 9 10 11 12)

(7 3 6 | 4 5 6 7 | 2 10 9 1 8)

Now some cities occur twice while others are missing in the new array. The crossing defines the mappings

11 -> 4   4 -> 5   12 -> 6   5 -> 7   (∗)

4 -> 11   5 -> 4   6 -> 12   7 -> 5   (∗∗)

Positions which must be fixed are indicated by X

(1 2 3| 11 4 12 5 | 8 9 10 X X)

(X 3 X | 4 5 6 7 | 2 10 9 1 8)

We fix the first array using the mapping (∗).

a) number 11 at position 10 must be fixed

i) map $11 \mapsto 4$ but 4 is in array $s2$

ii) map $4 \mapsto 5$ but 5 is in array $s2$

iii) map $5 \mapsto 7$ o.k. 7 is not in array $s2$

Thus replace number 11 at position 10 by number 7.

a) number 12 at position 11 must be fixed

i) map $12 \mapsto 6$ o.k. 6 is not in array s2

Thus replace number 12 at position 11 by number 6.

a) number 7 at position 0 must be fixed

i) map $7 \mapsto 5$ but 5 is in array $s1$

ii) map $5 \mapsto 4$ but 4 is in array $s1$

iii) map $4 \mapsto 11$ o.k. 11 is not in array $s1$

Thus replace number 7 at position 0 by number 11

b) number 6 at position 2 must be fixed

i) map $6 \mapsto 12$ o.k. 12 is not in array $s1$

Thus replace number 6 at position 2 by number 12. Consequently, the children are

(1 2 3 11 4 12 5 8 9 10 7 6)

(11 3 12 4 5 6 7 2 10 9 1 8)

Bac and Perov [4] proposed another operator of crossings using the *permutation group*. We illustrate the operator with an example and a C++ program. Let the parents be given by

```
(0 1 2 3 4 5 6 7 8 9) -> (8 7 3 4 5 6 0 2 1 9)   parent 1
(0 1 2 3 4 5 6 7 8 9) -> (7 6 0 1 2 9 8 4 3 5)   parent 2
```

The permutation map yields

```
0 -> 8 -> 3
1 -> 7 -> 4
2 -> 3 -> 1
```

etc. Thus the children are given by

```
(0 1 2 3 4 5 6 7 8 9) -> (3 4 1 2 9 8 7 0 6 5)
(0 1 2 3 4 5 6 7 8 9) -> (2 0 8 7 3 9 1 5 4 6)
```

The implementation of this permutation is straightforward.

```cpp
// tspperm.cpp
#include <iostream.h>

void crossing(int* a1,int* a2,int* a3,int* a4,int n)
{
    int i;
    for(i=0; i<n; i++)
    {
    int p = a1[i];
    a3[i] = a2[p];
    }

    for(i=0; i<n; i++)
    {
    int q = a2[i];
    a4[i] = a1[q];
    }
}

int main()
{
    int n = 10;
    int i;

    int* a1 = NULL; int* a2 = NULL;
    int* a3 = NULL; int* a4 = NULL;
```

```
a1 = new int[n]; a2 = new int[n];
a3 = new int[n]; a4 = new int[n];
a1[0] = 8; a1[1] = 7; a1[2] = 3; a1[3] = 4;
a1[4] = 5; a1[5] = 6; a1[6] = 0; a1[7] = 2;
a1[8] = 1; a1[9] = 9;
a2[0] = 7; a2[1] = 6; a2[2] = 0; a2[3] = 1;
a2[4] = 2; a2[5] = 9; a2[6] = 8; a2[7] = 4;
a2[8] = 3; a2[9] = 5;

crossing(a1,a2,a3,a4,n);

cout << endl;

for(i=0; i<n; i++)
{
cout << "a3[" << i << "] = " << a3[i] << "  ";
if(((i+1)%2) == 0) { cout << endl; }
}

cout << endl;

for(i=0; i<n; i++)
{
cout << "a4[" << i << "] = " << a4[i] << "  ";
if(((i+1)%2) == 0) { cout << endl; }
}

delete [] a1;  delete [] a2;
delete [] a3;  delete [] a4;

return 0;
}
```

In the following program we use these operators to find solutions to the traveling
salesman problem.

```cpp
// tsp.cpp
//
// traveling salesman problem

#include <fstream.h>
#include <stdlib.h>
#include <time.h>
#include "bitvect.h"

void readdist(char* filename,double **&dist,int &cities)
{
  int i,j;
  ifstream d(filename);
  d >> cities;
  dist = new double*[cities];
  for(i=0;i<cities;i++)
  dist[i] = new double[cities];
  for(i=0;i<cities;i++)
  for(j=i+1;j<cities;j++)
  {
  d >> dist[i][j];
  dist[j][i] = dist[i][j];
  }
  for(i=0;i<cities;i++) dist[i][i]=0;
  cout << "d[0][0] = " << dist[0][0] << endl;
  d.close();
}

void destroydist(double **dist,int cities)
{
  for(int i=0;i<cities;i++) delete[] dist[i];
  delete[] dist;
}

double distance(int *seq,int cities,double **dist)
{
  double sumdist = 0.0;
  for(int i=1;i<cities;i++)
  sumdist += dist[seq[i]][seq[i-1]];
  sumdist += dist[seq[0]][seq[cities-1]];
  return sumdist;
}

void setupfarm(int **farm,int n,int cities)
{
```

```
  BitVector used(cities);
  int city,i,j;
  srand(time(NULL));
  for(i=0;i<n;i++)
  {
  for(j=0;j<cities;j++) used.SetBit(j,0);
  for(j=0;j<cities;j++)
  {
  city = rand()%cities;
  if(!used.GetBit(city)) {farm[i][j]=city;used.SetBit(city);}
  else j--;
  }
  }
}


void mutate(int **farm,int n,int cities,double **dist)
{
  int i;
  int seq = rand()%n;
  int pos1 = rand()%cities;
  int pos2 = rand()%cities;
  while(pos2 == pos1) pos2 = rand()%cities;
  int *mutated = new int[cities];
  for(i=0;i<cities;i++) mutated[i] = farm[seq][i];
  mutated[pos1] = farm[seq][pos2];
  mutated[pos2] = farm[seq][pos1];
  if(distance(farm[seq],cities,dist) > distance(mutated,cities,dist))
  {
  delete farm[seq]; farm[seq] = mutated;
  }
  else delete mutated;
}


void permutate(int** farm,int n,int cities,double** dist)
{
  int i;
  int seq1 = rand()%n;
  int seq2 = rand()%n;
  int *result1, *result2, *result3, *result4;
  while(seq2 == seq1) seq2 = rand()%n;
  int *child1 = new int[cities];
  int *child2 = new int[cities];
  for(i=0;i<cities;i++)
  {
  child1[i] = farm[seq2][farm[seq1][i]];
  child2[i] = farm[seq1][farm[seq2][i]];
  }
  if(distance(farm[seq1],cities,dist)>distance(child1,cities,dist))
```

```
    result1 = child1;
    else result1 = farm[seq1];
    if(distance(farm[seq2],cities,dist) > distance(child2,cities,dist))
    result2 = child2;
    else result2 = farm[seq2];
    result3 = ((result1 == farm[seq1])?child1:farm[seq1]);
    result4 = ((result2 == farm[seq2])?child2:farm[seq2]);
    farm[seq1] = result1;
    farm[seq2] = result2;
    delete [] result3;
    delete [] result4;
}

int insequence(int el,int *seq,int p1,int p2)
{
  for(int i=p1;i<p2;i++)
  if(seq[i] == el) return i;
  return -1;
}

void pmx(int **farm,int n,int cities,double **dist)
{
  int i,pos;
  int seq1 = rand()%n;
  int seq2 = rand()%n;
  int *result1, *result2, *result3, *result4;
  while(seq2 == seq1) seq2 = rand()%n;
  int pos1 = rand()%cities;
  int pos2 = rand()%cities;
  while(pos2 == pos1) pos2=rand()%cities;
  if(pos2<pos1) { i = pos2; pos2 = pos1; pos1 = i; }
  int *child1 = new int[cities];
  int *child2 = new int[cities];
  for(i=0;i<cities;i++)
  {
  if((i<pos2) && (i>=pos1))
  {
  child1[i] = farm[seq2][i];
  child2[i] = farm[seq1][i];
  }
  else
  {
  child1[i] = farm[seq1][i];
  child2[i] = farm[seq2][i];
  }
  }

  for(i=0;i<cities;i++)
```

```
{
if((i<pos1) || (i>=pos2))
while((pos = insequence(child1[i],child1,pos1,pos2)) >= 0)
child1[i] = child2[pos];
if((i<pos1) || (i>=pos2))
while((pos=insequence(child2[i],child2,pos1,pos2))>=0)
child2[i] = child1[pos];
}

if(distance(farm[seq1],cities,dist) > distance(child1,cities,dist))
result1 = child1;
else result1 = farm[seq1];
if(distance(farm[seq2],cities,dist) > distance(child2,cities,dist))
result2 = child2;
else result2 = farm[seq2];
result3=((result1 == farm[seq1])?child1:farm[seq1]);
result4=((result2 == farm[seq2])?child2:farm[seq2]);
farm[seq1] = result1;
farm[seq2] = result2;
delete [] result3;
delete [] result4;
}

void main(void)
{
  int N = 20;                // number of animals/chromosomes
  int iterations = 300;
  cout << N << endl;
  int** farm = NULL;
  int i,j;
  double** dist = NULL; // array of distances
  int cities;               // number of cities
  readdist("tsp.dat",dist,cities);
  cout<<"Cities: "<<cities<<endl;
  farm = new int*[N];
  for(i=0;i<N;i++) farm[i]=new int[cities];
  setupfarm(farm,N,cities);
  for(i=0;i<iterations;i++)
  {
  mutate(farm,N,cities,dist);
  permutate(farm,N,cities,dist);
  pmx(farm,N,cities,dist);
  }
  for(i=0;i<N;i++)
  {
  for(j=0;j<cities;j++) cout<<farm[i][j]<<" ";
  cout << " distance:" << distance(farm[i],cities,dist) << endl;
  }
```

```
    destroydist(dist,cities);
}
```

The input file **tsp.dat** is:

```
8
14.5413
20.7663
13.5059
19.6041
10.4139
4.60977
14.5344
6.34114
5.09313
9.12195
5
12.0416
14.0357
8.70919
10.4938
11.2432
18.3742
18.8788
14.213
7.5326
12.6625
17.7071
9.72677
15.4729
10.5361
7.2111
10.198
10
```

A typical output is

```
Cities: 8
7 4 2 3 1 5 0 6     distance:64.8559
7 4 2 3 1 5 0 6     distance:64.8559
7 4 2 3 1 5 0 6     distance:64.8559
0 6 5 1 3 2 4 7     distance:66.1875
7 4 2 3 1 5 0 6     distance:64.8559
7 4 2 3 1 5 0 6     distance:64.8559
7 4 2 3 1 5 0 6     distance:64.8559
0 6 5 1 3 2 4 7     distance:66.1875
7 4 2 3 1 5 0 6     distance:64.8559
4 2 1 3 0 6 5 7     distance:67.9889
7 4 2 3 1 5 0 6     distance:64.8559
7 4 2 3 1 5 0 6     distance:64.8559
0 6 5 1 3 2 4 7     distance:66.1875
0 6 5 1 3 2 4 7     distance:66.1875
7 4 2 3 1 5 0 6     distance:64.8559
7 4 2 3 1 5 0 6     distance:64.8559
7 4 2 3 1 5 0 6     distance:64.8559
7 4 2 3 1 5 6 0     distance:66.1875
7 4 2 3 1 5 0 6     distance:64.8559
7 4 2 3 1 5 0 6     distance:64.8559
```

15.12   Other Applications for Genetic Algorithms

Many other optimization problems can be solved using genetic algorithms. Two of
them are the bin packing problem and Steiner's problem.

The *bin packing problem* is as follows. Given bins each of size S and m objects with
sizes

$$\{ s_0, s_1, \ldots, s_{m-1} \}$$

minimize the number of bins n

$$\text{subject to} \quad \sum_{i=0}^{n-1} s_i x_{i,j} \leq S, \quad \sum_{i=0}^{n-1} x_{i,j} = 1,$$

$$\text{where} \quad x_{i,j} \in \{0, 1\}.$$

We use $x_{i,j} = 1$ when object i is in bin j.

Genetic algorithms can also be applied to *Steiner's problem*. In this problem there
are n villages. Village j requires L_j phone lines to a station. A line costs c per
kilometer. Determine where to place a single station such that the total cost for the
phone lines is minimized. The set of positions of the villages is

$$\{ \mathbf{x}_0, \mathbf{x}_1, \ldots, \mathbf{x}_{n-1} \}.$$

Thus we must minimize

$$\sum_{i=0}^{n-1} c L_j |\mathbf{x}_j - \mathbf{s}|$$

where $\mathbf{s}$ is the location of the station.

15.13 Distributed Global Optimization

Distributed global optimization [172] is a technique which attempts to overcome some of the limitations of other techniques for global optimization such as gradient descent, Monte Carlo methods and genetic algorithms. It does not suffer the loss of speed due to extensive use of random numbers, the lack of a stopping condition and does not require the tuning of parameters (such as the population size in genetic algorithms) to aid effectiveness. Furthermore the function to be optimized does not have to be continuous or differentiable.

The algorithm starts with an initial guess $S(0)$. This can be randomly generated or specifically chosen. Each iteration of the algorithm produces a new point $S(i)$. We use the transformation $T(m_1, m_2, S)$ to generate two new points for every for every point S in a set. The transformations operate on the part of the point identified by m_1 and m_2, where m_1 and m_2 are bit positions. Each resolution must use a power of 2 as the number of bits of representation.

1. Set the resolution $n := 1$.

2. From $S(i)$ generate $2^{n+1} - 1$ points as follows.

   (a) For $j = 0, \ldots, n$ set

$$P(j) := \bigcup_{k=0}^{2^j-1} T(k2^{n-j}, (k+1)2^{n-j} - 1, S(i)).$$

   (b) The points are in $\bigcup P(j) \cup \{S(i)\}$.

3. Find the point k with minimum function value from the 2^{n+1} points.

4. If this is a new minimum set $S(i+1) := k$ and goto 2.

5. Increase the resolution, i.e. increment n.

6. If the resolution is less than the maximum resolution goto 2.

7. Stop.

Now we give an example program, to calculate the maximum of the function

$$\cos(x) - \sin(2x).$$

Since the discrete global optimization technique attempts to find a minimum function value, we use $-(\cos(x) - \sin(2x))$ in the program for function evaluation. For the transformation we use simple bit inversion (i.e. we apply NOT to the selected bits). Valafar [172] recommends using the Gray code to transform the selected bits. The chosen transform influences the effectiveness of the technique.

```cpp
// dgo.cpp

#include <iostream>
#include <math.h>
#include <stdlib.h>
#include <time.h>

using namespace std;

const double pi = 3.14115927;

// maximum bit resolution between 1 and 32
const int maxres = 20;

typedef double (*function)(unsigned int);
typedef unsigned int (*transform)(int,int,unsigned int);

unsigned int pow2(unsigned int y)
{
 static unsigned int pow2table[31];
 static int init = 0;
 unsigned int p2;

 if(!init)
 {
  pow2table[0]=1;
  for(p2=1;p2<32;p2++) pow2table[p2]=2*pow2table[p2-1];
 }
 if(y>31) return 0;
 return pow2table[y];
}

void dgo(unsigned int &S,function f,transform T)
{
 int n = 1,j,k,newmin;
 unsigned int *P,*q;
 double min;

 min = f(S);
 while(n < maxres)
 {
  q = P = new unsigned int[pow2(n+1)-1];
  newmin = 0;
  for(j=0;j <= n;j++)
   for(k=0;k < pow2(j);k++)
   {
    *q = T(k*pow2(n-j),(k+1)*pow2(n-j)-1,S);
    if((f(*q) < min)&&(S != *q))
```

```
     {
       min = f(*q);
       S = *q;
       newmin = 1;
       }
       q++;
     }
   delete[] P;
   if(!newmin) n++;
 }
}


double f(unsigned int a)
{
 double x = 2*pi*a/(pow2(8*sizeof(unsigned int))-1);
 return -cos(x)+sin(2*x);
}


unsigned int T(int m1,int m2,unsigned int x)
{
 unsigned int mask=pow2(m2+1)-pow2(m1);
 return x^mask;
}


void main(void)
{
 unsigned int S;
 double x;

 srand(time(NULL));
 S=rand();
 dgo(S,f,T);
 x = 2*pi*S/(pow2(8*sizeof(unsigned int))-1);
 cout << "For cos(x)-sin(2x) DGO gives " << -f(S)
      << " at " << x << endl;
}
```

The program output is

```
For cos(x)-sin(2x) DGO gives 1.76017 at 5.64887
```

15.14   Genetic Programming

Genetic programming [107] uses the techniques of genetic algorithms to generate
"programs". The programs are evaluated according to their effectiveness in solving
a certain problem. It may not be possible to find a program that solves a given
problem exactly. The constructs and instructions provided by a language may be
insufficient to solve a problem. In this case the genetic algorithm techniques attempt
to approximate a solution. A simple illustration of genetic programming is *symbolic
regression*, which attempts to find a function which best fits a given set of data
points. We construct the function symbolically, the function consists of the basic
operations +, - and * acting on polynomials of a single variable, or one of the
functions cos, sin or exp. We exclude division to avoid division by zero errors.
Obviously not all functions will be able to be simulated exactly without division.
Since each operation takes at most two arguments, the functions can be represented
as a binary tree. For mutation, a binary operation in the tree is selected and
randomly changed to one of the other operations, or a leaf node is replaced with
a randomly generated tree. For crossover, subtrees are swapped. We keep only
the fittest third of the population and replace the rest with new individuals. This
attempts to provide the variety of functions needed to find the best fit.

In the following program we apply the technique to the functions

$$\cos(x), \qquad x^2, \qquad \cos(2x)$$

using 10 data points, and

$$\frac{1}{1+x}$$

using 20 data points. We use SymbolicC++ [169] to create the expressions for the
functions and to evaluate the functions at the data points. The generation of a
tree for a symbolic expression is simple. The function type is randomly determined,
and if the function takes any parameters these must also be generated. For every
function parameter generated the probability that the next parameter is a leaf node
(a constant, or variable, but not a function) is increased. Thus we can ensure
that a randomly generated tree will not exceed a certain depth. The crossover
function randomly selects two individuals from the population, and then selects the
subtrees to be swapped. This is done by randomly determining if the current node
in the tree will be used as the subtree or if one of its branches will be used. The
process is repeated until a node is selected or a leaf node is found. An improvement
would be to use the roulette wheel method to select candidates for crossover. The
function fitness takes a symbolic tree, converts it to a symbolic expression and
then calculates the error for each data point. The sum of these errors is used as the
fitness. A fitness of zero is a perfect match.

The program below uses small populations and few iterations of the algorithm in order to reduce the execution time and memory requirements. To achieve better results large populations should be used to perform more extensive global search for the optimum.

```cpp
// sreg.cpp

#include <iostream.h>
#include <stdlib.h>
#include <time.h>
#include <math.h>
#include "Msymbol.h"

const double pmutate = 0.3;
const double pleaf = 1.7;
const int ops = 9;
const int ops1 = 3;
const int ops2 = 6;
const char types[ops]={'0','1','x','c','s','e','+','-','*'};

struct stree
{
 int type;
 struct stree *arg[2];
};

Sum<double> zero(0.0), one(1.0), x("x",1);

Sum<double> symbolic(struct stree st)
{
 switch(types[st.type])
 {
  case '0':  return zero;
  case '1':  return one;
  case 'x':  return x;
  case 'c':  return cos(symbolic(*(st.arg[0])));
  case 's':  return sin(symbolic(*(st.arg[0])));
  case 'e':  return exp(symbolic(*(st.arg[0])));
  case '+':  return symbolic(*(st.arg[0]))+symbolic(*(st.arg[1]));
  case '-':  return symbolic(*(st.arg[0]))-symbolic(*(st.arg[1]));
  case '*':  return symbolic(*(st.arg[0]))*symbolic(*(st.arg[1]));
  case '/':  return symbolic(*(st.arg[0]))/symbolic(*(st.arg[1]));
 }
 return zero;
}

void destroy(struct stree st)
```

```
{
 if(st.type >= ops1)
 {
  destroy(*(st.arg[0]));
  delete st.arg[0];
 }
 if(st.type >= ops2)
 {
  destroy(*(st.arg[1]));
  delete st.arg[1];
 }
}

struct stree generate(double p=0.1)
{
 double q = double(rand())/RAND_MAX;
 int type;
 struct stree st;

 if(q > p)
  type = rand()%ops;
 else
  type = rand()%ops1;

 st.type = type;
 if(type >= ops1)
  { st.arg[0] = new struct stree; *(st.arg[0]) = generate(p*pleaf);}
 if(type >= ops2)
  { st.arg[1] = new struct stree; *(st.arg[1]) = generate(p*pleaf);}
 if((symbolic(st) == zero) || (symbolic(st) == one))
 {
  destroy(st);
  return generate(p);
 }
 return st;
}

void mutate(struct stree *st,int n)
{
 int ind = rand()%n;
 int mut = rand();
 int branch = rand()%2;
 double p = double(rand())/RAND_MAX;

 if(n == 1) ind = 0;
 if(st[ind].type < ops1)
  { destroy(st[ind]); st[ind] = generate(); return; }
 else if(st[ind].type < ops2)
```

```
   { mut %= ops2-ops1; mut += ops1; branch = 0;}
 else
   { mut %= ops-ops2; mut += ops2; }

 if((p < pmutate) || (st[ind].type < ops2))
  st[ind].type = mut;
 else mutate(st[ind].arg[branch],1);
}

struct stree copy(struct stree st)
{
 stree str=st;
 if(st.type >= ops1)
 {
  str.arg[0]=new struct stree;
  *(str.arg[0])=copy(*(st.arg[0]));
 }
 if(st.type >= ops2)
 {
  str.arg[1]=new struct stree;
  *(str.arg[1])=copy(*(st.arg[1]));
 }
 return str;
}

void crossover(struct stree *st,int n)
{
 int ind1 = rand()%n;
 int ind2 = rand()%n;

 while(st[ind1].type < ops1) ind1 = rand()%n;
 while(st[ind2].type < ops1) ind2 = rand()%n;

 st[n] = copy(st[ind1]);
 st[n+1] = copy(st[ind2]);

 struct stree *stn1,*stn2,*stp1,*stp2;
 int d1 = 0,d2 = 0,arg1 = 0,arg2 = 0;

 stp1 = stn1 = &(st[ind1]); stp2 = stn2 = &(st[ind2]);
 while(!d1)
 {
  arg1 = rand()%2;
  double p = double(rand())/RAND_MAX;
  if (stn1->type < ops2) arg1 = 0;
  stp1 = stn1;
  stn1 = stn1->arg[arg1];
  if((stn1->type < ops1) || (p > 0.5))
```

```
    d1 = 1;
  }
  while(!d2)
  {
    arg2 = rand()%2;
    double p = double(rand())/RAND_MAX;
    if (stn2->type < ops2) arg2 = 0;
    stp2 = stn2;
    stn2 = stn2->arg[arg2];
    if((stn2->type < ops1) || (p > 0.5))
      d2 = 1;
  }

  struct stree *temp = stp1->arg[arg1];
  stp1->arg[arg1] = stp2->arg[arg2];
  stp2->arg[arg2] = temp;
}


double fitness(struct stree st,double* data,int n)
{
  double sum = 0.0;
  int i;
  for(i=0;i < n; i++)
  {
    x.set(*(data++));
    sum += fabs(symbolic(st).nvalue()-*(data++));
  }
  return sum;
}


void gp(int n,int m,double *data, int d)
{
  int i,j,k;
  struct stree *st = new struct stree[n+2];
  double minf;

  for(i=0;i < n+2;i++) st[i] = generate();
  for(i=0;i < m;i++)
  {
    cout << i+1 << "/" << m << " generations  \r";
    cout.flush();
    mutate(st,n);
    crossover(st,n);
    for(j=0; j< n+2;j++)
    {
      for(k=j+1;k < n+2; k++)
      if(fitness(st[j],data,d) > fitness(st[k],data,d))
      {
```

```
   struct stree temp = st[j];
    st[j] = st[k];
    st[k] = temp;
   }
  }
 for(j=n/3+1;j < n+2;j++)
  { destroy(st[j]); if (j < n) st[j] = generate();}
}
cout << endl;
Sum<double> f = symbolic(st[0]);
cout << f << endl;
for(i=0;i < 2*d;i+=2)
{
 x.set(data[i]);
  cout << data[i+1] << " " << f.nvalue() << endl;
}
for(i=0;i < n;i++) destroy(st[i]);
delete[] st;
}

void main(void)
{
 const double pi=3.1415297;
 double data[20*2];
 int i;

 srand(time(NULL));

 for(i=0;i < 20;i+=2) { data[i]=i*pi/10; data[i+1]=cos(data[i]); }
 gp(6,50,data,10);
 cout << endl;

 for(i=0;i < 20;i+=2) { data[i]=i; data[i+1]=i*i; }
 gp(6,50,data,10);
 cout << endl;

 for(i=0;i < 20;i+=2) { data[i]=i; data[i+1]=cos(2*i); }
 gp(8,60,data,10);
 cout << endl;

 for(i=0;i < 40;i+=2) { data[i]=i; data[i+1]=1.0/(i+1.0); }
 gp(3,100,data,20);
 cout << endl;
}
```

The program outputs the function which best fits the data points and the data point and function evaluation for the different input values. The left column of numbers are the desired values, and the right column the obtained values.

```
50/50 generations
cos(x)
1 1
0.809024 0.809024
0.309041 0.309041
-0.308981 -0.308981
-0.808987 -0.808987
-1 -1
-0.809061 -0.809061
-0.309101 -0.309101
0.308921 0.308921
0.80895 0.80895

50/50 generations
x^(2)
0 0
4 4
16 16
36 36
64 64
100 100
144 144
196 196
256 256
324 324

60/60 generations
sin(cos(2*x))
1 0.841471
-0.653644 -0.608083
-0.1455 -0.144987
0.843854 0.74721
-0.957659 -0.817847
0.408082 0.39685
0.424179 0.411573
-0.962606 -0.820683
0.834223 0.740775
-0.127964 -0.127615

100/100 generations
sin((exp(x)*x+x^(2))*sin(x)*x)
1 0
0.333333 0.396537
0.2 0.429845
0.142857 -0.199608
```

0.111111 0.976007
0.0909091 0.997236
0.0769231 -0.246261
0.0666667 0.327578
0.0588235 0.762098
0.0526316 -0.998154
0.047619 -0.677241
0.0434783 -0.998545
0.04 0.209575
0.037037 -0.863178
0.0344828 -0.987161
0.0322581 0.492514
0.030303 -0.326958
0.0285714 -0.801911
0.027027 -0.328644
0.025641 0.782954

15.15   Gene Expression Programming

Gene expression programming is a genome/phenome genetic algorithm [66] which combines the simplicity of genetic algorithms and the abilities of genetic programming. In a sense gene expression programming is a generalization of genetic algorithms and genetic programming.

A *gene* is a symbolic string with a head and a tail. Each symbol represents an operation. For example the operation "+" takes two arguments and adds them. The operation "x" would evaluate to the value of the variable x. The tail consists only of operations which take no arguments. The string represents expressions in prefix notation, i.e. $5 - 3$ would be stored as "$- 5\ 3$". The reason for the tail is to ensure that the expression is always complete. Suppose the string has h symbols in the head which is specified as an input to the algorithm, and t symbols in the tail which is determined from h. Thus if n is the maximum number of arguments for an operation we must have

$$h + t - 1 = hn.$$

The left-hand side is the total number of symbols except for the very first symbol. The right-hand side is the total number of arguments required for all operations. We assume, of course, that each operation requires the maximum number of arguments so that any string of this length is a valid string for the expression. Thus the equation states that there must be enough symbols to serve as arguments for all operations. Now we can determine the required length for the tail

$$t = h(n - 1) + 1.$$

Suppose we use $h = 8$, and $n = 2$ for arithmetic operations. Thus the tail length must be $t = 9$. So the total gene length is 17. We could then represent the expression

$$\cos(x^2 + 2) - \sin(x)$$

with the string

```
-c+*xx2s|x1x226x31
```

The vertical | is used to indicate the beginning of the tail. Here c represents cos() and s represents sin().

A *chromosome* is a series of genes. The genes combine to form an expression using some operation with the same number of arguments as genes in the chromosome. For example the expressions of genes of a chromosome may be added together.

A number of operations are applied to chromosomes.

- **Replication**. The chromosome is unchanged. The roulette wheel selection technique can be used to select chromosomes for replication.

- **Mutation**. Randomly change symbols in a chromosome. Symbols in the tail of a gene may not operate on any arguments. Typically 2 point mutations per chromosome is used.

- **Insertion**. A portion of a chromosome is chosen to be inserted in the head of a gene. The tail of the gene is unaffected. Thus symbols are removed from the end of the head to make room for the inserted string. Typically a probability of 0.1 of insertion is used. Suppose +x2 is to be inserted into

  ```
  -c+*xx2s|x1x226x31
  ```

  at the fourth position in the head. We obtain

  ```
  -c++x2*x|x1x226x31
  ```

  which represents
  $$\cos((x+2)+x^2) - 1.$$

- **Gene transposition**. One gene in a chromosome is randomly chosen to be the first gene. All other genes in the chromosome are shifted downwards in the chromosome to make place for the first gene.

- **Recombination**. The crossover operation. This can be one point (the chromosomes are split in two and corresponding sections are swapped), two point (chromosomes are split in three and the middle portion is swapped) or gene (one entire gene is swapped between chromosomes) recombination. Typically the sum of the probabilities of recombination is used as 0.7.

In the following program we implement these techniques. The example is the same as for genetic programming. This implementation is faster and more accurate than the implementation for genetic programming. This is a result of the relative simplicity of gene expression programming. For simplicity we use only one gene in a chromosome and only one point recombination.

```cpp
// gep.cpp

#include <iostream.h>
#include <stdlib.h>
#include <time.h>
#include <math.h>
#include <string.h>

const double pi=3.1415297;
const int nsymbols=8;
// 2 terminal symbols (no arguments) x and 1
const int terminals=2;
// terminal symbols first
const char symbols[nsymbols] = {'1','x','c','s','e','+','-','*'};
const int n = 2;    // for +,- and *
int h = 5;

double evalr(char *&e,double x)
{
 switch(*(e++))
 {
  case '1': return 1.0;
  case 'x': return x;
  case 'c': return cos(evalr(e,x));
  case 's': return sin(evalr(e,x));
  case 'e': return exp(evalr(e,x));
  case '+': return evalr(e,x)+evalr(e,x);
  case '-': return evalr(e,x)-evalr(e,x);
  case '*': return evalr(e,x)*evalr(e,x);
  default : return 0.0;
 }
}


double eval(char *e,double x)
{
 char *c=e;
 return evalr(c,x);
}


void printr(char *&e)
{
 switch(*(e++))
 {
  case '1': cout << '1';
            break;
  case 'x': cout << 'x';
            break;
  case 'c': cout << "cos(";printr(e);
```

```
                cout << ")";
                break;
case 's': cout << "sin(";printr(e);
                cout << ")";
                break;
case 'e': cout << "exp(";printr(e);cout<<")";
                break;
case '+': cout << '(';
                printr(e);
                cout << '+';
                printr(e);
                cout << ')';
                break;
case '-': cout << '(';
                printr(e);
                cout << '-';
                printr(e);
                cout << ')';
                break;
case '*': cout << '(';
                printr(e);
                cout << '*';
                printr(e);
                cout<<')';
                break;
  }
}


void print(char *e)
{
 char *c=e;
 printr(c);
}


double fitness(char *c,double *data,int N)
{
 int j;
 double sum=0.0;

 // the average error
 for(j=0;j<N;j++)
  sum+=fabs(eval(c,data[2*j])-data[2*j+1]);
 return sum;
}


// N number of data points
// population of size P
// eps = accuracy required
```

```
void gep(double *data,int N,int P, double eps)
{
 int i,j,k,replace,replace2,rlen,rp;
 int t = h*(n-1)+1;
 int gene_len = h+t;
 int pop_len = P*gene_len;
 int iterations=0;
 char *population = new char[pop_len];
 char *elim = new char[P];
 int toelim = P/2;
 double bestf,f;        // best fitness, fitness value
 double sumf = 0.0;     // sum of fitness values
 double pm = 0.1;       // probability of mutation
 double pi = 0.4;       // probability of insertion
 double pr = 0.7;       // probability of recombination
 double r,lastf;        // random numbers and roulette wheel selection
 char *best = (char*)NULL,*iter; // best gene, iteration variable

 // initialize the population
 for(i=0;i < pop_len;i++)
  if(i%gene_len < h)
   population[i] = symbols[rand()%nsymbols];
  else
   population[i] = symbols[rand()%terminals];

 // initial calculations
 bestf = fitness(population,data,N);
 best = population;
 for(i=0,sumf=0.0,iter=population;i < P;i++,iter+=gene_len)
 {
  f = fitness(iter,data,N);
  sumf += f;
  if(f<bestf)
  {
   bestf = f;
   best = population+i*gene_len;
  }
 }

 while(bestf >= eps)
 {
 // reproduction
 // roulette wheel selection
 for(i=0;i<P;i++)
  elim[i]=0;
 for(i=0;i < toelim;i++)
 {
  r = sumf*(double(rand())/RAND_MAX);
```

```
lastf = 0.0;
for(j=0;j < P;j++)
{
 f = fitness(population+j*gene_len,data,N);
 if((lastf<=r) && (r<f+lastf))
 {
  elim[j] = 1;
  j = P;
 }
 lastf += f;
}
}

for(i=0;i < pop_len;)
{
 if(population+i==best)
  i += gene_len; // never modify/replace best gene
 else for(j=0;j < gene_len;j++,i++)
 {
  // mutation or elimination due to failure in selection
  // for reproduction
  if((double(rand())/RAND_MAX<pm) || elim[i/gene_len])
  if(i%gene_len < h)
   population[i] = symbols[rand()%nsymbols];
  else
   population[i] = symbols[rand()%terminals];
 }

 // insertion
 if(double(rand())/RAND_MAX<pi)
 {
  // find a position in the head of this gene for insertion
  // -gene_len for the gene since we have already moved
  //  onto the next gene
  replace = i-gene_len;
  rp = rand()%h;
  // a random position for insertion source
  replace2 = rand()%pop_len;
  // a random length for insertion from the gene
  rlen = rand()%(h-rp);
  // create the new gene
  char *c = new char[gene_len];
  // copy the shifted portion of the head
  strncpy(c+rp+rlen,population+replace+rp,h-rp-rlen);
  // copy the tail
  strncpy(c+h,population+replace+h,t);
  // copy the segment to be inserted
  strncpy(c+rp,population+replace2,rlen);
```

```
    // if the gene is fitter use it
    if(fitness(c,data,N) < fitness(population+replace,data,N))
     strncpy(population+replace,c,h);
    delete[] c;
  }


  // recombination
  if(double(rand())/RAND_MAX < pr)
  {
   // find a position in the gene for one point recombination
   replace = i-gene_len;
   rlen = rand()%gene_len;
   // a random gene for recombination
   replace2 = (rand()%P)*gene_len;
   // create the new genes
   char *c[5];
   c[0] = population+replace;
   c[1] = population+replace2;
   c[2] = new char[gene_len];
   c[3] = new char[gene_len];
   c[4] = new char[gene_len];
   strncpy(c[2],c[0],rlen);
   strncpy(c[2]+rlen,c[1]+rlen,gene_len-rlen);
   strncpy(c[3],c[1],rlen);
   strncpy(c[3]+rlen,c[0]+rlen,gene_len-rlen);
   // take the fittest genes
   for(j=0;j < 4;j++)
   for(k=j+1;j < 4;j++)
    if(fitness(c[k],data,N) < fitness(c[j],data,N))
    {
     strncpy(c[4],c[j],gene_len);
     strncpy(c[j],c[k],gene_len);
     strncpy(c[k],c[4],gene_len);
    }
  delete[] c[2];
  delete[] c[3];
  delete[] c[4];
 }
}


// fitness
for(i=0,sumf=0.0,iter=population;i < P;i++,iter+=gene_len)
{
 f = fitness(iter,data,N);
 sumf += f;
 if(f < bestf)
 {
  bestf = f;
```

```
     best = population+i*gene_len;
    }
  }
  iterations++;
}

print(best);
cout << endl;
cout << "Fitness of " << bestf << " after "
     << iterations << " iterations." << endl;
for(i=0;i < N;i++)
  cout << data[2*i+1] << " " << eval(best,data[2*i]) << endl;

delete[] population;
delete[] elim;
}

void main(void)
{
  double data[20*2];
  int i;

  srand(time(NULL));

  for(i=0;i < 20;i+=2)
  {
    data[i] = i*pi/10;
    data[i+1] = cos(data[i]);
  }

  gep(data,10,50,0.001);
  cout << endl;

  for(i=0;i < 20;i+=2)
  {
    data[i] = i;
    data[i+1] = i*i;
  }

  gep(data,10,50,0.001);
  cout << endl;

  for(i=0;i < 20;i+=2)
  {
    data[i] = i;
    data[i+1] = cos(2*i);
  }
```

```
  gep(data,10,50,0.001);
  cout << endl;
}


/*
Results:

cos(x)
Fitness of 1.95888e-16 after 6 iterations.
1 1
0.809024 0.809024
0.309041 0.309041
-0.308981 -0.308981
-0.808987 -0.808987
-1 -1
-0.809061 -0.809061
-0.309101 -0.309101
0.308921 0.308921
0.80895 0.80895

(x*x)
Fitness of 0 after 0 iterations.
0 0
4 4
16 16
36 36
64 64
100 100
144 144
196 196
256 256
324 324

cos((((x+x)+x)-x))
Fitness of 1.59324e-16 after 191 iterations.
1 1
-0.653644 -0.653644
-0.1455 -0.1455
0.843854 0.843854
-0.957659 -0.957659
0.408082 0.408082
0.424179 0.424179
-0.962606 -0.962606
0.834223 0.834223
-0.127964 -0.127964
*/
```

Part II

Quantum Computing

Chapter 16

Quantum Mechanics

16.1 Hilbert Spaces

In this chapter we introduce the *Hilbert space* which plays the central rôle in quantum mechanics. For a more detailed discussion of this subject we refer to the books of Balakrishnan [5], Richtmyer [137], Sewell [146], Stakgold [152], Steeb [163], Weidmann [181], Yosida [185]. Moreover the proofs of the theorems given in this chapter can be found in these books. We assume that the reader is familiar with the notation of a linear space. First we introduce the pre-Hilbert space.

Definition. A linear space L is called a *pre-Hilbert space* if there is defined a numerical function called the *scalar product* (or *inner product*) which assigns to every f, g of vectors of L ($f, g \in L$) a complex number $\mathbf{C}$. The scalar product satisfies the conditions

$$(a) \qquad \langle f, f \rangle \geq 0, \qquad \langle f, f \rangle = 0 \quad \text{iff} \quad f = 0$$

$$(b) \qquad \langle f, g \rangle = \overline{\langle g, f \rangle}$$

$$(c) \qquad \langle cf, g \rangle = c \langle f, g \rangle \text{ where } c \text{ is an arbitrary complex number}$$

$$(d) \qquad \langle f_1 + f_2, g \rangle = \langle f_1, g \rangle + \langle f_2, g \rangle$$

where $\overline{\langle g, f \rangle}$ denotes the complex conjugate of $\langle g, f \rangle$.

It follows that

$$\langle f, g_1 + g_2 \rangle = \langle f, g_1 \rangle + \langle f, g_2 \rangle$$

and

$$\langle f, cg \rangle = \bar{c} \langle f, g \rangle.$$

Definition. A linear space E is called a *normed space*, if for every $f \in E$ there is associated a real number $\|f\|$, the norm of the vector f such that

$$(a) \qquad \|f\| \geq 0, \qquad \|f\| = 0 \quad \text{iff} \quad f = 0$$

$$(b) \qquad \|cf\| = |c| \|f\| \text{ where } c \text{ is an arbitrary complex number}$$

$$(c) \qquad \|f + g\| \leq \|f\| + \|g\|.$$

The conditions imply that

$$\|f - g\| \geq |\, \|f\| - \|g\|\, |\,.$$

This can be seen as follows. From

$$\|f - g\| + \|g\| \geq \|f\|$$

we obtain

$$\|f - g\| \geq \|f\| - \|g\|\,.$$

On the other hand

$$\|f - g\| = |-1|\,\|g - f\| \geq \|g\| - \|f\|\,.$$

The topology of a normed linear space E is thus defined by the *distance*

$$d(f, g) = \|f - g\|.$$

If a scalar product is given we can introduce a norm. The norm of f is defined by

$$\|f\| := \sqrt{\langle f, f \rangle}.$$

A vector $f \in L$ is called normalized if $\|f\| = 1$.

Definition. Two functions $f \in L$ and $g \in L$ are called *orthogonal* if

$$\langle f, g \rangle = 0.$$

Example. Consider the pre-Hilbert space $\mathbf{R}^4$ and

$$\mathbf{x} = \begin{pmatrix} -1 \\ 1 \\ -1 \\ 1 \end{pmatrix}, \qquad \mathbf{y} = \begin{pmatrix} 1 \\ -1 \\ -1 \\ 1 \end{pmatrix}.$$

Then $\mathbf{x}^T \mathbf{y} = 0$.

♣

Definition. A sequence $\{f_n\}$ ($n \in \mathbf{N}$) of elements in a normed space E is called a *Cauchy sequence* if, for every $\epsilon > 0$, there exists a number M_ϵ such that $\|f_p - f_q\| < \epsilon$ for $p, q > M_\epsilon$.

Definition. A normed space E is said to be *complete* if every Cauchy sequence of elements in E converges to an element in E.

Definition. A complete pre-Hilbert space is called a *Hilbert space*.

Definition. A complete normed space is called a *Banach space*.

Example. The vector space $C([a,b])$ of all continuous (real or complex valued) functions on an interval $[a,b]$ with the norm

$$\|f\| = \max_{[a,b]} |f(x)|$$

is a Banach space.   ♣

A Hilbert space will be denoted by $\mathcal{H}$ in the following. A Banach space will be denoted by $\mathcal{B}$ in the following.

Theorem. Every pre-Hilbert space L admits a completion $\mathcal{H}$ which is a Hilbert space.

Example. Let $L = \mathbf{Q}$. Then $\mathcal{H} = \mathbf{R}$.   ♣

Before we discuss some examples of Hilbert spaces we give the definitions of strong and weak convergence in Hilbert spaces.

Definition. A sequence $\{f_n\}$ of vectors in a Hilbert space $\mathcal{H}$ is said to *converge strongly* to f if

$$\|f_n - f\| \to 0$$

as $n \to \infty$. We write $s - \lim_{n \to \infty} f_n \to f$.

Definition. A sequence $\{f_n\}$ of vectors in a Hilbert space $\mathcal{H}$ is said to *converge weakly* to f if

$$\langle f_n, g \rangle \to \langle f, g \rangle$$

as $n \to \infty$, for any vector g in $\mathcal{H}$. We write $w - \lim_{n \to \infty} f_n \to f$.

It can be shown that strong convergence implies weak convergence. The converse is not generally true, however.

Example. Consider the sequence

$$f_n(x) := \sin(nx), \qquad n = 1, 2, \ldots$$

in the Hilbert space $L_2[0, \pi]$. The sequence does not tend to a limit in the sense of strong convergence. However, the sequence tends to 0 in the sense of weak convergence.   ♣

Let us now give several examples of Hilbert spaces which are important in quantum mechanics (and quantum computing). Although quantum computing is mainly discussed for finite dimensional Hilbert spaces (see chapter 17), infinite dimensional Hilbert spaces are now also discussed [29, 30, 31, 112, 175].

Example. Every finite dimensional vector space with an inner product is a Hilbert space. Let $\mathbf{C}^n$ be the linear space of n-tuples of complex numbers with the scalar product

$$\langle \mathbf{u}, \mathbf{v} \rangle := \sum_{j=1}^{n} u_j \bar{v}_j.$$

Then $\mathbf{C}^n$ is a Hilbert space. Let $\mathbf{u} \in \mathbf{C}^n$. We write the vector $\mathbf{u}$ as a column vector

$$\mathbf{u} = \begin{pmatrix} u_1 \\ u_2 \\ . \\ . \\ u_n \end{pmatrix}.$$

Thus we can write the scalar product in matrix notation

$$\langle \mathbf{u}, \mathbf{v} \rangle = \mathbf{u}^T \bar{\mathbf{v}}$$

where $\mathbf{u}^T$ is the transpose of $\mathbf{u}$.   ♣

Example. By $l_2(\mathbf{N})$ we mean the set of all infinite dimensional vectors (sequences) $\mathbf{u} = (u_1, u_2, \ldots)^T$ of complex numbers u_j such that

$$\sum_{j=1}^{\infty} |u_j|^2 < \infty.$$

Here $l_2(\mathbf{N})$ is a linear space with operations $(a \in \mathbf{C})$

$$a\mathbf{u} = (au_1, au_2, \ldots)^T$$
$$\mathbf{u} + \mathbf{v} = (u_1 + v_1, u_2 + v_2, \ldots)^T$$

with $\mathbf{v} = (v_1, v_2, \ldots)^T$ and $\sum_{j=1}^{\infty} |v_j|^2 < \infty$. One has

$$\sum_{j=1}^{\infty} |u_j + v_j|^2 \leq \sum_{j=1}^{\infty} (|u_j|^2 + |v_j|^2 + 2|u_j v_j|) \leq 2 \sum_{j=1}^{\infty} (|u_j|^2 + |v_j|^2) < \infty.$$

The scalar product is defined as

$$\langle \mathbf{u}, \mathbf{v} \rangle := \sum_{j=1}^{\infty} u_j \bar{v}_j = \mathbf{u}^T \bar{\mathbf{v}}.$$

It can also be proved that this pre-Hilbert space is complete. Therefore $l_2(\mathbf{N})$ is a Hilbert space. As an example, let us consider

$$\mathbf{u} = (1, \frac{1}{2}, \frac{1}{3}, \ldots, \frac{1}{n}, \ldots)^T.$$

Since

$$\sum_{n=1}^{\infty} \frac{1}{n^2} < \infty$$

we find that $\mathbf{u} \in l_2(\mathbf{N})$. Let

$$\mathbf{u} = (1, \frac{1}{\sqrt{2}}, \frac{1}{\sqrt{3}}, \ldots, \frac{1}{\sqrt{n}}, \ldots)^T.$$

Then $\mathbf{u} \notin l_2(\mathbf{N})$.

♣

Example. $L_2(M)$ is the space of *Lebesgue square-integrable functions* on M, where M is a Lebesgue measurable subset of $\mathbf{R}^n$, where $n \in \mathbf{N}$. If $f \in L_2(M)$, then

$$\int_M |f|^2 \, dm < \infty.$$

The integration is performed in the Lebesgue sense. The scalar product in $L_2(M)$ is defined as

$$\langle f, g \rangle := \int_M f(x)\bar{g}(x) \, dm$$

where $\bar{g}$ denotes the complex conjugate of g. It can be shown that this pre-Hilbert space is complete. Therefore $L_2(M)$ is a Hilbert space. Instead of dm we also write

dx in the following. If the Riemann integral exists then it is equal to the Lebesgue integral. However, the Lebesgue integral exists also in cases in which the Riemann integral does not exist.

♣

Example. Consider the linear space M^n of all $n \times n$ matrices over $\mathbf{C}$. The *trace* of an $n \times n$ matrix $A = (a_{jk})$ is given by

$$\mathrm{tr}A := \sum_{j=1}^{n} a_{jj}.$$

We define a scalar product by

$$\langle A, B \rangle := \mathrm{tr}(AB^*)$$

where tr denotes the trace and B^* denotes the conjugate transpose matrix of B. We recall that $\mathrm{tr}(C + D) = \mathrm{tr}C + \mathrm{tr}D$, where C and D are $n \times n$ matrices. For example, if A is the $n \times n$ unit matrix we find

$$\langle A, A \rangle = \mathrm{tr}(AA^*) = \mathrm{tr}(A) = n.$$

Thus $||A|| = \sqrt{n}$.

♣

Example. Consider the linear space of all infinite dimensional matrices $A = (a_{jk})$ over $\mathbf{C}$ such that

$$\sum_{j=1}^{\infty} \sum_{k=1}^{\infty} |a_{jk}|^2 < \infty.$$

We define a scalar product by

$$\langle A, B \rangle := \mathrm{tr}(AB^*)$$

where tr denotes the trace and B^* denotes the conjugate transpose matrix of B. We recall that $\mathrm{tr}(C + D) = \mathrm{tr}C + \mathrm{tr}D$ where C and D are infinite dimensional matrices. The infinite dimensional unit matrix does not belong to this Hilbert space.

♣

Example. Let D be an open set of the Euclidean space $\mathbf{R}^n$. Now $L_2(D)^{pq}$ denotes the space of all $q \times p$ matrix functions Lebesgue measurable on D such that

$$\int_D \mathrm{tr}f(x)f(x)^* dm < \infty$$

where m denotes the Lebesgue measure, * denotes the conjugate transpose, and tr is the trace of the $q \times q$ matrix. We define the scalar product as

$$\langle f, g \rangle := \int_D \mathrm{tr} f(x) g(x)^* dm.$$

Then $L_2(D)^{pq}$ is a Hilbert space. ♣

Theorem. All complex infinite dimensional Hilbert spaces are isomorphic to $l_2(\mathbf{N})$ and consequently are mutually isomorphic.

Definition. Let S be a subset of the Hilbert space $\mathcal{H}$. The subset S is dense in $\mathcal{H}$ if for every $f \in \mathcal{H}$ there exists a Cauchy sequence $\{f_j\}$ in S such that $f_j \to f$ as $j \to \infty$.

Definition. A Hilbert space is called *separable* if it contains a countable dense subset $\{f_1, f_2, \ldots\}$.

Example. The set of all $\mathbf{u} = (u_1, u_2, \ldots)^T$ in $l_2(\mathbf{N})$ with only finitely many nonzero components u_j is dense in $l_2(\mathbf{N})$. ♣

Example. Let $C^1_{(2)}(\mathbf{R})$ be the linear space of the once continuously differentiable functions that vanish at infinity together with their first derivative and which are square integrable. Then $C^1_{(2)}(\mathbf{R})$ is dense in $L_2(\mathbf{R})$. ♣

In almost all applications in quantum mechanics the underlying Hilbert space is separable.

Definition. A *subspace* $\mathcal{K}$ of a Hilbert space $\mathcal{H}$ is a subset of vectors which themselves form a Hilbert space.

It follows from this definition that, if $\mathcal{K}$ is a subspace of $\mathcal{H}$, then so too is the set $\mathcal{K}^\perp$ of vectors orthogonal to all those in $\mathcal{K}$. The subspace $\mathcal{K}^\perp$ is termed the *orthogonal complement* of $\mathcal{K}$ in $\mathcal{H}$. Moreover, any vector f in $\mathcal{H}$ may be uniquely decomposed into components $f_\mathcal{K}$ and $f_{\mathcal{K}^\perp}$, lying in $\mathcal{K}$ and $\mathcal{K}^\perp$, respectively, i.e.

$$f = f_\mathcal{K} + f_{\mathcal{K}^\perp}.$$

Example. Consider the Hilbert space $\mathcal{H} = l_2(\mathbf{N})$. Then the vectors

$$\mathbf{u}^T = (u_1, u_2, \ldots, u_N, 0, \ldots)$$

with $u_n = 0$ for $n > N$, form a subspace $\mathcal{K}$. The orthogonal complement $\mathcal{K}^\perp$ of $\mathcal{K}$ then consists of the vectors

$$(0, \ldots, 0, u_{N+1}, u_{N+2}, \ldots)$$

with $u_n = 0$ for $n \leq N$. ♣

Definition. A sequence $\{\phi_j\}$, $j \in I$ and $\phi_j \in \mathcal{H}$ is called an *orthonormal sequence* if

$$\langle \phi_j, \phi_k \rangle = \delta_{jk}$$

where I is a countable index set and δ_{jk} denotes the *Kronecker delta*, i.e.

$$\delta_{jk} := \begin{cases} 1 & \text{for } j = k \\ 0 & \text{for } j \neq k \end{cases}$$

Definition. An orthonormal sequence $\{\phi_j\}$ in $\mathcal{H}$ is an orthonormal basis if every $f \in \mathcal{H}$ can be expressed as

$$f = \sum_{j \in I} a_j \phi_j \qquad I : \text{ Index set}$$

for some constants $a_j \in \mathbf{C}$. The expansion coefficients a_j are given by

$$a_j := \langle f, \phi_j \rangle$$

Example. Consider the Hilbert space $\mathcal{H} = \mathbf{C}^4$. The scalar product is defined as

$$\langle \mathbf{u}, \mathbf{v} \rangle := \sum_{j=1}^{4} u_j \bar{v}_j.$$

The *Bell basis* for $\mathbf{C}^4$ is an orthonormal basis given by

$$\Phi^+ = \frac{1}{\sqrt{2}} \begin{pmatrix} 1 \\ 0 \\ 0 \\ 1 \end{pmatrix}, \qquad \Phi^- = \frac{1}{\sqrt{2}} \begin{pmatrix} 1 \\ 0 \\ 0 \\ -1 \end{pmatrix},$$

$$\Psi^+ = \frac{1}{\sqrt{2}} \begin{pmatrix} 0 \\ 1 \\ 1 \\ 0 \end{pmatrix}, \qquad \Psi^- = \frac{1}{\sqrt{2}} \begin{pmatrix} 0 \\ 1 \\ -1 \\ 0 \end{pmatrix}.$$

Let

$$\mathbf{u} = \frac{1}{2}\begin{pmatrix} 1 \\ 1 \\ 1 \\ 1 \end{pmatrix}$$

be a normalized state in $\mathbf{C}^4$. Then the expansion coefficients are given by

$$a_1 = \langle \mathbf{u}, \Phi^+ \rangle = \frac{1}{\sqrt{2}}, \qquad a_2 = \langle \mathbf{u}, \Phi^- \rangle = 0,$$

$$a_3 = \langle \mathbf{u}, \Psi^+ \rangle = \frac{1}{\sqrt{2}}, \qquad a_4 = \langle \mathbf{u}, \Psi^- \rangle = 0.$$

Consequently

$$\frac{1}{2}\begin{pmatrix} 1 \\ 1 \\ 1 \\ 1 \end{pmatrix} = \frac{1}{\sqrt{2}}\Phi^+ + \frac{1}{\sqrt{2}}\Psi^+.$$

♣

Example. Let $\mathcal{H} = L_2(-\pi, \pi)$. Then an orthonormal basis is given by

$$\left\{ \phi_k(x) := \frac{1}{\sqrt{2\pi}} \exp(ikx) \quad : \quad k \in \mathbf{Z} \right\}.$$

Let $f \in L_2(-\pi, \pi)$ with $f(x) = x$. Then the expansion coefficients are

$$a_k = \langle f, \phi_k \rangle = \int_{-\pi}^{\pi} f(x)\bar{\phi}_k(x)dx = \frac{1}{\sqrt{2\pi}} \int_{-\pi}^{\pi} x \exp(-ikx)dx.$$

♣

Remark. We call the expansion

$$f = \sum_{k \in \mathbf{Z}} \langle f, \phi_k \rangle \phi_k$$

the *Fourier expansion* of f.

Theorem. Every separable Hilbert space has at least one orthonormal basis.

Inequality of Schwarz. Let $f, g \in \mathcal{H}$. Then

$$|\langle f, g \rangle| \leq \|f\| \cdot \|g\|$$

Triangle inequality. Let $f, g \in \mathcal{H}$. Then

$$\|f + g\| \leq \|f\| + \|g\|$$

Let $B = \{ \phi_n : n \in I \}$ be an orthonormal basis in a Hilbert space $\mathcal{H}$. I is the countable index set. Then

$$(1) \qquad \langle \phi_n, \phi_m \rangle = \delta_{nm}$$

$$(2) \qquad \bigwedge_{f \in \mathcal{H}} \quad f = \sum_{n \in I} \langle f, \phi_n \rangle \phi_n$$

$$(3) \qquad \bigwedge_{f, g \in \mathcal{H}} \quad \langle f, g \rangle = \sum_{n \in I} \langle f, \phi_n \rangle \overline{\langle g, \phi_n \rangle}$$

$$(4) \qquad \left(\bigwedge_{\phi_n \in B} \langle f, \phi_n \rangle = 0 \right) \quad \Rightarrow \quad f = 0$$

$$(5) \qquad \bigwedge_{f \in \mathcal{H}} \|f\|^2 = \sum_{n \in I} |\langle f, \phi_n \rangle|^2$$

Equation (3) is called *Parseval's relation.*

Examples of orthonormal bases.

1) $\mathbf{R}^n$ or $\mathbf{C}^n$

$$B = \left\{ \begin{pmatrix} 1 \\ 0 \\ 0 \\ \vdots \\ 0 \end{pmatrix}, \begin{pmatrix} 0 \\ 1 \\ 0 \\ \vdots \\ 0 \end{pmatrix}, \quad \cdots \quad \begin{pmatrix} 0 \\ 0 \\ \vdots \\ 0 \\ 1 \end{pmatrix} \right\}$$

2) M^n

$$B = \{ (E_{jk}); \quad j, k = 1, 2, \ldots, n \}$$

where (E_{jk}) is the matrix with a one for the entry in row j, column k and zero everywhere else.

3) $l_2(\mathbf{N})$

$$B = \left\{ \begin{pmatrix} 1 \\ 0 \\ 0 \\ \vdots \end{pmatrix}, \begin{pmatrix} 0 \\ 1 \\ 0 \\ \vdots \end{pmatrix}, \begin{pmatrix} 0 \\ 0 \\ 1 \\ \vdots \end{pmatrix}, \cdots \right\}$$

4) $L_2(-\pi, \pi)$

$$B = \left\{ \frac{1}{\sqrt{2\pi}} e^{ikx} \quad : \quad |x| < \pi \quad : \quad k \in \mathbf{Z} \right\}$$

5) $L_2(-1, 1)$

$$B = \left\{ \frac{\sqrt{2l+1}}{\sqrt{2}} \frac{(-1)^l}{2^l l!} \frac{d^l}{dx^l} (1 - x^2)^l \quad : \quad l = 0, 1, 2, \dots \right\}$$

The polynomials are called the *Legendre polynomials*. For the first four Legendre polynomials we find $P_0(x) = 1$, $P_1(x) = x$, $P_2(x) = \frac{1}{2}(3x^2 - 1)$ and $P_3(x) = \frac{1}{2}(5x^3 - 3x)$.

6) $L_2[0, a]$ with $a > 0$

$$B_1 = \left\{ \frac{1}{\sqrt{a}} \exp(2\pi i x n / a) \quad : \quad n \in \mathbf{Z} \right\}$$

$$B_2 = \left\{ \frac{1}{\sqrt{a}}, \quad \sqrt{\frac{2}{a}} \cos\left(\frac{2\pi x n}{a}\right), \quad \sqrt{\frac{2}{a}} \sin\left(\frac{2\pi x n}{a}\right) \quad : \quad n \in \mathbf{N} \right\}$$

$$B_3 = \left\{ \sqrt{\frac{2}{a}} \sin\left(\frac{\pi x n}{a}\right) \quad : \quad n \in \mathbf{N} \right\}$$

$$B_4 = \left\{ \frac{1}{\sqrt{a}}, \quad \sqrt{\frac{2}{a}} \cos\left(\frac{\pi x n}{a}\right) \quad : \quad n \in \mathbf{N} \right\}$$

7) $L_2(\prod_{j=1}^{n}(-\pi,\pi))$

$$B = \left\{ \frac{1}{(2\pi)^{n/2}} \exp(i\mathbf{k}\cdot\mathbf{x}) \quad : \quad \mathbf{k}\cdot\mathbf{x} := k_1 x_1 + \ldots + k_n x_n \right\}$$

where $|x_j| < \pi$ and $k_j \in \mathbf{Z}$.

8) $L_2([0,a] \times [0,a] \times [0,a])$

$$B = \left\{ \frac{1}{a^{3/2}} e^{i2\pi \mathbf{n}\cdot\mathbf{x}/a} \quad : \quad \mathbf{n}\cdot\mathbf{x} = n_1 x_1 + n_2 x_2 + n_3 x_3 \right\}$$

where $a > 0$ and $n_j \in \mathbf{Z}$.

9) $L_2(S^2)$ where $S^2 := \{(x_1, x_2, x_3) : x_1^2 + x_2^2 + x_3^2 = 1\}$

$$Y_{lm}(\theta,\phi) := \frac{(-1)^{l+m}}{2^l l!} \sqrt{\frac{2l+1}{4\pi} \cdot \frac{(l-m)!}{(l+m)!}} \sin^m\theta \frac{d^{l+|m|}(\sin\theta)^{2l}}{d(\cos\theta)^{l+|m|}} e^{im\phi}$$

where

$$l = 0, 1, 2, 3, \ldots$$

$$m = -l, -l+1, \ldots, +l$$

and $0 \le \phi < 2\pi$, $0 \le \theta < \pi$. The functions Y_{lm} are called *spherical harmonics*.
The orthogonality relation is given by

$$\langle Y_{lm}, Y_{l'm'} \rangle := \int_{\theta=0}^{\pi} \int_{\phi=0}^{2\pi} Y_{lm}(\theta,\phi) \bar{Y}_{l'm'}(\theta,\phi) \overbrace{\sin\theta\, d\theta\, d\phi}^{d\Omega} = \delta_{ll'}\delta_{mm'}.$$

The first few spherical harmonics are given by

$$Y_{00}(\theta,\phi) = \frac{1}{\sqrt{4\pi}} \qquad\qquad Y_{11}(\theta,\phi) = -\sqrt{\frac{3}{8\pi}}\sin\theta e^{i\phi}$$

$$Y_{10}(\theta,\phi) = \sqrt{\frac{3}{4\pi}}\cos\theta \qquad\qquad Y_{1,-1}(\theta,\phi) = \sqrt{\frac{3}{8\pi}}\cos\theta e^{-i\phi}$$

10) $L_2(\mathbf{R})$

$$B = \left\{ \frac{(-1)^k}{2^{\frac{k}{2}} \sqrt{k!} \sqrt[4]{\pi}} e^{x^2/2} \frac{d^k}{dx^k} e^{-x^2} \quad : \quad k = 0, 1, 2, \dots \right\}$$

The functions

$$H_n(x) := (-1)^n e^{x^2} \frac{d^n}{dx^n} e^{-x^2}$$

are called the *Hermite polynomials*. For the first four Hermite polynomials we find
$H_0(x) = 1$, $H_1(x) = 2x$, $H_2(x) = 4x^2 - 2$, $H_3(x) = 8x^3 - 12x$.

11) $L_2(0, \infty)$

$$B = \left\{ e^{-x/2} L_n(x) \quad : \quad n = 0, 1, 2, \dots \right\}$$

where

$$L_n(x) := e^x \frac{d^n}{dx^n} (x^n e^{-x}).$$

The functions L_n are called *Laguerre polynomials*. For the first four Laguerre polynomials we find $L_0(x) = 1$, $L_1(x) = -x + 1$, $L_2(x) = x^2 - 4x + 2$, $L_3(x) = -x^3 + 9x^2 - 18x + 6$. ♣

In many applications in quantum mechanics such as spin-orbit coupling we need the *tensor product* of Hilbert spaces. The tensor product also plays a central role in quantum computing. Let $\mathcal{H}_1$ and $\mathcal{H}_2$ be two Hilbert spaces. We first consider the *algebraic tensor product* considering the spaces merely as linear spaces. The algebraic tensor product space is the linear space of all formal finite sums

$$h = \sum_{j=1}^{n} (f_j \otimes g_j), \quad f_j \in \mathcal{H}_1, \quad g_j \in \mathcal{H}_2$$

with the following expressions identified

$$c(f \otimes g) = (f \otimes cg) = (cf \otimes g)$$
$$(f_1 + f_2) \otimes g = (f_1 \otimes g) + (f_2 \otimes g)$$
$$f \otimes (g_1 + g_2) = (f \otimes g_1) + (f \otimes g_2)$$

where $c \in \mathbf{C}$. Let $f_j, h_l \in \mathcal{H}_1$ and $g_j, k_l \in \mathcal{H}_2$. We endow this linear space with the inner product

$$\left\langle \sum_{j=1}^{n_1} (f_j \otimes g_j), \sum_{l=1}^{n_2} (h_l \otimes k_l) \right\rangle := \sum_{j=1}^{n_1} \sum_{l=1}^{n_2} \langle f_j, h_l \rangle \langle g_j, k_l \rangle.$$

Thus we have a pre-Hilbert space. The completion of this space is denoted by $\mathcal{H}_1 \otimes \mathcal{H}_2$ and is called the tensor product Hilbert space.

As an example we consider the two Hilbert spaces $\mathcal{H}_1 = L_2(a, b)$ and $\mathcal{H}_2 = L_2(c, d)$. Then the tensor product Hilbert space $\mathcal{H}_1 \otimes \mathcal{H}_2$ is readily seen to be

$$L_2((a, b) \times (c, d))$$

the space of the functions $f(x_1, x_2)$ with $a < x_1 < b$, $c < x_2 < d$ and

$$\int_c^d \int_a^b |f(x_1, x_2)|^2 dx_1 dx_2 < \infty.$$

The inner product is defined by

$$\langle f, g \rangle := \int_c^d \int_a^b f(x_1, x_2) \bar{g}(x_1, x_2) dx_1 dx_2.$$

Let $\mathcal{H}_1 = L_2(a, b)$ and $\mathcal{H}_2 = L_2(c, d)$. Then we have the following

Theorem. Let

$$\{ \phi_n : n \in \mathbf{N} \}$$

be an orthonormal basis in the Hilbert space $L_2(a, b)$ and let

$$\{ \psi_n : n \in \mathbf{N} \}$$

be an orthonormal basis in the Hilbert space $L_2(c, d)$. Then the set

$$\{ \phi_n \psi_m \quad : \quad n \in \mathbf{N}, \quad m \in \mathbf{N} \}$$

is an orthonormal basis in the Hilbert space $L_2((a \leq x_1 \leq b) \times (c \leq x_2 \leq d))$.

It is easy to verify by integration that the set is an orthonormal set over the rectangle. To prove completeness it suffices to show that every continuous function f with

$$\int\limits_{c}^{d} \int\limits_{a}^{b} f(x_1, x_2) dx_1 dx_2 < \infty$$

whose Fourier coefficients with respect to the set are all zero, vanishes identically over the rectangle.

In some textbooks and articles the so-called *Dirac notation* is used to describe Hilbert space theory in quantum mechanics (Dirac [59]). Let $\mathcal{H}$ be a Hilbert space and $\mathcal{H}_*$ be the dual space endowed with the multiplication law of the form

$$(\lambda, \phi) = \bar{\lambda} \phi$$

where $\lambda \in \mathbf{C}$ and $\phi \in \mathcal{H}$. The inner product can be viewed as a bilinear form (duality)

$$\langle . | . \rangle : \mathcal{H}_* \times \mathcal{H} \to \mathbf{C}$$

such that the linear mappings

$$\langle \phi | : \psi \to \langle \phi | \psi \rangle, \quad \langle . | : \mathcal{H}_* \to \mathcal{H}'$$

$$| \psi \rangle : \phi \to \langle \phi | \psi \rangle, \quad | . \rangle : \mathcal{H} \to \mathcal{H}'_*$$

where prime denotes the space of linear continuous functionals on the corresponding space, are monomorphisms. The vectors $\langle \phi |$ and $| \psi \rangle$ are called *bra* and *ket* vectors, respectively. The ket vector $| \phi \rangle$ is uniquely determined by a vector $\phi \in \mathcal{H}$, therefore we write $| \phi \rangle \in \mathcal{H}$. A *dyadic product* of a bra vector $\langle \phi_2 |$ and a ket vector $| \phi_1 \rangle$ is a linear operator defined as

$$(| \phi_1 \rangle \langle \phi_2 |) | \psi \rangle = \langle \phi_2 | \psi \rangle | \phi_1 \rangle.$$

In some chapters we will adopt the Dirac notation.

16.2 Linear Operators in Hilbert Spaces

A linear operator, A, in a Hilbert space, $\mathcal{H}$, is a linear transformation of a linear manifold, $\mathcal{D}(A)$ ($\subset \mathcal{H}$), into $\mathcal{H}$. The manifold $\mathcal{D}(A)$ is termed the domain of definition, or simply the *domain*, of A. Throughout this section we consider linear operators.

Definition. The linear operator A is termed *bounded* if the set of numbers, $\|Af\|$, is bounded as f runs through the normalized vectors in $\mathcal{D}(A)$. In this case, we define

$\|A\|$, the *norm* of A, to be the supremum, i.e. the least upper bound, of $\|Af\|$, as f runs through these normalized vectors, i.e.

$$\|A\| := \sup_{\|f\|=1} \|Af\|.$$

Example. Let $\mathcal{H} = \mathbf{C}^n$. Then all $n \times n$ matrices over $\mathbf{C}$ are bounded linear operators. If I_n is the $n \times n$ identity matrix we have $\|I_n\| = 1$.   ♣

Example. Consider the Hilbert space $L_2(0, a)$ with $a > 0$. Let $Af(x) := xf(x)$. Then $\mathcal{D}(A) = L_2(0, a)$ and $\|A\| = a$.   ♣

It follows from this definition that

$$\|Af\| \leq \|A\| \|f\| \quad \text{for all vectors } f \text{ in } \mathcal{D}(A).$$

If A is bounded, we may take $\mathcal{D}(A)$ to be $\mathcal{H}$, since, even if this domain is originally defined to be a proper subset of $\mathcal{H}$, we can always extend it to the whole of this space as follows. Since

$$\|Af_m - Af_n\| \leq \|A\| \|f_m - f_n\|$$

we conclude that the convergence of a sequence of vectors $\{f_n\}$ in $\mathcal{D}(A)$ implies that of $\{Af_n\}$. Hence, we may extend the definition of A to $\overline{\mathcal{D}(A)}$, the closure of $\mathcal{D}(A)$, by defining

$$A \lim_{n \to \infty} f_n := \lim_{n \to \infty} Af_n.$$

We may then extend A to the full Hilbert space $\mathcal{H}$, by defining it to be zero on $\overline{\mathcal{D}(A)}^\perp$, the orthogonal complement of $\overline{\mathcal{D}(A)}$.

On the other hand, if A is unbounded, then in general, $\mathcal{D}(A)$ does not comprise the whole of $\mathcal{H}$ and cannot be extended to do so.

Example. Consider the differential operator d/dx acting on the Hilbert space, $\mathcal{H}$, of square-integrable functions of the real variable x. The domain of this operator consists of those functions $f(x)$ for which both $\int |f(x)|^2 dx$ and $\int |df(x)/dx|^2 dx$ are both finite, and this set of functions does not comprise the whole of $\mathcal{H}$.   ♣

Definition. Let A be a bounded operator in $\mathcal{H}$. We define A^*, the *adjoint operator* of A, by the formula

$$\langle f, A^*g \rangle := \langle Af, g \rangle \quad \text{for all} \quad f, g \in \mathcal{H}.$$

Definition. The operator A is termed *self-adjoint* if $A^* = A$ or, equivalently, if

$$\langle f, Ag \rangle = \langle Af, g \rangle \quad \text{for all} \quad f, g \in \mathcal{H}.$$

Example. Let $\mathcal{H} = \mathbf{C}^2$. Then

$$A = \begin{pmatrix} 0 & i \\ -i & 0 \end{pmatrix}$$

is a self-adjoint operator (hermitian matrix). ♣

In the case where A is an unbounded operator in $\mathcal{H}$ we again define its adjoint, A^*, by the same formula, except that f is confined to $\mathcal{D}(A)$ and g to the domain $\mathcal{D}(A^*)$, which is specified as follows: g belongs to $\mathcal{D}(A^*)$ if there is a vector g_A in $\mathcal{H}$ such that

$$\langle f, g_A \rangle = \langle Af, g \rangle \quad \text{for all } f \text{ in } \mathcal{D}(A)$$

in which case $g_A = A^*g$. The operator A is termed self-adjoint if $\mathcal{D}(A^*) = \mathcal{D}(A)$ and $A^* = A$. The coincidence of $\mathcal{D}(A^*)$ with $\mathcal{D}(A)$ is essential here. The domain of a self-adjoint operator is dense in $\mathcal{H}$.

Remark. If merely $\langle Af, g \rangle = \langle f, Ag \rangle$ for all $f, g \in \mathcal{D}(A)$, and if $\mathcal{D}(A)$ is dense in $\mathcal{H}$, i.e., if $A \subset A^*$, then A is called hermitian (symmetric); $\mathcal{D}(A^*)$ may be larger than $\mathcal{D}(A)$, in which case A^* is a proper extension of A.

Definition. Let A be a linear operator with dense domain. Then its *nullspace* is defined by

$$N(A) := \{ u \in \mathcal{H} \,:\, Au = 0 \}.$$

Definition. A self-adjoint operator A is termed *positive* if

$$\langle f, Af \rangle \geq 0$$

for all vectors f in $\mathcal{D}(A)$.

Example. Let B be a bounded operator. We define $A := B^*B$. Then A is a bounded self-adjoint operator and the operator A is positive. ♣

Remark. If B is unbounded, then B^*B need not be self-adjoint.

Remark. An operator product AB is defined on a domain

$$\mathcal{D}(AB) = \{ v \in \mathcal{D}(B) \,:\, Bv \in \mathcal{D}(A) \}$$

and then

$$(AB)v := A(Bv).$$

Therefore $\mathcal{D}(A^*A)$ may be smaller than $\mathcal{D}(A)$.

Next we summarize the algebraic properties of the operator norm. It follows from the definitions of the norm and the adjoint of a bounded operator, together with the triangular inequality that if A, B are bounded operators and $c \in \mathbf{C}$, then

$$\|cA\| = |c|\|A\|$$

$$\|A^*A\| = \|A\|^2$$

$$\|A + B\| \leq \|A\| + \|B\|$$

$$\|AB\| \leq \|A\|\|B\|.$$

Definition. Suppose that $\mathcal{K}$ is a subspace of $\mathcal{H}$. Then since any vector f in $\mathcal{H}$ may be resolved into unequally defined components $f_\mathcal{K}$ and $f_{\mathcal{K}^\perp}$ in $\mathcal{K}$ and $\mathcal{K}^\perp$, respectively, we may define a linear operator Π by the formula $\Pi f = f_\mathcal{K}$. This is termed the *projection operator* from $\mathcal{H}$ to $\mathcal{K}$, or simply the projection operator or projector for the subspace $\mathcal{K}$.

It follows from this definition and the orthogonality of $f_\mathcal{K}$ and $f_{\mathcal{K}^\perp}$ that

$$\|\Pi f\|^2 = \|f_\mathcal{K}\|^2 \leq \|f_\mathcal{K}\|^2 + \|f_{\mathcal{K}^\perp}\|^2 = \|f\|^2$$

and therefore that Π is bounded. It also follows from the definition of Π that

$$\Pi^2 = \Pi = \Pi^*.$$

This formula is generally employed as a definition of a projection operator, Π, since it implies that the set of elements $\{\Pi f\}$ form a subspace of $\mathcal{H}$, as f runs through the vectors in $\mathcal{H}$.

Example. Let $\mathcal{H} = \mathbf{R}^2$. Then

$$\Pi_1 = \frac{1}{2}\begin{pmatrix} 1 & 1 \\ 1 & 1 \end{pmatrix}, \qquad \Pi_2 = \frac{1}{2}\begin{pmatrix} 1 & -1 \\ -1 & 1 \end{pmatrix}$$

are projection operators (projection matrices). We have $\Pi_1\Pi_2 = 0$. ♣

Example. In the case where $\mathcal{K}$ is a one-dimensional subspace, consisting of the scalar multiples of a normalized vector ϕ, the projection operator $\Pi = \Pi(\phi)$ is given by

$$\Pi(\phi)f = \langle \phi, f \rangle \phi.$$ ♣

Definition. An operator, U, in a Hilbert space $\mathcal{H}$ is termed a *unitary operator* if

$$\langle Uf, Ug \rangle = \langle f, g \rangle$$

for all vectors f, g in $\mathcal{H}$, and if U has an inverse U^{-1}, i.e. $UU^{-1} = U^{-1}U = I$, where I is the identity operator, i.e. $If = f$ for all $f \in \mathcal{H}$.

In other words, a unitary operator is an invertible one which preserves the form of the scalar product in $\mathcal{H}$. The above definition of unitarity is equivalent to the condition that

$$U^*U = UU^* = I \qquad \text{i.e.} \quad U^* = U^{-1}.$$

A unitary mapping of $\mathcal{H}$ onto a second Hilbert space $\mathcal{H}'$ is an invertible transformation V, from $\mathcal{H}$ to $\mathcal{H}'$, such that

$$\langle f, g \rangle_{\mathcal{H}} = \langle Vf, Vg \rangle_{\mathcal{H}'}.$$

Example. Let $\mathcal{H} = \mathbf{C}^4$. Then the permutation matrix

$$U = \begin{pmatrix} 0 & 0 & 0 & 1 \\ 0 & 0 & 1 & 0 \\ 0 & 1 & 0 & 0 \\ 1 & 0 & 0 & 0 \end{pmatrix}$$

is a unitary operator (unitary matrix). ♣

Next we discuss operator convergence. Suppose that A and the sequence $\{A_n\}$ are bounded linear operators in $\mathcal{H}$.

Definition. The sequence of operators A_n is said to *converge uniformly*, or in norm, to A as $n \to \infty$ if

$$\|A_n - A\| \to 0 \quad \text{as} \quad n \to \infty.$$

Definition. The sequence of operators A_n is said to *converge strongly* to A if $A_n f$ tends strongly to Af for all vectors f in $\mathcal{H}$.

Definition. The sequence of operators A_n is said to *converge weakly* to A if $A_n f$ tends weakly to Af for all vectors f in $\mathcal{H}$.

Example. Consider the Hilbert space $L_2(\mathbf{R})$. Let A_n be the *translation operator*

$$(A_n f)(x) := f(x + 2n).$$

The operator A_n converges weakly to the zero operator. However, A_n does not converge strongly to anything, since if it did the limit would have to be zero, whereas, for any f, $\|A_n f\| = \|f\|$, which does not tend to zero. ♣

From these definitions, it follows that norm convergence implies strong operator convergence, which in turn implies weak operator convergence. The converse statements are applicable only if $\mathcal{H}$ is finite dimensional, i.e. $\mathcal{H} = \mathbf{C}^n$.

Definition. A *density matrix*, ρ, is an operator in $\mathcal{H}$ of the form

$$\rho := \sum_{n=1}^{\infty} w_n \Pi(\phi_n) = \operatorname{norm} \lim_{N \to \infty} \sum_{n=1}^{N} w_n \Pi(\phi_n)$$

where $\{\Pi(\phi_n)\}$ are the projection operators for an orthonormal sequence $\{\phi_n\}$ of vectors and $\{w_n\}$ is a sequence of non-negative numbers whose sum is unity. Thus, a density matrix is bounded and positive.

Definition. The *trace* of a positive operator B is defined to be

$$\operatorname{tr}(B) := \sum_{n \in I} \langle \phi_n, B\phi_n \rangle$$

where $\{\phi_n : n \in I\}$ is any orthonormal basis set. The value of $\operatorname{tr}(B)$, which is infinite for some operators, is independent of the choice of basis.

It follows from these definitions of density matrices and trace that a density matrix is a positive operator whose trace is equal to unity.

Example. Consider the Hilbert space $\mathcal{H} = \mathbf{C}^4$. The Bell states

$$\Phi^+ = \frac{1}{\sqrt{2}} \begin{pmatrix} 1 \\ 0 \\ 0 \\ 1 \end{pmatrix}, \qquad \Phi^- = \frac{1}{\sqrt{2}} \begin{pmatrix} 1 \\ 0 \\ 0 \\ -1 \end{pmatrix},$$

$$\Psi^+ = \frac{1}{\sqrt{2}} \begin{pmatrix} 0 \\ 1 \\ 1 \\ 0 \end{pmatrix}, \qquad \Psi^- = \frac{1}{\sqrt{2}} \begin{pmatrix} 0 \\ 1 \\ -1 \\ 0 \end{pmatrix}$$

form a basis in $\mathbf{C}^4$. Consider the density matrix (we apply the Dirac notation)

$$\rho := \frac{5}{8} |\Phi^+\rangle\langle\Phi^+| + \frac{1}{8} (|\Phi^-\rangle\langle\Phi^-| + |\Psi^+\rangle\langle\Psi^+| + |\Psi^-\rangle\langle\Psi^-|).$$

This density matrix describes the *Werner state*. This mixed state, a $\frac{5}{8}$ vs. $\frac{1}{8}$ singlet-triplet mixture, can be produced by mixing equal amounts of singlets and random uncorrelated spins, or equivalently by sending one spin of an initially pure singlet through a 50% depolarizing channel [19, 182]. Since

$$\rho^2 = \frac{25}{64}|\Phi^+\rangle\langle\Phi^+| + \frac{1}{64}(|\Phi^-\rangle\langle\Phi^-| + |\Psi^+\rangle\langle\Psi^+| + |\Psi^-\rangle\langle\Psi^-|)$$

we find

$$\rho^2 \neq \rho$$

and

$$\text{tr}(\rho^2) = \frac{25}{64} + \frac{1}{64} = \frac{13}{32} < 1.$$

We can characterize a *mixed state* by $\text{tr}(\rho^2) < 1$.

The *pure states* of a quantum mechanical system are given by normalized vectors in a Hilbert space. The expectation value of an observable A (self-adjoint operator), for the state represented by $|\psi\rangle$ is

$$\langle A\rangle_\psi := \langle\psi|A\psi\rangle \equiv \text{tr}(\rho A) \qquad \text{where} \qquad \rho = |\psi\rangle\langle\psi|.$$

We have

$$\rho = \rho^*, \qquad \text{tr}(\rho) = 1, \qquad \rho^2 = \rho.$$

For a statistical mixture of pure states, given by an orthonormal set of vectors

$$\{ \psi_n \;:\; n = 1, 2, \ldots, N \}$$

in a Hilbert space, with respective probabilities

$$\{ w_n \;:\; n = 1, 2, \ldots, N \}$$

where

$$\sum_{n=1}^{N} w_n = 1, \qquad w_i \geq 0 \text{ for } i = 1, 2, \ldots, N$$

the expectation value of an observable is

$$\langle A\rangle = \sum_{n=1}^{N} w_n\langle\psi_n|A|\psi_n\rangle.$$

Using the density matrix

$$\rho := \sum_{n=1}^{N} w_n |\psi_n\rangle\langle\psi_n|$$

we can also write

$$\langle A\rangle = \mathrm{tr}(\rho A)$$

where

$$\mathrm{tr}(\rho) = 1, \qquad \rho^* = \rho$$

and

$$\rho^2 \neq \rho \ \text{ and } \ \mathrm{tr}(\rho^2) < 1 \ \text{ if } \ w_n \neq 0 \text{ for more than one } n.$$

Definition. A *one-parameter group* of unitary transformations of $\mathcal{H}$ is a family $\{U_t\}$ of unitary operators in $\mathcal{H}$, with t running through the real numbers, such that

$$U_t U_s = U_{t+s}, \qquad U_0 = I.$$

The group is said to be continuous if U_t converges strongly to I as t tends to zero; or equivalently, if U_t converges strongly to U_{t_0} as t tends to t_0, for any real t_0. In this case, Stone's theorem tells us that there is a unique self-adjoint operator, K, in $\mathcal{H}$ such that

$$\frac{d}{dt}U_t f = iKU_t f = iU_t K f \text{ for all } f \text{ in } \mathcal{D}(K).$$

This equation is formally expressed as

$$U_t = e^{iKt}$$

and iK is termed the infinitesimal generator of the group $\{U_t\}$.

Example. Let

$$K = \begin{pmatrix} 0 & i \\ -i & 0 \end{pmatrix}.$$

Then

$$U_t = e^{iKt} = \begin{pmatrix} \cos t & -\sin t \\ \sin t & \cos t \end{pmatrix}. \qquad\qquad \clubsuit$$

Next we consider linear operators in tensor product space. Suppose that $\mathcal{H}_1$ and $\mathcal{H}_2$ are Hilbert spaces, and that $\mathcal{H}$ is a third Hilbert space, defined in terms of $\mathcal{H}_1$ and

$\mathcal{H}_2$ as follows. We recall that for each pair of vectors f_1, f_2 in $\mathcal{H}_1, \mathcal{H}_2$, respectively, there is a vector in $\mathcal{H}$, denoted by $f_1 \otimes f_2$, such that

$$\langle f_1 \otimes f_2, g_1 \otimes g_2 \rangle = \langle f_1, g_1 \rangle_{\mathcal{H}_1} \langle f_2, g_2 \rangle_{\mathcal{H}_2}.$$

If A_1 and A_2 are operators in $\mathcal{H}_1$ and $\mathcal{H}_2$, respectively, we define the operator $A_1 \otimes A_2$ in $\mathcal{H}_1 \otimes \mathcal{H}_2$ by the formula

$$(A_1 \otimes A_2)(f_1 \otimes f_2) := (A_1 f_1) \otimes (A_2 f_2).$$

$A_1 \otimes A_2$ is called the *tensor product* of A_1 and A_2.

Similarly, we may define the tensor product $\mathcal{H}_1 \otimes \mathcal{H}_2 \otimes \cdots \otimes \mathcal{H}_n$ as well as that, $A_1 \otimes A_2 \otimes \cdots \otimes A_n$, of operators $A_1, \cdots, A_n$. In standard notation, one writes

$$\bigotimes_{j=1}^{n} \mathcal{H}_j = \mathcal{H}_1 \otimes \mathcal{H}_2 \otimes \cdots \otimes \mathcal{H}_n$$

and

$$\bigotimes_{j=1}^{n} A_j = A_1 \otimes A_2 \otimes \cdots \otimes A_n.$$

Example. Let $\mathcal{H}_1 = \mathcal{H}_2 = \mathbf{C}^2$. Then $\mathcal{H}_1 \otimes \mathcal{H}_2$ can be identified with $\mathbf{C}^4$. If

$$A_1 = \begin{pmatrix} 0 & 1 \\ 1 & 0 \end{pmatrix}, \qquad A_2 = \begin{pmatrix} 1 & 0 \\ 0 & 1 \end{pmatrix}$$

we obtain

$$A_1 \otimes A_2 = \begin{pmatrix} 0 & 0 & 1 & 0 \\ 0 & 0 & 0 & 1 \\ 1 & 0 & 0 & 0 \\ 0 & 1 & 0 & 0 \end{pmatrix}, \qquad A_2 \otimes A_1 = \begin{pmatrix} 0 & 1 & 0 & 0 \\ 1 & 0 & 0 & 0 \\ 0 & 0 & 0 & 1 \\ 0 & 0 & 1 & 0 \end{pmatrix}.$$

We see that $A_1 \otimes A_2 \neq A_2 \otimes A_1$.

Let us now discuss the spectrum of a linear operator. Let T be a linear operator whose domain $\mathcal{D}(T)$ and range $R(T)$ both lie in the same complex linear topological space X. In our case X is a Hilbert space $\mathcal{H}$. We consider the linear operator

$$T_\lambda := \lambda I - T$$

where λ is a complex number and I the identity operator. The distribution of the values of λ for which T_λ has an inverse and the properties of the inverse when it exists, are called the *spectral theory* for the operator T. We discuss the general theory of the inverse of T_λ (Yosida [185]).

Definition. If λ_0 is such that the range $R(T_{\lambda_0})$ is dense in X and T_{λ_0} has a continuous inverse $(\lambda_0 I - T)^{-1}$, we say that λ_0 is in the *resolvent set* $\varrho(T)$ of T, and we denote this inverse $(\lambda_0 I - T)^{-1}$ by $R(\lambda_0; T)$ and call it the *resolvent* (at λ_0) of T. All complex numbers λ not in $\varrho(T)$ form a set $\sigma(T)$ called the *spectrum* of T. The spectrum $\sigma(T)$ is decomposed into disjoint sets $P_\sigma(T)$, $C_\sigma(T)$ and $R_\sigma(T)$ with the following properties:

$P_\sigma(T)$ is the totality of complex numbers λ for which T_λ does not have an inverse. $P_\sigma(T)$ is called the *point spectrum* of T. In other words the point spectrum $P_\sigma(T)$ is the set of eigenvalues of T; that is

$$P_\sigma(T) := \{\lambda \in \mathbf{C} : Tf = \lambda f \text{ for some nonzero } f \text{ in } X\}.$$

$C_\sigma(T)$ is the totality of complex numbers λ for which T_λ has a discontinuous inverse with domain dense in X. $C_\sigma(T)$ is called the *continuous spectrum* of T.

$R_\sigma(T)$ is the totality of complex numbers λ for which T_λ has an inverse whose domain is not dense in X. $R_\sigma(T)$ is called the *residual spectrum* of T.

For these definitions and the linearity of the operator T we find the

Proposition. A necessary and sufficient condition for $\lambda_0 \in P_\sigma(T)$ is that the equation

$$Tf = \lambda_0 f$$

has a solution $f \neq 0$ ($f \in X$). In this case λ_0 is called an *eigenvalue* of T, and f the corresponding *eigenvector*. The null space $N(\lambda_0 I - T)$ of T_{λ_0} is called the *eigenspace* of T corresponding to the eigenvalue λ_0 of T. It consists of the vector 0 and the totality of eigenvectors corresponding to λ_0. The dimension of the eigenspace corresponding to λ_0 is called the *multiplicity* of the eigenvalue λ_0.

Theorem. Let X be a complex Banach-space, and T a closed linear operator with its domain $\mathcal{D}(T)$ and range $R(T)$ both in X. Then, for any $\lambda_0 \in \varrho(T)$, the resolvent $(\lambda_0 I - T)^{-1}$ is an everywhere defined continuous linear operator. For the proof we refer to Yosida [185].

Example. If the linear space X is of finite dimension, then any bounded linear operator T is represented by a matrix (t_{ij}). The eigenvalues of T are obtained as the roots of the algebraic equation, the so-called *secular* or *characteristic equation* of the matrix (t_{ij}):

$$\det(\lambda \delta_{ij} - t_{ij}) = 0$$

where $\det(.)$ denotes the determinant of the matrix. ♣

Example. Consider the Hilbert space $\mathcal{H} = L_2(\mathbf{R})$. Let T be defined by

$$Tf(x) := xf(x)$$

that is,

$$\mathcal{D}(T) = \{ f(x) \ : \ f(x) \text{ and } xf(x) \in L_2(\mathbf{R}) \}$$

and $Tf(x) = xf(x)$ for $f(x) \in \mathcal{D}(T)$. Then every real number λ_0 is in $C_\sigma(T)$, i.e. T has a purely continuous spectrum consisting of the entire real axis. For the proof we refer to Yosida [185]. ♣

Example. Let X be the Hilbert space $l_2(\mathbf{N})$. Let T be defined by

$$T(u_1, u_2, \cdots)^T := (0, u_1, u_2, \cdots)^T.$$

Then 0 is in the residual spectrum of T, since $R(T)$ is not dense in $l_2(\mathbf{N})$. ♣

Example. Let H be a self-adjoint operator in a Hilbert space $\mathcal{H}$. The spectrum $\sigma(H)$ lies on the real axis. The resolvent set $\varrho(H)$ of H comprises all the complex numbers λ with $\Im(\lambda) \neq 0$, and the resolvent $R(\lambda; H)$ is a bounded linear operator with the estimate

$$\|R(\lambda; H)\| \leq \frac{1}{|\Im(\lambda)|}.$$

Moreover,

$$\Im\langle(\lambda I - H)f, f\rangle = \Im(\lambda)\|f\|^2, \qquad f \in \mathcal{D}(H). \qquad ♣$$

Example. Let U be a unitary operator. The spectrum lies on the unit circle $|\lambda| = 1$; i.e. the interior and exterior of the unit circle are the resolvent set $\varrho(U)$. The residual spectrum is empty. ♣

Example. Consider the linear bounded self-adjoint operator in a Hilbert space $l_2(\mathbf{N})$

$$A = \begin{pmatrix} 0 & 1 & 0 & 0 & \cdots \\ 1 & 0 & 1 & 0 & \cdots \\ 0 & 1 & 0 & 1 & \cdots \\ & & \ddots & & \ddots \\ & & & \ddots & & \ddots \\ & & & & \ddots & & \ddots \end{pmatrix}.$$

In other words

$$A_{ij} = \begin{cases} 1 & \text{if } i = j + 1 \\ 1 & \text{if } i = j - 1 \\ 0 & \text{otherwise} \end{cases}$$

with $i, j \in \mathbf{N}$. We find $\operatorname{spec} A = [-2, 2]$, i.e. we have a continuous spectrum [163].

Example. The operator $-d^2/dx^2$, with a suitably chosen domain in $L_2(\mathbf{R})$ has a purely continuous spectrum consisting of the nonnegative real axis. The negative real axis belongs to the resolvent set.
♣

Example. The operator $-d^2/dx^2 + x^2$, with a suitably chosen domain in $L_2(\mathbf{R})$, has a pure point spectrum consisting of the positive odd integers, each of which is a simple eigenvalue.
♣

Example. Let $\mathcal{H} = l_2(\mathbf{N})$. Let A be the unitary operator that maps

$$\mathbf{u} = (u_1, u_2, u_3, u_4, u_5, \ldots, u_{2n}, u_{2n+1}, \ldots)^T$$

onto

$$A\mathbf{u} = (u_2, u_4, u_1, u_6, u_3, \ldots, u_{2n+2}, u_{2n-1}, \ldots)^T.$$

The point spectrum is empty and the continuous spectrum is the entire unit circle in the λ plane.
♣

Example. In the Hilbert space $\mathcal{H} = l_2(\mathbf{N} \cup \{0\})$, *annihilation* and *creation operators* denoted by b and b^* are defined as follows. They have a common domain

$$\mathcal{D}_1 = \mathcal{D}(b) = \mathcal{D}(b^*) = \{\mathbf{u} = (u_0, u_1, u_2, \ldots) : \sum_{n=0}^{\infty} n|u_n|^2 < \infty\}.$$

Then $b\mathbf{u}$ and $b^*\mathbf{u}$ are given by

$$b(u_0, u_1, u_2, \ldots)^T := (u_1, \sqrt{2}u_2, \sqrt{3}u_3, \ldots)^T$$
$$b^*(u_0, u_1, u_2, \ldots)^T := (0, u_0, \sqrt{2}u_1, \sqrt{3}u_2, \ldots)^T.$$

The physical interpretation for a simple model is that the vector

$$\varphi_n = (0, 0, \ldots, 0, u_n = 1, 0, \ldots)^T$$

represents a state of a physical system in which n particles are present. In particular, φ_0 represents the vacuum state, i.e.

$$b\varphi_0 = b(1, 0, 0, \ldots)^T = (0, 0, \ldots)^T.$$

The action of the operators b and b^* on these states is given by

$$b\varphi_n = \sqrt{n}\varphi_{n-1}$$
$$b^*\varphi_n = \sqrt{n+1}\varphi_{n+1}.$$

We find that b^* is the adjoint of b. We can show that

$$b^*b - bb^* = -I$$

in the sense that for all $\mathbf{u}$ in a certain domain $\mathcal{D}_2(\subset \mathcal{D}_1)$

$$b^*b\mathbf{u} - bb^*\mathbf{u} = -\mathbf{u}.$$

The operator denotes the identity operator. The operator $\hat{N} = b^*b$ with domain $\mathcal{D}_2$ is called the *particle-number operator*. Its action on the states φ_n is given by

$$\hat{N}\varphi_n = N\varphi_n$$

where $N = 0, 1, 2, \dots$. Thus the eigenvalues of $\hat{N}$ are $N = 0, 1, 2, \dots$. The point spectrum of b is the entire complex plane. The point spectrum of b^* is empty. The equation

$$b^*\mathbf{u} = \lambda\mathbf{u}$$

implies $\mathbf{u} = 0$. We can show that the residual spectrum of b^* is the entire complex plane. ♣

Remark. Instead of the notation φ_n the notation $|n\rangle$ is used in physics, where $n = 0, 1, 2, \dots$.

Remark. A point λ_0 in the spectrum $\sigma(A)$ of a self-adjoint operator A is a limit point of $\sigma(A)$ if it is either an eigenvalue of infinite multiplicity or an accumulation point of $\sigma(A)$. The set $\sigma_l(A)$ of all limit points of $\sigma(A)$ is called the essential spectrum of A, and its complement $\sigma_d(A) = \sigma(A) \setminus \sigma_l(A)$, i.e., the set of all isolated eigenvalues of finite multiplicity, is called the *discrete spectrum* of A.

Now we discuss the spectral analysis of a self-adjoint operator. Suppose that A is a self-adjoint, possibly unbounded, operator in $\mathcal{H}$ and ϕ is a vector in $\mathcal{H}$, such that

$$A\phi = \lambda\phi$$

where λ is a number, then ϕ is termed an eigenvector and λ the corresponding eigenvalue of A. The self-adjointness of A ensures that λ is real. The self-adjoint operator A is said to have a discrete spectrum if it has a set of eigenvectors $\{\phi_n : n \in I\}$ which form an orthonormal basis in $\mathcal{H}$. In this case, A may be expressed in the form

$$A = \sum_{n \in I} \lambda_n \Pi(\phi_n)$$

where $\Pi(\phi_n)$ is the projection operator and λ_n the eigenvalue for ϕ_n.

In general, even when the operator A does not have a discrete spectrum, it may still be resolved into a linear combination of projection operators according to the *spectral theorem* [137] which serves to express A as a Stieltjes integral

$$A = \int \lambda dE(\lambda)$$

where $\{E(\lambda)\}$ is a family of intercommuting projectors such that

$$
\begin{aligned}
E(-\infty) &= 0 \\
E(\infty) &= I \\
E(\lambda) &\leq E(\lambda') \quad \text{if} \quad \lambda < \lambda'
\end{aligned}
$$

and $E(\lambda')$ converges strongly to $E(\lambda)$ as λ' tends to λ from above. Here $E(\lambda)$ is a function of A, i.e. $\chi_\lambda(A)$, where

$$
\chi_\lambda(x) = \begin{cases} 1 & \text{for } x < \lambda \\ 0 & \text{for } x \geq \lambda. \end{cases}
$$

In the particular case where A has a discrete spectrum, i.e.

$$A = \sum_{n \in I} \lambda_n \Pi(\phi_n),$$

then

$$E(\lambda) = \sum_{\lambda_n < \lambda} \Pi(\phi_n).$$

In general, it follows from the spectral theorem that, for any positive N, we may express A in the form

$$A = A_N + A'_N$$

where

$$A_N = \int\limits_{-N-0}^{N+0} \lambda dE(\lambda) \qquad \text{and} \qquad A'_N = \int\limits_{-\infty}^{-N-0} \lambda dE(\lambda) + \int\limits_{N+0}^{\infty} \lambda dE(\lambda).$$

Thus, A is decomposed into parts, A_N and A'_N, whose spectra lie inside and outside the interval $[-N, N]$, respectively, and

$$A = \lim_{N \to \infty} A_N$$

on the domain of A. This last formula expresses unbounded operators as limits of bounded ones.

Let us consider two self-adjoint operators

$$A := \int_{\mathbf{R}} \lambda dE(\lambda), \qquad B := \int_{\mathbf{R}} \lambda dF(\lambda).$$

They are said to commute if

$$E(\lambda)F(\mu) = F(\mu)E(\lambda) \qquad \text{for all} \quad \lambda, \mu.$$

Since A and B are generally unbounded, one cannot say that $AB = BA$ unless the domains of AB and BA happen to be the same, whereas $E(\lambda)$ and $F(\lambda)$ are defined on all $\mathcal{H}$; however $ABu = BAu$ for all u (if any) such that both sides of the equation are meaningful. Commuting operators A and B are said to have a simple joint spectrum or to form a complete set of commuting observables if there is an element χ in $\mathcal{H}$ such that the closed linear span of the elements

$$\{ E(\lambda)F(\mu)\chi : -\infty < \mu, \lambda < \infty \}$$

is all of $\mathcal{H}$. If A and B are two bounded operators in a Hilbert space we can define the *commutator* $[A, B] := AB - BA$ in the sense that for all $u \in \mathcal{H}$ we have

$$[A, B]u = (AB)u - (BA)u = A(Bu) - B(Au).$$

Important special cases are discussed in Steeb [163].

16.3 Schmidt Decomposition

Let $\mathcal{H}_1$ and $\mathcal{H}_2$ be two finite dimensional Hilbert spaces with the underlying field $\mathbf{C}$. We define for the coupled system $\mathcal{H}_1 \otimes \mathcal{H}_2$ (described by the density operator $\rho_{\mathcal{H}_1 \otimes \mathcal{H}_2}$)

$$\rho_{\mathcal{H}_1} := \text{tr}_{\mathcal{H}_2}(\rho_{\mathcal{H}_1 \otimes \mathcal{H}_2})$$

where $\text{tr}_{\mathcal{H}_2}$ denotes the *partial trace* over $\mathcal{H}_2$, i.e we use $I_m \otimes |\beta_j\rangle$ as the basis for the trace where $|\beta_j\rangle$ is an orthonormal basis in $\mathcal{H}_2$. Let $\dim(\mathcal{H}_1) = m$ and $\dim(\mathcal{H}_2) = n$.

Thus $\dim(\mathcal{H}_1 \otimes \mathcal{H}_2) = m \cdot n$. An arbitrary normalized vector $|\psi\rangle_{12}$ in the tensor product space $\mathcal{H}_1 \otimes \mathcal{H}_2$ can be expanded as

$$|\psi\rangle_{12} = \sum_{i=1}^{m} \sum_{j=1}^{n} a_{ij} |i\rangle_1 \otimes |j\rangle_2$$

where $a_{ij} \in \mathbf{C}$ and $\{ |i\rangle_1 \mid i = 1, \ldots, m \}$ and $\{ |j\rangle_2 \mid j = 1, \ldots, n \}$ are orthogonal basis for $\mathcal{H}_1$ and $\mathcal{H}_2$, respectively. Let

$$|\tilde{\phi}_i\rangle := \sum_{j=1}^{n} a_{ij} |j\rangle_2.$$

We notice that the $|\tilde{\phi}_i\rangle$ need not be mutually orthogonal or normalized. Thus $|\psi\rangle_{12}$ can be written as

$$|\psi\rangle_{12} = \sum_{i=1}^{m} |i\rangle_1 \otimes |\tilde{\phi}_i\rangle.$$

Let

$$\rho := |\psi\rangle_{12}\, {}_{12}\langle\psi|$$

and let

$$\rho_1 = \mathrm{tr}_2(\rho), \qquad \rho_2 = \mathrm{tr}_1(\rho)$$

be the partial traces.

Theorem.

1. ρ_1 and ρ_2 have the same nonzero eigenvalues $\lambda_1, \ldots, \lambda_k$ (with the same multiplicities) and any extra dimensions are made up with zero eigenvalues, where $k \leq \min(m, n)$. There is no need for $\mathcal{H}_1$ and $\mathcal{H}_2$ to have the same dimension, so the number of zero eigenvalues of ρ_1 and ρ_2 can differ.

2. The state $|\psi\rangle_{12}$ can be written as

$$|\psi\rangle_{12} = \sum_{i=1}^{k} \sqrt{\lambda_i} |i\rangle_1 \otimes |\phi_i\rangle$$

where $|i\rangle_1$ (respectively $|\phi_i\rangle$) are orthonormal eigenvectors of ρ_1 in $\mathcal{H}_1$ (respectively ρ_2 in $\mathcal{H}_2$) belonging to λ_i. This expression is called the *Schmidt polar form* or *Schmidt decomposition*.

Proof. As stated above $|\psi\rangle_{12}$ can be written as

$$|\psi\rangle_{12} = \sum_{i=1}^{m} |i\rangle_1 \otimes |\tilde{\phi}_i\rangle$$

where the $|\tilde{\phi}_i\rangle$ are (not necessarily orthogonal) states in $\mathcal{H}_2$. Taking the partial trace of $|\psi\rangle_{12}\,_{12}\langle\psi|$ over $\mathcal{H}_2$ and equating to

$$\rho_1 = \sum_{i=1}^{m} \lambda_i |i\rangle_1\,_1\langle i| \qquad \text{gives} \qquad \langle\tilde{\phi}_i|\tilde{\phi}_j\rangle = \lambda_i \delta_{ij}.$$

Hence it turns out that the $\{\,|\tilde{\phi}_i\rangle\,\}$ are orthogonal after all. Thus at most $\min(m,n)$ eigenvalues are non-zero. Consequently, the set of states

$$\{\,|\phi_i\rangle = \frac{1}{\sqrt{\lambda_i}}|\tilde{\phi}_i\rangle\,\}$$

is an orthonormal set in the Hilbert space $\mathcal{H}_2$, where we exclude the zero eigenvalues. It follows that

$$|\psi\rangle_{12} = \sum_{i=1}^{k} \sqrt{\lambda_i}|i\rangle_1 \otimes |\phi_i\rangle$$

and taking the partial trace over $\mathcal{H}_1$ gives

$$\rho_2 = \sum_{i=1}^{k} \lambda_i |\phi_i\rangle\langle\phi_i|.$$

♠

Example. Consider $\dim(\mathcal{H}_1) = 3$, $\dim(\mathcal{H}_2) = 2$ and

$$|\psi\rangle := \frac{1}{\sqrt{2}}(\,1\ \ 0\ \ 0\ \ 0\ \ 0\ \ 1\,)^T.$$

We have

$$\rho = |\psi\rangle\langle\psi|, \quad \rho_1 = \frac{1}{2}\begin{pmatrix} 1 & 0 & 0 \\ 0 & 0 & 0 \\ 0 & 0 & 1 \end{pmatrix}, \quad \rho_2 = \frac{1}{2}\begin{pmatrix} 1 & 0 \\ 0 & 1 \end{pmatrix}.$$

The eigenvalues of ρ_1 are $\frac{1}{2}, \frac{1}{2}$ and 0. The eigenvalues of ρ_2 are $\frac{1}{2}$ and $\frac{1}{2}$. Thus

$$\rho_1 = \frac{1}{2}\begin{pmatrix} 1 \\ 0 \\ 0 \end{pmatrix}(\,1\ \ 0\ \ 0\,) + \frac{1}{2}\begin{pmatrix} 0 \\ 0 \\ 1 \end{pmatrix}(\,0\ \ 0\ \ 1\,)$$

$$\rho_2 = \frac{1}{2}\begin{pmatrix} 1 \\ 0 \end{pmatrix}(\,1\ \ 0\,) + \frac{1}{2}\begin{pmatrix} 0 \\ 1 \end{pmatrix}(\,0\ \ 1\,)$$

$$|\psi\rangle = \frac{1}{\sqrt{2}}\begin{pmatrix} 1 \\ 0 \\ 0 \end{pmatrix} \otimes \begin{pmatrix} 1 \\ 0 \end{pmatrix} + \frac{1}{\sqrt{2}}\begin{pmatrix} 0 \\ 0 \\ 1 \end{pmatrix} \otimes \begin{pmatrix} 0 \\ 1 \end{pmatrix}.$$

♣

16.4 Spin Matrices and Kronecker Product

In this section we study spin systems. In Pauli's nonrelativistic theory of spin certain spin wave functions, vectors, or spinor functions – along with spin operators, or matrices – are introduced to facilitate computation. We define

$$|\uparrow\rangle := \begin{pmatrix} 1 \\ 0 \end{pmatrix} \qquad \text{spin-up vector}$$

$$|\downarrow\rangle := \begin{pmatrix} 0 \\ 1 \end{pmatrix} \qquad \text{spin-down vector}$$

and

$$\sigma_x := \begin{pmatrix} 0 & 1 \\ 1 & 0 \end{pmatrix}, \qquad \sigma_y := \begin{pmatrix} 0 & -i \\ i & 0 \end{pmatrix}, \qquad \sigma_z := \begin{pmatrix} 1 & 0 \\ 0 & -1 \end{pmatrix}.$$

The matrices σ_x, σ_y, σ_z, are called the *Pauli spin matrices*. Let I be the 2×2 unit matrix. We find the following relationships. After squaring the spin matrices, we have

$$\sigma_x^2 = I, \qquad \sigma_y^2 = I, \qquad \sigma_z^2 = I.$$

Since the squares of the spin matrices are the 2×2 unit matrix, their eigenvalues are ± 1. The anticommutators are given by

$$\sigma_x \sigma_y + \sigma_y \sigma_x = 0, \qquad \sigma_y \sigma_z + \sigma_z \sigma_y = 0, \qquad \sigma_z \sigma_x + \sigma_x \sigma_z = 0$$

where 0 is the 2×2 zero matrix. Summarizing these results we have

$$\sigma_i \sigma_j + \sigma_j \sigma_i = 2\delta_{ij} I$$

where i and j may independently be x, y, or z and δ_{ij} is the Kronecker delta. The matrices I, σ_x, σ_y and σ_z form an orthogonal basis in the Hilbert space M^2. This means every 2×2 matrix can be written as

$$M = c_x \sigma_x + c_y \sigma_y + c_z \sigma_z + c_1 I$$

where $c_x, c_y, c_z, c_1 \in \mathbf{C}$. Another orthonormal basis (standard basis) is given by

$$\begin{pmatrix} 1 & 0 \\ 0 & 0 \end{pmatrix}, \qquad \begin{pmatrix} 0 & 1 \\ 0 & 0 \end{pmatrix}, \qquad \begin{pmatrix} 0 & 0 \\ 1 & 0 \end{pmatrix}, \qquad \begin{pmatrix} 0 & 0 \\ 0 & 1 \end{pmatrix}.$$

The trace of a matrix is the sum of the diagonal terms. For all three Pauli spin matrices the trace is zero. The Pauli spin matrices are self-adjoint operators (hermitian matrices) and therefore have real eigenvalues. The commutators are given by

$$
\begin{aligned}
[\sigma_x, \sigma_y] = \sigma_x \sigma_y - \sigma_y \sigma_x &= 2i\sigma_z \\
[\sigma_y, \sigma_z] = \sigma_y \sigma_z - \sigma_z \sigma_y &= 2i\sigma_x \\
[\sigma_z, \sigma_x] = \sigma_z \sigma_x - \sigma_x \sigma_z &= 2i\sigma_y.
\end{aligned}
$$

These three relationships may be combined in a single equation $\sigma \times \sigma = 2i\sigma$, where $\times$ denotes the vector product and $\sigma = (\sigma_x, \sigma_y, \sigma_z)^T$. We define

$$\sigma_+ := \frac{1}{2}(\sigma_x + i\sigma_y) = \begin{pmatrix} 0 & 1 \\ 0 & 0 \end{pmatrix}, \qquad \sigma_- := \frac{1}{2}(\sigma_x - i\sigma_y) = \begin{pmatrix} 0 & 0 \\ 1 & 0 \end{pmatrix}.$$

These are the spin-flip operators. We define

$$\Lambda_+ := \frac{1}{2}(I + \sigma_z) = \begin{pmatrix} 1 & 0 \\ 0 & 0 \end{pmatrix}, \qquad \Lambda_- := \frac{1}{2}(I - \sigma_z) = \begin{pmatrix} 0 & 0 \\ 0 & 1 \end{pmatrix}.$$

The two matrices are projection matrices. As mentioned above the four matrices $\sigma_\pm$, $\Lambda_\pm$ form an orthonormal basis in the Hilbert space M^2.

Let us now study the action of spin matrices on spin vectors. A vector $\mathbf{u} \in \mathbf{C}^2$ can be written as

$$\mathbf{u} = u_1 \begin{pmatrix} 1 \\ 0 \end{pmatrix} + u_2 \begin{pmatrix} 0 \\ 1 \end{pmatrix} = \begin{pmatrix} u_1 \\ u_2 \end{pmatrix}$$

where $u_1, u_2 \in \mathbf{C}$. We find the following relations

$$\sigma_x| \uparrow\rangle = |\downarrow\rangle, \qquad \sigma_x| \downarrow\rangle = |\uparrow\rangle$$
$$\sigma_y| \uparrow\rangle = i|\downarrow\rangle, \qquad \sigma_y| \downarrow\rangle = -i|\uparrow\rangle,$$
$$\sigma_z| \uparrow\rangle = |\uparrow\rangle, \qquad \sigma_z| \downarrow\rangle = -|\downarrow\rangle$$

and

$$\sigma_+| \uparrow\rangle = \begin{pmatrix} 0 \\ 0 \end{pmatrix}, \qquad \sigma_-| \downarrow\rangle = \begin{pmatrix} 0 \\ 0 \end{pmatrix}, \qquad \sigma_+| \downarrow\rangle = |\uparrow\rangle, \qquad \sigma_-| \uparrow\rangle = |\downarrow\rangle.$$

Furthermore we find

$$\Lambda_+| \uparrow\rangle = |\uparrow\rangle, \qquad \Lambda_-| \downarrow\rangle = |\downarrow\rangle, \qquad \Lambda_-| \uparrow\rangle = \begin{pmatrix} 0 \\ 0 \end{pmatrix}, \qquad \Lambda_+| \downarrow\rangle = \begin{pmatrix} 0 \\ 0 \end{pmatrix}.$$

The projection operators $\Lambda_\pm$ select the positive or negative spin components of a vector

$$\Lambda_+\mathbf{u} = \begin{pmatrix} 1 & 0 \\ 0 & 0 \end{pmatrix}\begin{pmatrix} u_1 \\ u_2 \end{pmatrix} = \begin{pmatrix} u_1 \\ 0 \end{pmatrix} = u_1 \begin{pmatrix} 1 \\ 0 \end{pmatrix}$$

and

$$\Lambda_-\mathbf{u} = \begin{pmatrix} 0 & 0 \\ 0 & 1 \end{pmatrix}\begin{pmatrix} u_1 \\ u_2 \end{pmatrix} = \begin{pmatrix} 0 \\ u_2 \end{pmatrix} = u_2 \begin{pmatrix} 0 \\ 1 \end{pmatrix}.$$

The matrices $\sigma_\pm$ and $\Lambda_\pm$ obey

$$\sigma_\pm^2 = 0, \qquad \Lambda_\pm^2 = \Lambda_\pm.$$

In studying spin systems such as the Heisenberg model, the XY model and the Dirac spin matrices we have to introduce the Kronecker product (Steeb [162]). Also in the spectral representation of hermitian matrices the Kronecker product plays an important role.

Definition. Let A be an $m \times n$ matrix and let B be a $p \times q$ matrix. Then

$$A \otimes B := \begin{pmatrix} a_{11}B & a_{12}B & \cdots & a_{1n}B \\ a_{21}B & a_{22}B & \cdots & a_{2n}B \\ \cdots & \cdots & \cdots & \cdots \\ a_{m1}B & a_{m2}B & \cdots & a_{mn}B \end{pmatrix}.$$

$A \otimes B$ is an $(mp) \times (nq)$ matrix and $\otimes$ is the *Kronecker product* (sometimes also called tensor product or direct product).

We have the following properties. Let A be an $m \times n$ matrix, B be a $p \times q$ matrix, C be an $n \times r$ matrix and D be an $r \times s$ matrix. Then

$$(A \otimes B)(C \otimes D) = (AC) \otimes (BD)$$

where AC and BD denote the ordinary matrix product. An extension is

$$(A_1 \otimes B_1)(A_2 \otimes B_2)(A_3 \otimes B_3) = (A_1 A_2 A_3) \otimes (B_1 B_2 B_3).$$

The size of the matrices must be such that the matrix products exist. Further rules are

$$\begin{aligned} A \otimes (B + C) &= A \otimes B + A \otimes C \\ (A \otimes B)^T &= A^T \otimes B^T \\ B \otimes A &= P(A \otimes B)Q \end{aligned}$$

where P and Q are certain permutation matrices. Let A be an $m \times m$ matrix and let B be a $p \times p$ matrix. Then

$$\begin{aligned} \operatorname{tr}(A \otimes B) &= (\operatorname{tr}A)(\operatorname{tr}B) \\ (A \otimes B)^{-1} &= A^{-1} \otimes B^{-1} \quad \text{if } A^{-1} \text{ and } B^{-1} \text{ exist} \\ \det(A \otimes B) &= (\det A)^p (\det B)^m \end{aligned}$$

where tr denotes the trace and det the determinant. The Kronecker product of two orthogonal matrices is again an orthogonal matrix.

Theorem. Let A be an $m \times m$ matrix and B be a $p \times p$ matrix. Let $\lambda_1, \lambda_2, \ldots, \lambda_m$ be the eigenvalues of A. Let $\mu_1, \mu_2, \ldots, \mu_p$ be the eigenvalues of B. Then $\lambda_j \mu_k$

$(j = 1, \ldots, m; k = 1, \ldots, p)$ are the eigenvalues of $A \otimes B$. Let $\mathbf{u}_j$ $(j = 1, \ldots, m)$ be the eigenvectors of A. Let $\mathbf{v}_k$ $(k = 1, \ldots, p)$ be the eigenvectors of B. Then $\mathbf{u}_j \otimes \mathbf{v}_k$ $(j = 1, \ldots, m; k = 1, \ldots, p)$ are the eigenvectors of $A \otimes B$.

Theorem. Let A be an $m \times m$ matrix and B be a $p \times p$ matrix. Let $\lambda_1, \lambda_2, \ldots, \lambda_m$ be the eigenvalues of A. Let $\mu_1, \mu_2, \ldots, \mu_p$ be the eigenvalues of B. Then $\lambda_j + \mu_k$ $(j = 1, \ldots, m; k = 1, \ldots, p)$ are the eigenvalues of $A \otimes I_p + I_m \otimes B$. Let $\mathbf{u}_j$ $(j = 1, \ldots, m)$ be the eigenvectors of A. Let $\mathbf{v}_k$ $(k = 1, \ldots, p)$ be the eigenvectors of B. Then $\mathbf{u}_j \otimes \mathbf{v}_k$ $(j = 1, \ldots, m; k = 1, \ldots, p)$ are the eigenvectors of $A \otimes I_p + I_m \otimes B$.

For the proofs we refer to Steeb [162].

With the help of the eigenvalues and eigenvectors of a hermitian matrix A we can reconstruct the matrix A using the Kronecker product.

Theorem. Let A be an $n \times n$ hermitian matrix. Assume that the eigenvalues $\lambda_1, \ldots, \lambda_n$ are all distinct. Then the normalized eigenvectors $\mathbf{u}_1$, $\mathbf{u}_2$, ..., $\mathbf{u}_n$ are orthonormal and form an orthonormal basis in the Hilbert space $\mathbf{C}^n$. Then

$$A = \sum_{j=1}^{n} \lambda_j \mathbf{u}_j^* \otimes \mathbf{u}_j.$$

Example. We consider the matrix

$$A = \begin{pmatrix} 0 & 1 \\ 1 & 0 \end{pmatrix}$$

with eigenvalues $\lambda_1 = +1$, $\lambda_2 = -1$. The normalized eigenvectors are given by

$$\mathbf{u}_1 = \frac{1}{\sqrt{2}} \begin{pmatrix} 1 \\ 1 \end{pmatrix}, \qquad \mathbf{u}_2 = \frac{1}{\sqrt{2}} \begin{pmatrix} 1 \\ -1 \end{pmatrix}.$$

Since

$$(a, b) \otimes \begin{pmatrix} c \\ d \end{pmatrix} = \begin{pmatrix} ac & bc \\ ad & bd \end{pmatrix}$$

we find that the spectral representation of the matrix A is given by

$$A = \sum_{j=1}^{2} \lambda_j \mathbf{u}_j^* \otimes \mathbf{u}_j = \frac{1}{\sqrt{2}}(1,1) \otimes \frac{1}{\sqrt{2}}\begin{pmatrix} 1 \\ 1 \end{pmatrix} - \frac{1}{\sqrt{2}}(1,-1) \otimes \frac{1}{\sqrt{2}}\begin{pmatrix} 1 \\ -1 \end{pmatrix}$$

$$= \frac{1}{2}\begin{pmatrix} 1 & 1 \\ 1 & 1 \end{pmatrix} - \frac{1}{2}\begin{pmatrix} 1 & -1 \\ -1 & 1 \end{pmatrix} = \begin{pmatrix} 0 & 1 \\ 1 & 0 \end{pmatrix}.$$

Moreover we have

$$\Pi_1 = \frac{1}{\sqrt{2}}(1,1) \otimes \frac{1}{\sqrt{2}}\begin{pmatrix} 1 \\ 1 \end{pmatrix} = \frac{1}{2}\begin{pmatrix} 1 & 1 \\ 1 & 1 \end{pmatrix}$$

and

$$\Pi_2 = \frac{1}{\sqrt{2}}(1,-1) \otimes \frac{1}{\sqrt{2}}\begin{pmatrix} 1 \\ -1 \end{pmatrix} = \frac{1}{2}\begin{pmatrix} 1 & -1 \\ -1 & 1 \end{pmatrix}$$

where Π_1 and Π_2 are projection matrices. Thus $\Pi_1^2 = \Pi_1$, $\Pi_2^2 = \Pi_2$ and $\Pi_1\Pi_2 = 0$. ♣

The *Dirac spin matrices* $\alpha_1, \alpha_2, \alpha_3$, and β can be constructed using the Pauli spin matrices and the Kronecker product. These matrices play a central role in the description of the electron. We define

$$\alpha_1 := \sigma_x \otimes \sigma_x = \begin{pmatrix} 0 & 0 & 0 & 1 \\ 0 & 0 & 1 & 0 \\ 0 & 1 & 0 & 0 \\ 1 & 0 & 0 & 0 \end{pmatrix}, \qquad \alpha_2 := \sigma_x \otimes \sigma_y = \begin{pmatrix} 0 & 0 & 0 & -i \\ 0 & 0 & i & 0 \\ 0 & -i & 0 & 0 \\ i & 0 & 0 & 0 \end{pmatrix}$$

$$\alpha_3 := \sigma_x \otimes \sigma_z = \begin{pmatrix} 0 & 0 & 1 & 0 \\ 0 & 0 & 0 & -1 \\ 1 & 0 & 0 & 0 \\ 0 & -1 & 0 & 0 \end{pmatrix}, \qquad \beta := \sigma_z \otimes I_2 = \begin{pmatrix} 1 & 0 & 0 & 0 \\ 0 & 1 & 0 & 0 \\ 0 & 0 & -1 & 0 \\ 0 & 0 & 0 & -1 \end{pmatrix}.$$

The 4×4 matrices $\alpha_1, \alpha_2, \alpha_3$ and β satisfy the rules

$$\beta^2 = I, \qquad \alpha_i\alpha_j + \alpha_j\alpha_i = 2\delta_{ij}I, \qquad \alpha_i\beta + \beta\alpha_i = 0$$

where I is the 4×4 unit matrix.

The *spin matrices* are defined by

$$S_x := \frac{1}{2}\sigma_x, \qquad S_y := \frac{1}{2}\sigma_y, \qquad S_z := \frac{1}{2}\sigma_z.$$

Definition. Let $j = 1, 2, \ldots, N$. We define

$$\sigma_{\alpha,j} := I \otimes \ldots \otimes I \otimes \sigma_\alpha \otimes I \otimes \ldots \otimes I$$

where I is the 2×2 unit matrix, $\alpha = x, y, z$ and σ_α is the α-th Pauli matrix in the j-th location. Thus $\sigma_{\alpha,j}$ is a $2^N \times 2^N$ matrix. Analogously, we define

$$S_{\alpha,j} := I \otimes \ldots \otimes I \otimes S_\alpha \otimes I \otimes \ldots \otimes I.$$

In the following we set

$$\mathbf{S}_j := (S_{x,j}, S_{y,j}, S_{z,j})^T.$$

We calculate the eigenvalues and eigenvectors for the two-point *Heisenberg model*. The Heisenberg model is used to describe interacting spin-systems [162]. The model is given by

$$\hat{H} = J \sum_{j=1}^{2} \mathbf{S}_j \cdot \mathbf{S}_{j+1}$$

where J is the so-called exchange constant ($J > 0$ or $J < 0$) and $\cdot$ denotes the scalar product. We impose cyclic boundary conditions, i.e. $\mathbf{S}_3 \equiv \mathbf{S}_1$. It follows that

$$\hat{H} = J(\mathbf{S}_1 \cdot \mathbf{S}_2 + \mathbf{S}_2 \cdot \mathbf{S}_3) \equiv J(\mathbf{S}_1 \cdot \mathbf{S}_2 + \mathbf{S}_2 \cdot \mathbf{S}_1).$$

Therefore

$$\hat{H} = J(S_{x,1}S_{x,2} + S_{y,1}S_{y,2} + S_{z,1}S_{z,2} + S_{x,2}S_{x,1} + S_{y,2}S_{y,1} + S_{z,2}S_{z,1}).$$

Since

$$S_{x,1} = S_x \otimes I, \qquad S_{x,2} = I \otimes S_x$$

etc. where I is the 2×2 unit matrix, it follows that

$$\begin{aligned}
\hat{H} = {} & J[(S_x \otimes I)(I \otimes S_x) + (S_y \otimes I)(I \otimes S_y) + (S_z \otimes I)(I \otimes S_z) \\
& + (I \otimes S_x)(S_x \otimes I) + (I \otimes S_y)(S_y \otimes I) + (I \otimes S_z)(S_z \otimes I)].
\end{aligned}$$

Thus we obtain

$$\hat{H} = 2J[(S_x \otimes S_x) + (S_y \otimes S_y) + (S_z \otimes S_z)].$$

Since

$$S_x := \frac{1}{2}\sigma_x, \qquad S_y := \frac{1}{2}\sigma_y, \qquad S_z := \frac{1}{2}\sigma_z$$

we obtain

$$S_x \otimes S_x = \frac{1}{4}\begin{pmatrix} 0 & 1 \\ 1 & 0 \end{pmatrix} \otimes \begin{pmatrix} 0 & 1 \\ 1 & 0 \end{pmatrix} = \frac{1}{4}\begin{pmatrix} 0 & 0 & 0 & 1 \\ 0 & 0 & 1 & 0 \\ 0 & 1 & 0 & 0 \\ 1 & 0 & 0 & 0 \end{pmatrix}$$

etc. Then the Hamilton operator $\hat{H}$ is given by the 4×4 symmetric matrix

$$\hat{H} = \frac{J}{2}\begin{pmatrix} 1 & 0 & 0 & 0 \\ 0 & -1 & 2 & 0 \\ 0 & 2 & -1 & 0 \\ 0 & 0 & 0 & 1 \end{pmatrix} \equiv \frac{J}{2}\left[(1) \oplus \begin{pmatrix} -1 & 2 \\ 2 & -1 \end{pmatrix} \oplus (1)\right]$$

where $\oplus$ denotes the direct sum of matrices. The eigenvalues and eigenvectors can now easily be calculated. We define

$$|\uparrow\uparrow\rangle := |\uparrow\rangle \otimes |\uparrow\rangle, \quad |\uparrow\downarrow\rangle := |\uparrow\rangle \otimes |\downarrow\rangle, \quad |\downarrow\uparrow\rangle := |\downarrow\rangle \otimes |\uparrow\rangle, \quad |\downarrow\downarrow\rangle := |\downarrow\rangle \otimes |\downarrow\rangle$$

where $|\uparrow\rangle$ and $|\downarrow\rangle$ have been given above. Consequently,

$$|\uparrow\uparrow\rangle = \begin{pmatrix} 1 \\ 0 \\ 0 \\ 0 \end{pmatrix}, \qquad |\uparrow\downarrow\rangle = \begin{pmatrix} 0 \\ 1 \\ 0 \\ 0 \end{pmatrix}, \qquad |\downarrow\uparrow\rangle = \begin{pmatrix} 0 \\ 0 \\ 1 \\ 0 \end{pmatrix}, \qquad |\downarrow\downarrow\rangle = \begin{pmatrix} 0 \\ 0 \\ 0 \\ 1 \end{pmatrix}.$$

Obviously these vectors form the standard basis in $\mathbf{C}^4$. One sees at once that $|\uparrow\uparrow\rangle$ and $|\downarrow\downarrow\rangle$ are eigenvectors of the Hamilton operator with eigenvalues $J/2$ and $J/2$, respectively. This means the eigenvalue $J/2$ is degenerate. The eigenvalues of the matrix

$$\frac{J}{2}\begin{pmatrix} -1 & 2 \\ 2 & -1 \end{pmatrix}$$

are given by $J/2$ and $-3J/2$. The corresponding eigenvectors are given by

$$\frac{1}{2}(|\uparrow\downarrow\rangle + |\downarrow\uparrow\rangle), \qquad \frac{1}{2}(|\uparrow\downarrow\rangle - |\downarrow\uparrow\rangle).$$

16.5 Postulates of Quantum Mechanics

Quantum mechanics, as opposed to classical mechanics, gives a probabilistic description of nature. The probabilistic interpretation of measurement is contained in one of the standard postulates of quantum mechanics (Glimm and Jaffe [71], Prugovečki [135], Schommers [142]).

Remark. More than sixty years after the formulation of quantum mechanics the interpretation of this formalism is by far the most controversial problem of current research in the foundations of physics and divides the community of physicists into numerous opposing schools of thought. There is an immense diversity of opinions and a huge variety of interpretations. A more detailed discussion of the interpretation of the measurement in quantum mechanics is given in chapter 18.

The standard postulates of quantum mechanics are

PI. The pure states of a quantum system, S, are described by normalized vectors ψ which are elements of a Hilbert space, $\mathcal{H}$, that describes S. The pure states of a quantum mechanical system are rays in a Hilbert space $\mathcal{H}$ (i.e., unit vectors, with an arbitrary phase). Specifying a pure state in quantum mechanics is the most that can be said about a physical system. In this respect, it is analogous to a classical pure state. The concept of a state as a ray in a Hilbert space leads to the probability interpretation in quantum mechanics. Given a physical system in the state ψ, the probability that it is in the state χ is $|\langle\psi,\chi\rangle|^2$. Clearly

$$0 \le |\langle\psi,\chi\rangle|^2 \le 1.$$

While the phase of a vector ψ has no physical significance, the relative phase of two vectors does. This means for $|\alpha| = 1$, $|\langle\alpha\psi,\chi\rangle|$ is independent of α, but $|\langle\psi_1 + \alpha\psi_2,\chi\rangle|$ is not. It is most convenient to regard pure states ψ simply as vectors in $\mathcal{H}$, and to normalize them in an appropriate calculation.

PII. The states evolve in time according to

$$i\hbar\frac{\partial\psi}{\partial t} = \hat{H}\psi$$

where $\hat{H}$ is a self-adjoint operator which specifies the dynamics of the system S. This equation is called the *Schrödinger equation*. The formal solution takes the form

$$\psi(t) = \exp(-i\hat{H}t/\hbar)\psi(0)$$

where $\psi(0) \equiv \psi(t = 0)$ with $\langle\psi(0), \psi(0)\rangle = 1$. It follows that $\langle\psi(t), \psi(t)\rangle = 1$.

Example. Consider the Hamilton operator

$$\hat{H} = \omega S_x \equiv \frac{\hbar\omega}{\sqrt{2}} \begin{pmatrix} 0 & 1 & 0 \\ 1 & 0 & 1 \\ 0 & 1 & 0 \end{pmatrix}$$

in the Hilbert space $\mathbf{C}^3$, where ω is the constant frequency. Then we find

$$\exp(-i\hat{H}t/\hbar) = \begin{pmatrix} \frac{1}{2} + \frac{1}{2}\cos(\omega t) & -\frac{i}{\sqrt{2}}\sin(\omega t) & \frac{1}{2}\cos(\omega t) - \frac{1}{2} \\ -\frac{i}{\sqrt{2}}\sin(\omega t) & \cos(\omega t) & -\frac{i}{\sqrt{2}}\sin(\omega t) \\ \frac{1}{2}\cos(\omega t) - \frac{1}{2} & -\frac{i}{\sqrt{2}}\sin(\omega t) & \frac{1}{2} + \frac{1}{2}\cos(\omega t) \end{pmatrix}.$$

Let

$$\psi(0) = \frac{1}{\sqrt{3}}(1, 1, 1)^T$$

be the initial state. Then

$$\exp(-i\hat{H}t/\hbar)\psi(0) = \frac{1}{\sqrt{3}} \begin{pmatrix} \cos(\omega t) - \frac{i}{\sqrt{2}}\sin(\omega t) \\ \cos(\omega t) - \sqrt{2}i\sin(\omega t) \\ \cos(\omega t) - \frac{i}{\sqrt{2}}\sin(\omega t) \end{pmatrix}.$$

The probability

$$p(t) = |\langle\psi(t), \psi(0)\rangle|^2$$

is given by

$$p(t) = 1 - \frac{1}{9}\sin^2(\omega t).$$

♣

PIII. Every observable, a, is associated with a self-adjoint operator $\hat{A}$. The only possible outcome of a measurement of a is an eigenvalue A_j of $\hat{A}$, i.e.

$$\hat{A}\phi_j = A_j\phi_j \qquad \langle\phi_j,\phi_k\rangle = \delta_{jk}$$

where ϕ_j is an eigenfunction.

PIV. If the state of the system is described by the normalized vector ψ, then a measurement of a will yield the eigenvalue A_j with probability

$$p_j = |\langle\phi_j,\psi\rangle|^2.$$

Notice that $\langle\phi_j,\psi\rangle$ can be complex. It is obvious that $0 \le p_j \le 1$.

In order for successive measurements of a to yield the same value A_j it is necessary to have the projection postulate:

PV. Immediately after a measurement which yields the value A_j the state of the system is described by $\Pi_j\psi$, where Π_j is the projection operator which projects onto the eigenspace of the eigenvalue A_j.

The type of time evolution implied by PV is incompatible with the unitary time evolution implied by PII. PIV can be replaced by the weaker postulate:

PIV'. If a quantum system is described by the state ϕ_j then a measurement of a will yield the value A_j.

Clearly PIV' is a special case of PIV but it is not a statement about probabilities. The replacement of PIV by PIV' eliminates the immediate need for PV since the state is ϕ_j before and after the measurement.

PVI. Quantum mechanical observables are self-adjoint operators on $\mathcal{H}$. The expected (average) value of the observable b with the corresponding self-adjoint operator B in the normalized state ψ is

$$E_\psi(B) := \langle\psi, B\psi\rangle.$$

Examples of observables are the Hamiltonian (energy) observable, the momentum observable, and the position observable.

The statistical mixtures in quantum mechanics lead to quantum statistical mechanics. The usual statistical mixture is described by a positive trace class operator ρ, yielding the expectation

$$\rho(B) = \frac{\operatorname{tr}(\rho B)}{\operatorname{tr}\rho}$$

where tr denotes the trace. If ρ has rank 1, then $\rho(B)$ is a pure state with $\rho/\text{tr}\rho$ the projection onto ψ. Otherwise, $\rho(B)$ is a convex linear combination of pure states,

$$\rho(B) = \sum_j \alpha_j \langle \phi_j, B\phi_j \rangle$$

where the ϕ_j are the (orthonormal) eigenvectors of ρ and $\sum_j \alpha_j = 1$.

PVII. The Hamilton operator $\hat{H}$ is the infinitesimal generator of the unitary group

$$U(t) := \exp(-it\hat{H}/\hbar)$$

of time translations. The unit of action $\hbar$ $(h/2\pi)$ has the same dimension as pq, where p is the momentum and q is the position [59].

The momentum operator $\hat{\mathbf{p}}$ is the infinitesimal generator of the unitary space translation group

$$\exp(i\mathbf{q} \cdot \hat{\mathbf{p}}/\hbar)$$

where

$$\mathbf{q} \cdot \hat{\mathbf{p}} := \sum_{k=1}^{N} \sum_{j=1}^{3} q_{kj}\hat{p}_{kj}.$$

We recall that

$$\exp(\mathbf{a} \cdot \nabla)u(\mathbf{q}) = u(\mathbf{q} + \mathbf{a})$$

where u is a smooth function and

$$\mathbf{a} \cdot \nabla := \sum_{k=1}^{N} \sum_{j=1}^{3} a_{kj}\frac{\partial}{\partial q_{kj}}$$

and $\mathbf{q} = (q_{11}, q_{12}, q_{13}, q_{21}, \ldots, q_{N3})$.

The *angular momentum operator* $\hat{\mathbf{J}}$ is the infinitesimal generator for the unitary space rotation group

$$\exp(-i\boldsymbol{\theta} \cdot \hat{\mathbf{J}}).$$

Remark. This leads to the *quantization*. Consider the energy conservation equation

$$E = \sum_{k=1}^{N} \sum_{j=1}^{3} \frac{p_{kj}^2}{2m_k} + V(\mathbf{q}).$$

We make the formal substitution

$$E \to i\hbar \frac{\partial}{\partial t}, \qquad p_{kj} \to -i\hbar \frac{\partial}{\partial q_{kj}}.$$

We arrive at the formal operator relation

$$i\hbar \frac{\partial}{\partial t} = -\sum_{k=1}^{N} \sum_{j=1}^{3} \frac{\hbar^2}{2m_k} \frac{\partial^2}{\partial q_{kj}^2} + V(\mathbf{q}).$$

Applying this operator relation to a wave function $\psi(\mathbf{q}, t)$ we obtain the Schrödinger equation.

The time translation group $U(t)$ determines the dynamics. There are two standard descriptions: the *Schrödinger picture* and the *Heisenberg picture*. In the Schrödinger picture, the states $\psi \in \mathcal{H}$ evolve in time according to the Schrödinger equation, while the observables do not evolve. The vectors satisfy the Schrödinger equation. The time-dependent normalized state $\psi(t)$ yields the expectation

$$E_{\psi(t)}(B) = \langle \psi(t), B\psi(t) \rangle.$$

The second description of dynamics is the Heisenberg picture, in which the states remain fixed, and the observables evolve in time according to the automorphism group

$$B \to B(t) = e^{it\hat{H}/\hbar} B e^{-it\hat{H}/\hbar} = U(t)^* B U(t).$$

Obviously we assume that the Hamilton operator does not depend explicitly on t. Thus the observables B satisfy the dynamical equation (*Heisenberg equation of motion*)

$$-i\hbar \frac{dB(t)}{dt} = [\hat{H}, B(t)]$$

with the formal solution

$$B(t) = \sum_{n=0}^{\infty} \frac{(it/\hbar)^n}{n!} [\hat{H}, [\hat{H}, \ldots, [\hat{H}, B], \ldots]] = \exp(i\hat{H}t/\hbar) B \exp(-i\hat{H}t/\hbar).$$

The relation between the Heisenberg and Schrödinger pictures is given by

$$\langle \psi(t), B\psi(t) \rangle = \langle \psi, B(t)\psi \rangle$$

where $\psi = \psi(t = 0)$.

Postulate PVII ensures that the results of an experiment, i.e., inner products $\langle \psi, \chi \rangle$, are independent of the time at which the experiment is performed. This means

$$|\langle \psi, \chi \rangle| = |\langle \psi(t), \chi(t) \rangle|.$$

Theorem. Every symmetry of $\mathcal{H}$ can be implemented either by a unitary transformation U on $\mathcal{H}$,

$$\psi' = U\psi$$

or by an antiunitary operator A on $\mathcal{H}$

$$\psi' = A\psi.$$

The interpretation of this result is that every symmetry of $\mathcal{H}$ can be regarded as a coordinate transformation. In particular, the group of time translations is implemented by a unitary group of operators $U(t)$. Only certain discrete symmetries (e.g., time inversion in nonrelativistic quantum mechanics) are implemented by antiunitary transformations.

Example. In nonrelativistic quantum mechanics one usual representation for a system of N particles moving in a potential V is

$$\mathcal{H} = L_2(\mathbf{R}^{3N}).$$

This choice is called the Schrödinger representation (as distinct from the Schrödinger picture). The function $\psi(\mathbf{q}) \in \mathcal{H}$ has the interpretation of giving the probability distribution

$$\rho(\mathbf{q}) = |\psi(\mathbf{q})|^2$$

for the position of the particles in $\mathbf{R}^{3N}$. Using postulate PVII, we find

$$p_{kj} \to \hat{p}_{kj} = -i\hbar \frac{\partial}{\partial q_{kj}}$$

and a nonrelativistic Hamilton function of the form

$$H = \sum_{k=1}^{N} \sum_{j=1}^{3} \frac{p_{kj}^2}{2m_k} + V(\mathbf{q})$$

becomes the elliptic differential operator

$$\hat{H} = -\sum_{k=1}^{N}\sum_{j=1}^{3}\frac{\hbar^2}{2m_k}\frac{\partial^2}{\partial q_{kj}^2} + \hat{V}(\mathbf{q}).$$

In other words the Hamilton operator $\hat{H}$ follows from the Hamilton function H via the quantization

$$p_{kj} \rightarrow -i\hbar\frac{\partial}{\partial q_{kj}} \qquad q_{kj} \rightarrow \hat{q}_{kj}.$$

The operator $\hat{q}_{kj}$ is defined by $\hat{q}_{kj}f(\mathbf{q}) := q_{kj}f(\mathbf{q})$. We find for the (canonical) commutation relations

$$\begin{aligned}
[\hat{q}_{kj}, \hat{q}_{k'j'}] &= 0 \\
[\hat{p}_{kj}, \hat{p}_{k'j'}] &= 0 \\
[\hat{p}_{kj}, \hat{q}_{k'j'}] &= -i\hbar\delta_{kk'}\delta_{jj'}I
\end{aligned}$$

They are preserved by the Heisenberg equation of motion.   ♣

Thus far the spin of the particle is not taken into account. We have spin 0 for π mesons, spin $\frac{1}{2}$ for electrons, muons, protons, or neutrons, spin 1 for photons, and higher spins for other particles or nuclei. To consider spin-dependent forces (for example the coupling of the spin magnetic moment to a magnetic field) we have to extend the Hilbert space $L_2(\mathbf{R}^{3N})$ to the N-fold tensor product

$$\mathcal{H} = \otimes L_2(\mathbf{R}^3, S).$$

Here $L_2(\mathbf{R}^3, S)$ denotes functions defined on $\mathbf{R}^{3N}$ with values in the finite dimensional spin space S. For spin zero particles we have $S = \mathbf{C}$, and we are reduced to $L_2(\mathbf{R}^{3N})$. For nonzero spin s, we have $S = \mathbf{C}^{2s+1}$. We write $\psi(\mathbf{q})$ as a vector with components $\psi(\mathbf{q}, \zeta)$. A space rotation (generated by the angular momentum observable $\mathbf{J}$) will rotate both $\mathbf{q}$ and ζ, the latter by a linear transformation of the ζ coordinates according to an N-fold tensor product of a representation of the spin group $SU(2, \mathbf{R})$. The group $SU(2, \mathbf{R})$ consists of all 2×2 matrices with

$$UU^* = I \quad \text{and} \quad \det U = 1$$

Particles of a given type are indistinguishable. To obtain indistinguishable particles, we restrict ourselves to a subset of $\otimes L_2(\mathbf{R}^3, S)$ invariant under an irreducible representation of the symmetric group (permutation group) of the N particle coordinates

$(\mathbf{q}_k, \zeta_k)$, $k = 1, 2, \ldots, N$. The standard choices are the totally symmetric representation for integer spin particles and the totally antisymmetric representation for half-integer spin particles.

The choice of antisymmetry for atomic and molecular problems with spin $\frac{1}{2}$ is known as the Pauli exclusion principle. One can prove that integer spin particles cannot be antisymmetrized and half-integer spin particles cannot be symmetrized. Particles with integer spin are called bosons. Those with half-integer spin are called fermions.

Postulate VIII. A quantum mechanical state is symmetric under the permutation of identical bosons, and antisymmetric under the permutation of identical fermions.

Chapter 17
Quantum Bits and Quantum Computation

17.1 Introduction

Digital computers are based on devices that can take on only two states, one of which is denoted by 0 and the other by 1. By concatenating several 0s and 1s together, 0-1 combinations can be formed to represent as many different entities as desired. A combination containing a single 0 or 1 is called a bit. In general, n bits can be used to distinguish among 2^n distinct entities and each addition of a bit doubles the number of possible combinations. Computers use strings of bits to represent numbers, letters, punctuation marks, and any other useful pieces of information. In a classical computer, the processing of information is done by logic gate. A logic gate maps the state of its input bits into another state according to a *truth table*. Quantum computers require quantum logic, something fundamentally different to classical Boolean logic. This difference leads to a greater efficiency of quantum computation over its classical counterpart.

In the last few years a large number of authors have studied quantum computing ([122], [8]). The most exciting development in quantum information processing has been the discovery of quantum algorithms – for integer factorization and the discrete logarithm – that run exponentially faster than the best known classical algorithms. These algorithms take classical input (such as the number to be factored) and yield classical outputs (the factors), but obtain their speedup by using quantum interference computation paths during the intermediate steps. A *quantum network* is a quantum computing device consisting of quantum logic gates whose computational steps are synchronised in time. Quantum computation is defined as a unitary evolution of the network which takes its initial state input into some final state output.

17.2   Quantum Bits and Quantum Registers

17.2.1   Quantum Bits

In a quantum computer the *quantum bit* [8, 18, 21, 132, 138, 156, 178] or simply *qubit* is the natural extension of the classical notion of bit. A qubit is a quantum two-level system, that in addition to the two pairwise orthonormal states $|0\rangle$ and $|1\rangle$ in the Hilbert space $\mathbf{C}^2$ can be set in any superposition of the form

$$|\psi\rangle = c_0|0\rangle + c_1|1\rangle, \qquad c_0, c_1 \in \mathbf{C}.$$

Since $|\psi\rangle$ is normalized, i.e. $\langle\psi|\psi\rangle = 1$, $\langle 1|1\rangle = 1$, $\langle 0|0\rangle = 0$, and $\langle 0|1\rangle = 0$ we have

$$|c_0|^2 + |c_1|^2 = 1.$$

Any quantum two-level system is a potential candidate for a qubit. Examples are the polarization of a photon, the polarization of a spin-1/2 particle (electron), the relative phase and intensity of a single photon in two arms of an interferometer, or an arbirary superposition of two atomic states. Thus the classical Boolean states, 0 and 1 can be represented by a fixed pair of orthogonal states of the qubit. In the following we set

$$|0\rangle := \begin{pmatrix} 0 \\ 1 \end{pmatrix}, \qquad |1\rangle := \begin{pmatrix} 1 \\ 0 \end{pmatrix}.$$

Often the representations of $|0\rangle$ and $|1\rangle$ are reversed, this changes the matrix representation of operators but all computations and results are equivalent. In the following we think of a qubit as a spin-1/2 particle. The states $|0\rangle$ and $|1\rangle$ will correspond respectively to the spin-down and spin-up eigenstates along a pre-arranged axis of quantization, for example set by an external constant magnetic field. Although a qubit can be prepared in an infinite number of different quantum states (by choosing different complex coefficient c_i's) it cannot be used to transmit more than one bit of information. This is because no detection process can reliably differentiate between non-orthogonal states. However, qubits (and more generally information encoded in quantum systems) can be used in systems developed for quantum cryptography, quantum teleportation or quantum dense coding. The problem of measuring a quantum system is a central one in quantum theory. In a classical computer, it is possible in principle to inquire at any time and without disturbing the computer) about the state of any bit in the memory. In a quantum computer, the situation is different. Qubits can be in superposed states, or can even be entangled with each other, and the mere act of measuring the quantum computer alters its state. Performing a measurement on a qubit in a state given above will return 0 with probability $|c_0|^2$ and 1 with probability $|c_1|^2$. The state of the qubit after the measurement (post-measurement state) will be $|0\rangle$ or $|1\rangle$ (depending on the outcome), and not $c_0|0\rangle + c_1|1\rangle$. We think of the measuring apparatus as a Stern-Gerlach device [23, 68] into which the qubits (spins) are sent when we want to measure them. When measuring a state of outcomes 0 and 1 will be recorded with a probability $|c_0|^2$ and $|c_1|^2$ on the respective detector plate.

17.2.2 Quantum Registers

To study quantum computing we need a collection of qubits (a *quantum register*). Thus we call a collection of qubits a quantum register. This leads to the tensor product (product Hilbert space) of the Hilbert space $\mathbf{C}^2$. Since we consider finite dimensional Hilbert spaces over $\mathbf{C}$ we can identify the tensor product with the Kronecker product. As in the classical case, it can be used to encode more complicated information.

For instance, the binary form of 9 (decimal) is 1001 and loading a quantum register with this value is done by preparing four qubits in the state

$$|9\rangle \equiv |1001\rangle \equiv |1\rangle \otimes |0\rangle \otimes |0\rangle \otimes |1\rangle .$$

Thus the state $|9\rangle \equiv |1001\rangle$ is an element in the Hilbert space $\mathbf{C}^{16}$. In the literature the notation

$$|1\rangle|0\rangle|0\rangle|1\rangle$$

is sometimes used, i.e. the symbol $\otimes$ is omitted. Consider first the case with two quantum bits. Then we have the basis

$$
\begin{aligned}
|00\rangle &\equiv |0\rangle \otimes |0\rangle, \\
|01\rangle &\equiv |0\rangle \otimes |1\rangle, \\
|10\rangle &\equiv |1\rangle \otimes |0\rangle, \\
|11\rangle &\equiv |1\rangle \otimes |1\rangle
\end{aligned}
$$

in the Hilbert space $\mathbf{C}^4$. Thus the number of states is $2^2 = 4$.

Consider now the n qubit case. We use the notion

$$|a\rangle := |a_{n-1}\rangle \otimes |a_{n-2}\rangle \otimes \cdots \otimes |a_1\rangle \otimes |a_0\rangle$$

which denotes a quantum register prepared with the value

$$a := 2^0 a_0 + 2^1 a_1 + \cdots + 2^{n-1} a_{n-1} .$$

From $|a\rangle$ we find

$$\langle a| \equiv \langle a_{n-1}| \otimes \langle a_{n-1}| \otimes \ldots \otimes \langle a_1| \otimes \langle a_0|.$$

Thus the scalar product in the product space is

$$\langle a|b\rangle := \langle a_0|b_0\rangle \langle a_1|b_1\rangle \cdots \langle a_{n-1}|b_{n-1}\rangle .$$

Two states $|a\rangle$ and $|b\rangle$ are orthogonal if $a_j \neq b_j$ for at least one j. For an n-bit register, the most general state can be written as

$$|\psi\rangle = \sum_{x=0}^{2^n-1} c_x|x\rangle$$

where

$$\sum_{x=0}^{2^n-1} |c_x|^2 = 1.$$

Thus $|\psi\rangle$ is a state in the Hilbert space $\mathbf{C}^{2^n}$.

Quantum data processing consists of applying a sequence of unitary transformations to the state vector $|\psi\rangle$. This state describes the situation in which several different values of the register are present simultaneously; just as in the case of the qubit, there is no classical counterpart to this situation, and there is no way to gain a complete knowledge of the state of a register through a single measurement.

Measuring the state of a register is done by passing, one by one, the various spins that form the register into a Stern-Gerlach apparatus and recording the results. For instance a two-bit register initially prepared in the state

$$|\psi\rangle = \frac{1}{\sqrt{2}}(|0\rangle \otimes |0\rangle + |1\rangle \otimes |1\rangle)$$

will, with equal probability, result in either two successive clicks in the down-detector or two successive clicks in the up-detector. The post measurement state will be either

$$|0\rangle \otimes |0\rangle \quad \text{or} \quad |1\rangle \otimes |1\rangle,$$

depending on the outcome. A record of a click-up followed by a click-down, or the opposite (click-down followed by click-up), signals an experimental or a preparation error, because neither

$$|1\rangle \otimes |0\rangle \quad \text{nor} \quad |0\rangle \otimes |1\rangle$$

appear in the state $|\psi\rangle$.

17.3   Entangled States

Entangled quantum states [165] are an important component of quantum computing techniques such as quantum error-correction, dense coding and quantum teleportation. Entanglement is the characteristic trait of quantum mechanics which enforces its entire departure from classical lines of thought. We consider entanglement of pure states. Thus a basic question in quantum computing is as follows: given a normalized state $|\mathbf{u}\rangle$ in the Hilbert space $\mathbf{C}^4$, can two normalized states $|\mathbf{x}\rangle$ and $|\mathbf{y}\rangle$ in the Hilbert space $\mathbf{C}^2$ be found such that

$$|\mathbf{x}\rangle \otimes |\mathbf{y}\rangle = |\mathbf{u}\rangle$$

where $\otimes$ denotes the Kronecker product [162, 163]. In other words, what is the condition on $|\mathbf{u}\rangle$ such that $|\mathbf{x}\rangle$ and $|\mathbf{y}\rangle$ exist? If no such $|\mathbf{x}\rangle$ and $|\mathbf{y}\rangle$ exist then $|\mathbf{u}\rangle$ is said to be *entangled*. If $|\mathbf{x}\rangle$ and $|\mathbf{y}\rangle$ do exist we say that $|\mathbf{u}\rangle$ is not entangled. As an example the state

$$\frac{1}{2}\begin{pmatrix} 1 \\ -1 \\ 1 \\ -1 \end{pmatrix} = \frac{1}{\sqrt{2}}\begin{pmatrix} 1 \\ 1 \end{pmatrix} \otimes \frac{1}{\sqrt{2}}\begin{pmatrix} 1 \\ -1 \end{pmatrix}$$

is not entangled. The *Bell basis states* [132] (which form a basis in $\mathbf{C}^4$) are given by

$$\Phi^{\pm} = \frac{1}{\sqrt{2}}|\uparrow\uparrow\rangle \pm \frac{1}{\sqrt{2}}|\downarrow\downarrow\rangle, \qquad \Psi^{\pm} = \frac{1}{\sqrt{2}}|\uparrow\downarrow\rangle \pm \frac{1}{\sqrt{2}}|\downarrow\uparrow\rangle \qquad (17.1)$$

where $|ab\rangle = |a\rangle \otimes |b\rangle$ and

$$|\uparrow\rangle := \begin{pmatrix} 1 \\ 0 \end{pmatrix}, \qquad |\downarrow\rangle := \begin{pmatrix} 0 \\ 1 \end{pmatrix}.$$

Consider an arbitrary orthonormal basis $\{|\alpha\rangle, |\beta\rangle\}$ in $\mathbf{C}^2$. The state $|\alpha\rangle$ can be expressed as

$$|\alpha\rangle = a|\uparrow\rangle + b|\downarrow\rangle,$$

with $a, b \in \mathbf{C}$ and $|a|^2 + |b|^2 = 1$. Thus

$$|\beta\rangle = e^{i\theta}\left(\bar{b}|\uparrow\rangle - \bar{a}|\downarrow\rangle\right)$$

where $\theta \in \mathbf{R}$, and $\bar{a}$ and $\bar{b}$ denote the complex conjugate of a and b respectively. Now consider the state

$$\frac{1}{\sqrt{2}}\left(|\alpha\beta\rangle - |\beta\alpha\rangle\right) = \frac{e^{i\theta}}{\sqrt{2}}(a|\uparrow\rangle + b|\downarrow\rangle) \otimes (\bar{b}|\uparrow\rangle - \bar{a}|\downarrow\rangle)$$

$$-\frac{e^{i\theta}}{\sqrt{2}}(\bar{b}|\uparrow\rangle - \bar{a}|\downarrow\rangle) \otimes (a|\uparrow\rangle + b|\downarrow\rangle)$$

$$= \frac{e^{i\theta}}{\sqrt{2}}\left(-(|a|^2 + |b|^2)|\uparrow\downarrow\rangle + (|a|^2 + |b|^2)|\downarrow\uparrow\rangle\right)$$

$$= \frac{e^{i\theta}}{\sqrt{2}}\left(|\downarrow\uparrow\rangle - |\uparrow\downarrow\rangle\right)$$

$$= -e^{i\theta}\Psi^-.$$

Thus measurement of Ψ^- always yields opposite outcomes for the two qubits, independent of the basis. The Bell states

$$\Phi^+ = \frac{1}{\sqrt{2}}\begin{pmatrix} 1 \\ 0 \\ 0 \\ 1 \end{pmatrix}, \qquad \Phi^- = \frac{1}{\sqrt{2}}\begin{pmatrix} 1 \\ 0 \\ 0 \\ -1 \end{pmatrix},$$

$$\Psi^+ = \frac{1}{\sqrt{2}}\begin{pmatrix} 0 \\ 1 \\ 1 \\ 0 \end{pmatrix}, \qquad \Psi^- = \frac{1}{\sqrt{2}}\begin{pmatrix} 0 \\ 1 \\ -1 \\ 0 \end{pmatrix}$$

are entangled. The entangled state Ψ^+ is also called the *EPR state*, after Einstein, Podolsky and Rosen [60]. Entangled states exhibit nonlocal correlations. This means that two entangled systems which have interacted in the past and are no longer interacting still show correlations. These correlations are used for example in dense coding and quantum error-correction techniques [19, 132]. The Bell states can be characterized as the simultaneous eigenvectors of the 4×4 matrices

$$\sigma_1 \otimes \sigma_1, \qquad \sigma_3 \otimes \sigma_3$$

where σ_1 and σ_3 are the Pauli spin matrices

$$\sigma_1 := \begin{pmatrix} 0 & 1 \\ 1 & 0 \end{pmatrix}, \qquad \sigma_3 := \begin{pmatrix} 1 & 0 \\ 0 & -1 \end{pmatrix}.$$

The measure for entanglement for pure states $E(\mathbf{u})$ is defined as follows [19, 132]

$$E(\mathbf{u}) := S(\rho_A) = S(\rho_B)$$

where the density matrices are defined as

$$\rho_A := \operatorname{tr}_B |\mathbf{u}\rangle\langle\mathbf{u}|, \qquad \rho_B := \operatorname{tr}_A |\mathbf{u}\rangle\langle\mathbf{u}|$$

and

$$S(\rho) := -\operatorname{tr}\rho \log_2 \rho.$$

Thus $0 \le E \le 1$. If $E = 1$ we call the pure state maximally entangled. If $E = 0$, the pure state is not entangled.

As an example consider Bohm's singlet state

$$|\psi\rangle := \frac{1}{\sqrt{2}}(|\uparrow\downarrow\rangle - |\downarrow\uparrow\rangle)$$

which is Ψ^- in the Bell basis. Since

$$|\psi\rangle\langle\psi| = \frac{1}{2}\begin{pmatrix} 0 & 0 & 0 & 0 \\ 0 & 1 & -1 & 0 \\ 0 & -1 & 1 & 0 \\ 0 & 0 & 0 & 0 \end{pmatrix}$$

we find

$$\begin{aligned} \rho_B &:= \operatorname{tr}_A(|\psi\rangle\langle\psi|) \\ &= ((\langle\uparrow| \otimes I_2)|\psi\rangle\langle\psi|(|\uparrow\rangle \otimes I_2) + ((\langle\downarrow| \otimes I_2)|\psi\rangle\langle\psi|(|\downarrow\rangle \otimes I_2) \\ &= \tfrac{1}{2}I_2 \end{aligned}$$

where I_2 is the 2×2 unit matrix. Thus

$$E_\psi = S(\rho_B) = -\operatorname{tr}(\frac{1}{2}\log_2 \frac{1}{2}I_2) = 1.$$

This state is maximally entangled. As another example consider

$$|\phi\rangle := |\uparrow\downarrow\rangle \equiv |\uparrow\rangle \otimes |\downarrow\rangle.$$

Since

$$|\phi\rangle\langle\phi| = \begin{pmatrix} 0 & 0 & 0 & 0 \\ 0 & 1 & 0 & 0 \\ 0 & 0 & 0 & 0 \\ 0 & 0 & 0 & 0 \end{pmatrix}$$

we find

$$\rho_B = \begin{pmatrix} 1 & 0 \\ 0 & 0 \end{pmatrix}.$$

Thus

$$E_\phi = S(\rho_B) = -\mathrm{tr}\rho_B \log_2 \rho_B = 0.$$

This state is not entangled.

The *Schmidt number* can also be used to characterize entanglement. The Schmidt number is the number of nonzero eigenvalues of ρ_A and ρ_B. A pure state is entangled if its Schmidt number is greater than one. In this case we have $E > 0$. Otherwise the pure state is not entangled and we have $E = 0$.

Next we derive the requirement for a state in $\mathbf{C}^4$ to be entangled. We use the representation

$$|\mathbf{u}\rangle = \begin{pmatrix} u_1 \\ u_2 \\ u_3 \\ u_4 \end{pmatrix}, \qquad |\mathbf{x}\rangle = \begin{pmatrix} x_1 \\ x_2 \end{pmatrix}, \qquad |\mathbf{y}\rangle = \begin{pmatrix} y_1 \\ y_2 \end{pmatrix}.$$

Since $|\mathbf{u}\rangle$ is normalized at least one of u_1, u_2, u_3, u_4 is nonzero. From the normalization conditions and (1) we find

$$|u_1|^2 + |u_2|^2 + |u_3|^2 + |u_4|^2 \;=\; 1 \tag{17.2}$$
$$|x_1|^2 + |x_2|^2 \;=\; 1 \tag{17.3}$$
$$|y_1|^2 + |y_2|^2 \;=\; 1 \tag{17.4}$$
$$x_1 y_1 \;=\; u_1 \tag{17.5}$$
$$x_1 y_2 \;=\; u_2 \tag{17.6}$$
$$x_2 y_1 \;=\; u_3 \tag{17.7}$$
$$x_2 y_2 \;=\; u_4 . \tag{17.8}$$

From (17.5)–(17.8) we find that the condition on $|\mathbf{u}\rangle$ is given by

$$u_1 u_4 = u_2 u_3 \tag{17.9}$$

From (17.3)–(17.8) we obtain

$$|x_1|^2 = |u_1|^2 + |u_2|^2 \tag{17.10}$$
$$|x_2|^2 = |u_3|^2 + |u_4|^2 \tag{17.11}$$
$$|y_1|^2 = |u_1|^2 + |u_3|^2 \tag{17.12}$$
$$|y_2|^2 = |u_2|^2 + |u_4|^2 \tag{17.13}$$

Let

$$\alpha_1 := \arg(x_1), \qquad \alpha_2 := \arg(x_2), \qquad \beta_1 := \arg(y_1), \qquad \beta_2 := \arg(y_2).$$

Now equations (17.5)–(17.8) become

$$\alpha_1 + \beta_1 \;=\; \arg(u_1) \quad \text{mod } 2\pi \tag{17.14}$$
$$\alpha_1 + \beta_2 \;=\; \arg(u_2) \quad \text{mod } 2\pi \tag{17.15}$$
$$\alpha_2 + \beta_1 \;=\; \arg(u_3) \quad \text{mod } 2\pi \tag{17.16}$$
$$\alpha_2 + \beta_2 \;=\; \arg(u_4) \quad \text{mod } 2\pi. \tag{17.17}$$

Suppose that (17.9) holds, then a solution is given by

$$x_1 = \left(\sqrt{|u_1|^2 + |u_2|^2}\right)e^{i\alpha_1}, \qquad x_2 = \left(\sqrt{|u_3|^2 + |u_4|^2}\right)e^{i\alpha_2} \tag{17.18}$$

$$y_1 = \left(\sqrt{|u_1|^2 + |u_3|^2}\right)e^{i\beta_1}, \qquad y_2 = \left(\sqrt{|u_2|^2 + |u_4|^2}\right)e^{i\beta_2} \tag{17.19}$$
$$\alpha_1 = 0, \qquad \alpha_2 = \arg(u_3) - \beta_1 \tag{17.20}$$
$$\beta_1 = \arg(u_1), \qquad \beta_2 = \arg(u_2) \tag{17.21}$$

The decomposition (if possible) of $|\mathbf{u}\rangle$ is not unique. For example

$$\frac{1}{\sqrt{2}}\begin{pmatrix} 1 \\ 1 \end{pmatrix} \otimes \frac{1}{\sqrt{2}}\begin{pmatrix} -1 \\ 1 \end{pmatrix} = \frac{1}{2}\begin{pmatrix} -1 \\ 1 \\ -1 \\ 1 \end{pmatrix}$$

and

$$\frac{1}{\sqrt{2}}\begin{pmatrix} i \\ i \end{pmatrix} \otimes \frac{1}{\sqrt{2}}\begin{pmatrix} i \\ -i \end{pmatrix} = \frac{1}{2}\begin{pmatrix} -1 \\ 1 \\ -1 \\ 1 \end{pmatrix}.$$

This follows from the fact that if $|\mathbf{u}\rangle = |\mathbf{x}\rangle \otimes |\mathbf{y}\rangle$ is a decomposition of $|\mathbf{u}\rangle$ then

$$|\mathbf{u}\rangle = (e^{i\theta}|\mathbf{x}\rangle) \otimes (e^{-i\theta}|\mathbf{y}\rangle), \qquad \theta \in \mathbf{R}$$

is also a decomposition of $|\mathbf{u}\rangle$.

Suppose $|\mathbf{u}\rangle = |\mathbf{x}_1\rangle \otimes |\mathbf{y}_1\rangle$ and $|\mathbf{u}\rangle = |\mathbf{x}_2\rangle \otimes |\mathbf{y}_2\rangle$. Suppose u_j is nonzero, then

$$x_{1k}, y_{1l}, x_{2k}, y_{2l} \neq 0 \tag{17.22}$$

$$x_{1k}y_{1l} = u_j \tag{17.23}$$

$$x_{2k}y_{2l} = u_j \tag{17.24}$$

for some $k, l \in \{1, 2\}$. x_{1k} can be written as $x_{1k} = cx_{2k}$, $c \in \mathbf{C}$ which gives $cy_{1l} = y_{2l}$. Let $k' := 3 - k$ and $l' := 3 - l$. If $x_{1k'}$ is nonzero then $x_{2k'}$ is nonzero and $x_{1k'}y_{1l} = x_{2k'}y_{2l}$ so that $x_{1k'} = cx_{2k'}$. Similarly if $y_{1l'}$ is nonzero then $cy_{1l'} = y_{2l'}$. Thus decomposition is unique up to a phase factor.

Next we describe the relation between condition (17.9) and the measure of entanglement introduced above. Since

$$|\mathbf{u}\rangle\langle\mathbf{u}| = \begin{pmatrix} u_1\overline{u_1} & u_1\overline{u_2} & u_1\overline{u_3} & u_1\overline{u_4} \\ u_2\overline{u_1} & u_2\overline{u_2} & u_2\overline{u_3} & u_2\overline{u_4} \\ u_3\overline{u_1} & u_3\overline{u_2} & u_3\overline{u_3} & u_3\overline{u_4} \\ u_4\overline{u_1} & u_4\overline{u_2} & u_4\overline{u_3} & u_4\overline{u_4} \end{pmatrix} \tag{17.25}$$

we find

$$\rho_A = \text{tr}_B(|\mathbf{u}\rangle\langle\mathbf{u}|) = \begin{pmatrix} u_1\overline{u_1} + u_2\overline{u_2} & u_1\overline{u_3} + u_2\overline{u_4} \\ u_3\overline{u_1} + u_4\overline{u_2} & u_3\overline{u_3} + u_4\overline{u_4} \end{pmatrix} \tag{17.26}$$

$$\rho_B = \text{tr}_A(|\mathbf{u}\rangle\langle\mathbf{u}|) = \begin{pmatrix} u_1\overline{u_1} + u_3\overline{u_3} & u_1\overline{u_2} + u_3\overline{u_4} \\ u_2\overline{u_1} + u_4\overline{u_3} & u_2\overline{u_2} + u_4\overline{u_4} \end{pmatrix}. \tag{17.27}$$

The 2×2 density matrices ρ_A and ρ_B given by (17.26) and (17.27) are hermitian and have the same eigenvalues. Thus the eigenvalues λ_1 and λ_2 are real. The matrices are also positive semi-definite i.e. for all $|\mathbf{a}\rangle \in \mathbf{C}^2$ we have $\langle\mathbf{a}|\rho_{A,B}|\mathbf{a}\rangle \geq 0$. Thus the eigenvalues are non-negative. The eigenvalues are given by

$$\lambda_1 = \frac{1 + \sqrt{1 - 4|u_1u_4 - u_2u_3|^2}}{2}, \qquad \lambda_2 = \frac{1 - \sqrt{1 - 4|u_1u_4 - u_2u_3|^2}}{2}. \tag{17.28}$$

Since $|\mathbf{u}\rangle$ is normalized we have

$$\text{tr}(\text{tr}_A|\mathbf{u}\rangle\langle\mathbf{u}|) = 1 \tag{17.29}$$

$$\text{tr}(\text{tr}_B|\mathbf{u}\rangle\langle\mathbf{u}|) = 1 \tag{17.30}$$

and therefore

$$\lambda_1 + \lambda_2 = 1 \tag{17.31}$$

where we used the fact that the trace of an $n \times n$ matrix is the sum of its eigenvalues. This can also be seen from (17.28). Thus $0 \le \lambda_1, \lambda_2 \le 1$. Now we have

$$
\begin{aligned}
\det(\text{tr}_A(|\mathbf{u}\rangle\langle\mathbf{u}|)) &= \det(\text{tr}_B(|\mathbf{u}\rangle\langle\mathbf{u}|)) \\
&= (u_1 u_4 - u_2 u_3)(\overline{u_1}\,\overline{u_4} - \overline{u_2}\,\overline{u_3}) \\
&= |u_1 u_4 - u_2 u_3|^2 .
\end{aligned}
\tag{17.32}
$$

Thus if $u_1 u_4 = u_2 u_3$ the determinant is equal to 0. Since the determinant of an $n \times n$ matrix is the product of the eigenvalues we find that one eigenvalue is equal to 0 and owing to (17.31) the other eigenvalue is 1. Obviously the entanglement can be written as

$$E(\mathbf{u}) = -(\lambda \log_2 \lambda + (1 - \lambda) \log_2(1 - \lambda)) \tag{17.33}$$

where $\lambda \in \{\lambda_1, \lambda_2\}$ is one of the eigenvalues given above. Using these facts and

$$\log_2 1 = 0, \quad 0 \log_2 0 = 0$$

we find that $E(\mathbf{u}) = 0$ if condition (17.9) is satisfied. Thus if $\lambda = 0$ or $\lambda = 1$ we have $E(\mathbf{u}) = 0$. For $\lambda = \frac{1}{2}$ the entanglement $E(\mathbf{u})$ has a maximum and we find $E(\mathbf{u}) = 1$. Vice versa we can prove that if $E(\mathbf{u}) = 0$ the condition (17.9) follows. The squares of the density operators are given by

$$
\rho_A^2 = \begin{pmatrix} (|u_1|^2 + |u_2|^2)^2 + |u_1\overline{u_3} + u_2\overline{u_4}|^2 & u_1\overline{u_3} + u_2\overline{u_4} \\ u_3\overline{u_1} + u_4\overline{u_2} & (|u_3|^2 + |u_4|^2)^2 + |u_1\overline{u_3} + u_2\overline{u_4}|^2 \end{pmatrix}
$$

$$
\rho_B^2 = \begin{pmatrix} (|u_1|^2 + |u_3|^2)^2 + |u_1\overline{u_2} + u_3\overline{u_4}|^2 & u_1\overline{u_2} + u_3\overline{u_4} \\ u_2\overline{u_1} + u_4\overline{u_3} & (|u_2|^2 + |u_4|^2)^2 + |u_1\overline{u_2} + u_3\overline{u_4}|^2 \end{pmatrix}
$$

When (17.9) holds we find that

$$\rho_A^2 = \rho_A \quad \text{and} \quad \rho_B^2 = \rho_B.$$

Using the computer algebra system SymbolicC++ [169] the expression $u_1 u_4 - u_2 u_3$ can be evaluated symbolically and compared against 0 which then provides the information whether the state is entangled or not. SymbolicC++ includes among other classes a template class `Complex` and a `Sum` class to do the symbolic manipulations. If the state is entangled, then we can use equations (38) and (33) to find the entanglement E.

A remark is in order about the precision of the numerical calculations of the condition $u_1u_4 = u_2u_3$ and the entanglement E to test for non-entanglement. To test the condition $u_1u_4 = u_2u_3$ has the advantage that it consists of only multiplication of complex numbers and the normalization factor of the vector $|u\rangle$ must not be taken into account. On the other hand if the difference $|u_1u_4 - u_2u_3|$ is of order $O(10^{-15})$ the term

$$\log(1 + 4|u_1u_4 - u_2u_3|^2)$$

can be taken as

$$\log(1 + O(10^{-30})).$$

Therefore $1 + O(10^{-30})$ is rounded to $\log 1$ for data type double. Thus the entanglement E is less affected by the problem of the floating point comparison. However in calculating E we have to take into account the normalization factor of the vector $|u\rangle$. Warnings should be issued if E or $|u_1u_4 - u_2u_3|$ are close to zero when we use the data type double. Java and a number of computer algebra systems admit a data type of arbitrary precision of floating point numbers. For example, Java has the abstract data type `BigDecimal`. Then we can work with higher precision. An important special case arises when one of the components of the vector $|u\rangle$ is equal to zero. For example, say $u_4 = 0$. If the state $|u\rangle$ is non-entangled then u_2 or u_3 must be zero.

The analysis of separability can be extended to higher dimensions, for example Steeb and Hardy [167] consider when states in $\mathbf{C}^9$ can be separated into a product of two states in $\mathbf{C}^3$. They have only considered separability of pure states.

The more general question of the separability of mixed states has been considered in [90, 91, 127].

17.4 Quantum Gates

17.4.1 Introduction

Quantum computation is a unitary transformation, where a measurement is performed at the end to extract the result. A unitary transformation is itself reversible; therefore, we have to use reversible gates in order to be able to implement quantum gates. A unitary transformation may operate on a single qubit or multiple qubits. Some transformations on multiple qubits cannot be expressed as a sequence of operations on single qubits.

States evolve according to the Schrödinger equation

$$i\hbar\frac{\partial\psi}{\partial t} = \hat{H}\psi$$

where $\hat{H}$ is a linear self-adjoint operator. The formal solution is given by

$$\psi(t) = e^{-it\hat{H}/\hbar}\psi(0)$$

and since $\hat{H}$ is self-adjoint $\exp(-it\hat{H}/\hbar)$ is unitary. Thus the evolution of states in quantum computation is described by unitary operations.

For example, a general unitary transformation in the two-dimensional space $\mathbf{C}^2$ can be defined as follows

$$U(\theta, \delta, \sigma, \tau) = \begin{pmatrix} e^{i(\delta+\sigma+\tau)}\cos(\theta/2) & e^{-i(\delta+\sigma-\tau)}\sin(\theta/2) \\ -e^{i(\delta-\sigma+\tau)}\sin(\theta/2) & e^{i(\delta-\sigma-\tau)}\cos(\theta/2) \end{pmatrix}$$

with $\theta, \delta, \sigma, \tau \in \mathbf{R}$.

Thus any quantum gate operating on a single qubit is given by an appropriate choice of θ, δ, σ, and τ. Single qubit operations are not sufficient to implement arbitrary unitary transforms required by quantum algorithms. Thus it is important to determine if some basic set of unitary operations are sufficient to implement any unitary transform.

Definition. A unitary transformation on n qubits is called *simple* if $n - 2$ of the qubits always remain unchanged by the transformation.

Theorem. Given any unitary transformation U and $\epsilon > 0$ there exists simple unitary transformations $U_1, U_2, \ldots, U_k$ such that

$$\|U - U_1 U_2 \ldots U_k\| < \epsilon,$$

where k is a polynomial function of 2^n and $\log_2 \frac{1}{\epsilon}$ and

$$\|A\| := \max_{\||x\rangle\|=1} \| A|x\rangle \|.$$

This theorem is important for the discussion of universality (see section 17.4.6).

Next we discuss some important quantum gates.

17.4.2 NOT Gate

The corresponding quantum gate of the classical NOT gate is implemented via a unitary matrix U_{NOT} operation that evolves the basis states into the corresponding states according to the same truth table. The quantum version of the classical NOT gate is the unitary operation U_{NOT} such that

$$U_{NOT}|0\rangle := |1\rangle, \qquad U_{NOT}|1\rangle := |0\rangle.$$

Since $|0\rangle = (0\ 1)^T$ and $|1\rangle = (1\ 0)^T$ we find the unitary matrix

$$U_{NOT} := \begin{pmatrix} 0 & 1 \\ 1 & 0 \end{pmatrix}.$$

The quantum NOT gate for the two quantum bit case would be then the unitary 4×4 matrix

$$U_{NOT} = \begin{pmatrix} 0 & 1 \\ 1 & 0 \end{pmatrix} \otimes \begin{pmatrix} 0 & 1 \\ 1 & 0 \end{pmatrix} \equiv \begin{pmatrix} 0 & 0 & 0 & 1 \\ 0 & 0 & 1 & 0 \\ 0 & 1 & 0 & 0 \\ 1 & 0 & 0 & 0 \end{pmatrix}$$

since

$$U_{NOT}|00\rangle = |11\rangle, \quad U_{NOT}|10\rangle = |01\rangle, \quad U_{NOT}|01\rangle = |10\rangle, \quad U_{NOT}|11\rangle = |00\rangle.$$

This can be extended to any dimension. The unitary matrix U_{NOT} is a permutation matrix. The NOT gate is a special case of the unitary matrix

$$U(\alpha) = \frac{1}{2} \begin{pmatrix} 1 + e^{i\pi\alpha} & 1 - e^{i\pi\alpha} \\ 1 - e^{i\pi\alpha} & 1 + e^{i\pi\alpha} \end{pmatrix}$$

if $\alpha = 1$.

The NOT gate is denoted as

Figure 17.1: NOT Gate

17.4.3 Walsh-Hadamard Gate

In quantum mechanics, the notation of gates can be extended to operations that have no classical counterpart. For instance, the operation U_H (*Walsh-Hadamard gate*) that evolves according to

$$U_H|0\rangle := \frac{1}{\sqrt{2}}(|0\rangle + |1\rangle)$$

$$U_H|1\rangle := \frac{1}{\sqrt{2}}(|0\rangle - |1\rangle).$$

Note that it evolves classical states into superpositions and therefore cannot be regarded as classical. Thus U_H is given by the 2×2 unitary matrix

$$U_H \equiv \frac{1}{\sqrt{2}} \begin{pmatrix} 1 & 1 \\ -1 & 1 \end{pmatrix}$$

since

$$U_H|0\rangle \equiv \frac{1}{\sqrt{2}} \begin{pmatrix} 1 & 1 \\ -1 & 1 \end{pmatrix} \begin{pmatrix} 0 \\ 1 \end{pmatrix} = \frac{1}{\sqrt{2}} \begin{pmatrix} 1 \\ 1 \end{pmatrix} = \frac{1}{\sqrt{2}}(|0\rangle + |1\rangle)$$

$$U_H|1\rangle \equiv \frac{1}{\sqrt{2}} \begin{pmatrix} 1 & 1 \\ -1 & 1 \end{pmatrix} \begin{pmatrix} 1 \\ 0 \end{pmatrix} = \frac{1}{\sqrt{2}} \begin{pmatrix} 1 \\ -1 \end{pmatrix} = \frac{1}{\sqrt{2}}(|0\rangle - |1\rangle).$$

The unitary operation represented by the unitary matrix U_H corresponds to a 45° rotation of the polarization. This is intrinsically nonclassical because it transforms Boolean states into superpositions. The inverse matrix of U_H is given by

$$U_H^{-1} \equiv U_H^T = \frac{1}{\sqrt{2}} \begin{pmatrix} 1 & -1 \\ 1 & 1 \end{pmatrix}.$$

The Walsh-Hadamard gate is a special case of the rotation matrix when $\theta = -\frac{\pi}{4}$,

$$U_R(\theta) := \begin{pmatrix} \cos\theta & -\sin\theta \\ \sin\theta & \cos\theta \end{pmatrix}.$$

The Walsh-Hadamard gate is denoted as

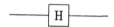

Figure 17.2: Walsh-Hadamard Gate

The Walsh-Hadamard gate is quite useful when extended using the Kronecker product. If we take an n-bit quantum register initially in the state

$$|00\ldots0\rangle$$

and apply U_H to every single qubit of the register. The resulting state is

$$|\psi\rangle = (U_H \otimes U_H \otimes \cdots \otimes U_H)|00\cdots0\rangle = \frac{1}{2^{n/2}}(|00\cdots0\rangle + |00\cdots1\rangle + \cdots + |11\cdots1\rangle).$$

Thus we can write

$$|\psi\rangle = \frac{1}{2^{n/2}} \sum_{x=0}^{2^n-1} |x\rangle.$$

When the initial configuration of the qubits is

$$y = y_0 + y_1 2 + \ldots + y_{n-1} 2^{n-1},$$

in other words the register is prepared as

$$|y_{n-1} \cdots y_1 y_0\rangle,$$

applying the Walsh-Hadamard transform to each qubit yields

$$(U_H \otimes U_H \otimes \cdots \otimes U_H)|y\rangle = \frac{1}{2^{n/2}} \sum_{x=0}^{2^n-1} (-1)^{x*y} |x\rangle.$$

where

$$x * y = (x_0 \cdot y_0) \oplus (x_1 \cdot y_1) \oplus \ldots \oplus (x_{n-1} \cdot y_{n-1}).$$

This means with a linear number of operations (i.e. n applications of U_H) we have generated a register state that contains an exponential (2^n) number of distinct terms. Using quantum registers, n elementary operations can generate a state containing all 2^n possible numerical values of the register. In contrast, in classical registers n elementary operations can only prepare one state of the register representing one specific number. It is this ability of creating quantum superpositions which makes the quantum parallel processing possible. If after preparing the register in a coherent superposition of several numbers all subsequent computational operations are unitary and linear (i.e. preserve the superpositions of states) then with each computational step the computation is performed simultaneously on all the numbers present in the superposition.

17.4.4 XOR and the Controlled NOT Gate

The most important two-qubit quantum gate is the *XOR gate*. It is defined as

$$U_{XOR}|a, b\rangle := |a, a \oplus b\rangle.$$

Consequently

$$U_{XOR}|00\rangle = |00\rangle, \quad U_{XOR}|01\rangle = |01\rangle, \quad U_{XOR}|10\rangle = |11\rangle, \quad U_{XOR}|11\rangle = |10\rangle.$$

The vectors $|00\rangle$, $|01\rangle$, $|10\rangle$ and $|11\rangle$ form an orthonormal basis in $\mathbf{C}^4$. If we consider the basis in this order, the matrix representation of U_{XOR} is

$$U_{XOR} = \begin{pmatrix} 1 & 0 & 0 & 0 \\ 0 & 1 & 0 & 0 \\ 0 & 0 & 0 & 1 \\ 0 & 0 & 1 & 0 \end{pmatrix}.$$

If we consider the order $|11\rangle$, $|10\rangle$, $|01\rangle$, $|00\rangle$, the matrix representation is

$$U_{XOR} = \begin{pmatrix} 0 & 1 & 0 & 0 \\ 1 & 0 & 0 & 0 \\ 0 & 0 & 1 & 0 \\ 0 & 0 & 0 & 1 \end{pmatrix}.$$

Both matrices are permutation matrices. Sometimes in the literature the definition

$$U_{XOR}|a, b\rangle := |a \oplus b, b\rangle$$

is used. Furthermore the XOR gate is also called the *controlled NOT* gate (*CNOT gate*). The name comes from the fact that the gate effects a logical NOT on the second qubit (target bit), if and only if the first qubit (control bit) is in state 1. We see that U_{XOR} cannot be written as a Kronecker product of 2×2 matrices. Two interacting magnetic dipoles sufficiently close to each other can be used to implement this operation. The XOR gate is denoted by

Figure 17.3: XOR Gate

17.4.5 Other Quantum Gates

The *exchange gate* simply swaps two bits, i.e. it applies the transform

$$|00\rangle \mapsto |00\rangle, \quad |01\rangle \mapsto |10\rangle, \quad |10\rangle \mapsto |01\rangle, \quad |11\rangle \mapsto |11\rangle.$$

We have

$$U_{EXCH} := |00\rangle\langle00| + |10\rangle\langle01| + |01\rangle\langle10| + |11\rangle\langle11|.$$

The matrix representation is

$$U_{EXCH} := \begin{pmatrix} 1 & 0 & 0 & 0 \\ 0 & 0 & 1 & 0 \\ 0 & 1 & 0 & 0 \\ 0 & 0 & 0 & 1 \end{pmatrix}$$

which is a permutation matrix.

The *phase shift gate* is defined on two qubits as

$$U_{PS}(\phi)|ab\rangle := e^{i(a \cdot b)\phi}|ab\rangle$$

where $a, b \in \{0, 1\}$ and $\cdot$ denotes the classical AND operation. Thus we have

$$U_{PS}(\phi)|00\rangle = |00\rangle,$$
$$U_{PS}(\phi)|01\rangle = |01\rangle,$$
$$U_{PS}(\phi)|10\rangle = |10\rangle,$$
$$U_{PS}(\phi)|11\rangle = e^{i\phi}|11\rangle.$$

The gate performs a conditional phase shift, i.e. a multiplication by a phase factor $e^{i\phi}$ only if the two qubits are both in their $|1\rangle$ state. The three other basis states are unaffected. An important special case is if $\phi = \pi$. The phase shift gate is denoted by

Figure 17.4: Phase Shift Gate

The phase shift gate which acts on one qubit is defined as (in matrix notation)

$$\begin{pmatrix} e^{-i\phi} & 0 \\ 0 & e^{i\phi} \end{pmatrix}.$$

The *Toffoli gate* is a classically universal, reversible, 3 input, 3 output gate. It is sometimes also called the *controlled controlled NOT gate*. It transforms a state according to

$$|a, b, c\rangle \mapsto |a, b, (a \cdot b) \oplus c\rangle.$$

The NOT gate can be constructed as

$$|1, 1, a\rangle \mapsto |1, 1, \neg a\rangle,$$

and the AND gate can be implemented as

$$|a, b, 0\rangle \mapsto |a, b, a \cdot b\rangle.$$

The gate can be described in terms of U_{XOR}

$$U_{TOFFOLI} := |0\rangle\langle 0| \otimes I_2 \otimes I_2 + |1\rangle\langle 1| \otimes U_{XOR}.$$

We can also describe the gate by

$$(U_{TOFFOLI})_{x_1'x_2'x_3'}^{x_1x_2x_3} = \delta_{x_1'}^{x_1}\delta_{x_2'}^{x_2}\delta_{x_3'}^{x_3\oplus(x_1\cdot x_2)}$$

which is a special case of Deutsch's gate (given below) for $\alpha = 1$. Thus we obtain the matrix representation

$$U_{TOFFOLI} = \begin{pmatrix} 1 & 0 & 0 & 0 & 0 & 0 & 0 & 0 \\ 0 & 1 & 0 & 0 & 0 & 0 & 0 & 0 \\ 0 & 0 & 1 & 0 & 0 & 0 & 0 & 0 \\ 0 & 0 & 0 & 1 & 0 & 0 & 0 & 0 \\ 0 & 0 & 0 & 0 & 1 & 0 & 0 & 0 \\ 0 & 0 & 0 & 0 & 0 & 1 & 0 & 0 \\ 0 & 0 & 0 & 0 & 0 & 0 & 0 & 1 \\ 0 & 0 & 0 & 0 & 0 & 0 & 1 & 0 \end{pmatrix}.$$

The Toffoli gate is denoted by

Figure 17.5: Toffoli Gate

The *Fredkin gate* is described by

$$U_{FREDKIN} := |0\rangle\langle 0| \otimes I_2 \otimes I_2 + |1\rangle\langle 1| \otimes U_{EXCH}.$$

The gate is also called the *controlled exchange gate*. It is also possible to construct AND and NOT gates from the Fredkin gate.

Deutsch's gate acts on elements of the Hilbert space $\mathbf{C}^8$ and is given by

$$(U_D)_{a_1'a_2'a_3'}^{a_1a_2a_3}(\alpha) := \delta_{a_1'}^{a_1}\delta_{a_2'}^{a_2}\left[(1 - a_1\cdot a_2)\delta_{a_3'}^{a_3} + ia_1\cdot a_2 e^{-\frac{i}{2}\pi\alpha}S(\alpha)_{a_3'}^{a_3}\right]$$

where $a_1, a_2, a_3 \in \{0,1\}$ and

$$S(\alpha) = \frac{1}{2}\begin{pmatrix} 1 + e^{i\pi\alpha} & 1 - e^{i\pi\alpha} \\ 1 - e^{i\pi\alpha} & 1 + e^{i\pi\alpha} \end{pmatrix}$$

is a 2×2 unitary matrix. Here δ_i^j denotes the Kronecker symbol. Thus U_D is a unitary 8×8 matrix given by

$$
U_D(\alpha) = \begin{pmatrix}
1 & 0 & 0 & 0 & 0 & 0 & 0 & 0 \\
0 & 1 & 0 & 0 & 0 & 0 & 0 & 0 \\
0 & 0 & 1 & 0 & 0 & 0 & 0 & 0 \\
0 & 0 & 0 & 1 & 0 & 0 & 0 & 0 \\
0 & 0 & 0 & 0 & 1 & 0 & 0 & 0 \\
0 & 0 & 0 & 0 & 0 & 1 & 0 & 0 \\
0 & 0 & 0 & 0 & 0 & 0 & i\cos(\pi\alpha/2) & \sin(\pi\alpha/2) \\
0 & 0 & 0 & 0 & 0 & 0 & \sin(\pi\alpha/2) & i\cos(\pi\alpha/2)
\end{pmatrix}
$$

where we used the following ordering of $a_1 a_2 a_3$

$$000, \ 001, \ 010, \ 011, \ 100, \ 101, \ 110, \ 111.$$

17.4.6 Universal Sets of Quantum Gates

In classical computing we described how any Boolean function can be expressed in terms of the NAND (or NOR) operation. Is their a single quantum gate which can be used to implement any other quantum gate ?

Deutch's gate, described in the previous section is one such gate [57]. It is a class of gates described by a real parameter. Thus to prove a set of quantum gates is a universal set, all that is required is to show that the set can implement Deutch's gate. For example U_{XOR} together with the set of all single qubit transformations described by the matrix

$$
U(\theta, \delta, \sigma, \tau) = \begin{pmatrix}
e^{i(\delta+\sigma+\tau)} \cos(\theta/2) & e^{-i(\delta+\sigma-\tau)} \sin(\theta/2) \\
-e^{i(\delta-\sigma+\tau)} \sin(\theta/2) & e^{i(\delta-\sigma-\tau)} \cos(\theta/2)
\end{pmatrix}
$$

is a universal set of quantum gates [7, 138].

It has also been shown that a combination of single and double qubit operations [6, 57] can also form a universal set.

17.4.7   Functions

Now we illustrate how to construct a simple transformation implementing a classical function. We consider only the case where one qubit is changed, i.e. the function we compute over the input gives the value 0 or 1. A simple permutation, which is unitary, and its inverse allows functions with a greater number of output qubits to be computed. Suppose the input consists of n qubits, and the function to be calculated is $f := \{0, 1, \ldots, 2^n - 1\} \to \{0, 1\}$. A unitary transform given by

$$U_f := \sum_{j=0}^{2^n-1} \sum_{k=0}^{1} |j\rangle\langle j| \otimes |k \oplus f(j)\rangle\langle k|$$

which is a permutation. The terms in the sum consist of the mapping f

$$\Big(|j\rangle \otimes |f(j)\rangle\Big)\Big(\langle j| \otimes \langle 0|\Big)$$

and

$$\Big(|j\rangle \otimes |\overline{f(j)}\rangle\Big)\Big(\langle j| \otimes \langle 1|\Big)$$

to ensure unitarity.

For example, the sum bit of the full adder would be implemented as

$$
\begin{aligned}
U_{SUM} \ := \ & |0000\rangle\langle 0000| + |0011\rangle\langle 0010| + |0101\rangle\langle 0100| + |0110\rangle\langle 0110| + \\
& |1001\rangle\langle 1000| + |1010\rangle\langle 1010| + |1100\rangle\langle 1100| + |1111\rangle\langle 1110| + \\
& |0001\rangle\langle 0001| + |0010\rangle\langle 0011| + |0100\rangle\langle 0101| + |0111\rangle\langle 0111| + \\
& |1000\rangle\langle 1001| + |1011\rangle\langle 1011| + |1101\rangle\langle 1101| + |1110\rangle\langle 1111|
\end{aligned}
$$

and the carry bit would be implemented as

$$
\begin{aligned}
U_{CARRY} \ := \ & |0000\rangle\langle 0000| + |0010\rangle\langle 0010| + |0100\rangle\langle 0100| + |0111\rangle\langle 0110| + \\
& |1000\rangle\langle 1000| + |1011\rangle\langle 1010| + |1101\rangle\langle 1100| + |1111\rangle\langle 1110| + \\
& |0001\rangle\langle 0001| + |0011\rangle\langle 0011| + |0101\rangle\langle 0101| + |0110\rangle\langle 0111| + \\
& |1001\rangle\langle 1001| + |1010\rangle\langle 1011| + |1100\rangle\langle 1101| + |1110\rangle\langle 1111|.
\end{aligned}
$$

To compute a more complex function which maps to n bits, we describe each n functions which map to a single bit. This is possible since each bit value can be viewed as a function of the input separate from the other $n - 1$ bits.

Next we describe how quantum computers deal with functions ([8], [123]). Consider
a function

$$f : \{0, 1, \ldots, 2^m - 1\} \rightarrow \{0, 1, \ldots, 2^n - 1\}$$

where m and n are positive integers. A classical device computes f by evolving each
labelled input

$$0, 1, \ldots, 2^m - 1$$

into its respective labelled output

$$f(0), f(1), \ldots, f(2^m - 1).$$

Quantum computers, due to the unitary (and therefore reversible) nature of their
evolution, compute functions in a slightly different way. It is not directly possible
to compute a function f by a unitary operation that evolves $|x\rangle$ into $|f(x)\rangle$. If f is
not a one-to-one mapping (i.e. if $f(x) = f(y)$ for some $x \neq y$), then two orthogonal
kets $|x\rangle$ and $|y\rangle$ can be evolved into the same state

$$|f(x)\rangle = |f(y)\rangle.$$

Thus this violates unitarity. One way to compute functions which are not one-to-
one mappings, while preserving the reversibility of computation, is by keeping the
record of the input. To achieve this, a quantum computer uses two registers; the first
register to store the input data, the second one for the output data. Each possible
input x is represented by the state $|x\rangle$, the quantum state of the first register.
Analogously, each possible output $y = f(x)$ is represented by $|y\rangle$, the quantum
state of the second register. States corresponding to different inputs and different
outputs are orthogonal,

$$
\begin{aligned}
\langle x | x' \rangle &= \delta_{xx'}, \\
\langle y | y' \rangle &= \delta_{yy'}.
\end{aligned}
$$

Thus

$$((\langle y'| \otimes \langle x'|)(|x\rangle \otimes |y\rangle)) = \delta_{xx'}\delta_{yy'}.$$

The function evaluation is then determined by a unitary evolution operator U_f that
acts on both registers

$$U_f|x\rangle \otimes |0\rangle = |x\rangle \otimes |f(x)\rangle.$$

A reversible function evaluation, i.e. the one that keeps track of the input, is as
good as a regular, irreversible evaluation. This means that if a given function can
be computed in polynomial time, it can also be computed in polynomial time using
a reversible computation. The computations we are considering here are not only
reversible but also quantum, and we can do much more than computing values of
$f(x)$ one by one. We can prepare a superposition of all input values as a single state

and by running the computation U_f only once, we can compute all of the 2^m values
$f(0), \ldots, f(2^m - 1)$,

$$|\psi\rangle = U_f\left(\frac{1}{2^{m/2}} \sum_{x=0}^{2^m-1} |x\rangle\right) \otimes |0\rangle = \frac{1}{2^{m/2}} \sum_{x=0}^{2^m-1} |x\rangle \otimes |f(x)\rangle.$$

How much information about f does the state $|\psi\rangle$ contain? No quantum measurement can extract all of the 2^m values

$$f(0), \quad f(1), \ldots, f(2^m - 1)$$

from $|\psi\rangle$. Imagine, for instance, performing a measurement on the first register of $|\psi\rangle$. Quantum mechanics enables us to infer several facts. Since each value x appears with the same complex amplitude in the first register of state of $|\psi\rangle$, the outcome of the measurement is equiprobable and can be any value ranging from 0 to $2^m - 1$. Assuming that the result of the measurement is $|j\rangle$, the post-measurement state of the two registers (i.e. the state of the registers after the measurement) is

$$\widetilde{|\psi\rangle} = |j\rangle \otimes |f(j)\rangle.$$

Thus a subsequent measurement on the second register would yield with certainty the result $f(j)$, and no additional information about f can be gained.

Nielsen and Chuang [123] use a different notation. The initial state is assumed to be of the form

$$|d\rangle \otimes |P\rangle$$

where $|d\rangle$ is the state of the m-qubit data register, and $|P\rangle$ is a state of the n-qubit program register. The two registers are not entangled. The dynamics of the gate array is given by

$$|d\rangle \otimes |P\rangle \rightarrow G(|d\rangle \otimes |P\rangle)$$

where G is a unitary operator. This operation is implemented by some fixed quantum gate array. A unitary operator, U, acting on m qubits, is said to be implemented by this gate array if there exists a state $|P_U\rangle$ of the program register such that

$$G(|d\rangle \otimes |P_U\rangle) = (U|d\rangle) \otimes |P_U'\rangle$$

for all states $|d\rangle$ of the data register, and some state $|P_U'\rangle$ of the program register. To see that $|P_U'\rangle$ does not depend on $|d\rangle$, suppose that

$$G|d_1\rangle \otimes |P\rangle = (U|d_1\rangle) \otimes |P_1'\rangle$$

$$G|d\rangle \otimes |P\rangle = (U|d_2\rangle) \otimes |P_2'\rangle.$$

Taking the inner product of these two equations, we find that

$$\langle P_1' | P_2' \rangle = 1$$

provided

$$\langle d_1 | d_2 \rangle \neq 0.$$

Thus

$$|P_1'\rangle = |P_2'\rangle$$

and therefore there is no $|d\rangle$ dependence of $|P_U'\rangle$. The case $\langle d_1 | d_2 \rangle = 0$ follows by similar reasoning. Nielsen and Chuang [123] show how to construct quantum gate arrays that can be programmed to perform different unitary operations on a data register, depending on the input to some program register. Furthermore, they show that a universal quantum register gate array – a gate array which can be programmed to perform any unitary operation – exists only if one allows the gate array to operate in a probabilistic fashion. Since the number of possible unitary operations on m qubits is infinite, it follows that a universal gate array would require an infinite number of qubits in the program register, and thus no such array exists.

Suppose distinct (up to a global phase) unitary operators $U_1, \ldots U_N$ are implemented by some programmable quantum gate array. Nielsen and Chuang [123] showed that the program register is at least N dimensional, that is, contains at least $\log_2 N$ qubits. Moreover, the corresponding programs $|P_1\rangle, \ldots |P_N\rangle$ are mutually orthogonal. A deterministic programmable gate array must have as many Hilbert space dimensions in the program register as the number of programs implemented.

17.5   Garbage Disposal

In performing a calculation, an algorithm may use a number of temporary registers for storing intermediate results. Since operations in quantum computing involve unitary (reversable) operations, these temporary registers cannot be forced into some initial state independent of the register contents. The unitarity of operations does however provide a mechanism to return temporary registers to some initial state.

Suppose an algorithm requires the application of the sequence of unitary operators

$$U_1, U_2, \ldots, U_n$$

where each operator U_i places the result in register (of appropriate size) i. The final result is placed in register n. We assume each register i is in an initial state $|0\rangle_i$. The register indicated by 0 is in the initial state $|a\rangle$ which serves as a parameter to the algorithm. Each U_i successively places the result $f_i(a)$ of the computation in register i, given the values $a, f_1(a), \ldots, f_{i-1}(a)$ as parameters. Thus application of the operators gives

$$U_n U_{n-1} \ldots U_1(|a\rangle \otimes |0\rangle_1 \otimes \ldots \otimes |0\rangle_n) = |a\rangle \otimes |f_1(a)\rangle \otimes |f_2(a)\rangle \otimes \ldots \otimes |f_n(a)\rangle.$$

Since we only require the result $f_n(a)$ we can apply the inverse operations

$$U_{n-1}^*, U_{n-2}^*, \ldots, U_1^*$$

to return the temporary registers to their initial states

$$U_1^* U_2^* \ldots U_{n-1}^*(|a\rangle \otimes |f_1(a)\rangle \otimes |f_2(a)\rangle \otimes \ldots \otimes f_n(a)) = |a\rangle \otimes |0\rangle_1 \otimes \ldots |0\rangle_{n-1} \otimes |f_n\rangle.$$

This can be understood by examining the register content after each unitary operation

$$U_1(|a\rangle \otimes |0\rangle_1 \otimes \ldots \otimes |0\rangle_n) = |a\rangle \otimes |f_1(a)\rangle \otimes |0\rangle_2 \otimes \ldots \otimes |0\rangle_n$$
$$U_2(|a\rangle \otimes |f_1(a)\rangle \otimes |0\rangle_2 \otimes \ldots \otimes |0\rangle_n) = |a\rangle \otimes |f_1(a)\rangle \otimes |f_2(a)\rangle \otimes |0\rangle_3 \otimes \ldots \otimes |0\rangle_n$$

$$U_i\left(|a\rangle \otimes \left(\bigotimes_{k=1}^{i-1} |f_k(a)\rangle\right) \otimes \left(\bigotimes_{k=i}^{n} |0\rangle_k\right)\right) = \left(|a\rangle \otimes \left(\bigotimes_{k=1}^{i} |f_k(a)\rangle\right) \otimes \left(\bigotimes_{k=i+1}^{n} |0\rangle_k\right)\right)$$

Each step depends only on previously calculated values, thus reversing each computation from U_{n-1} to U_1 does not destroy the final result. This method of regaining the use of temporary registers is termed *garbage disposal*, as it eliminates "garbage" in temporary registers which is no longer useful.

17.6 Quantum Copying

In this section we consider the problems associated with duplicating information in quantum computing. Copying is an extensively used operation in classical algorithms. In this section we show that copying of arbitrary states in quantum computing is not possible, and then consider limited copying processes.

Theorem. Given an arbitrary state, $|\psi\rangle$ no unitary matrix U exists such that $U|\psi, 0\rangle = |\psi, \psi\rangle$.

Proof. Suppose U does exist. Then for a state $|\mathbf{a}\rangle$ and a different state $|\mathbf{b}\rangle$ we have

$$U|\mathbf{a}, 0\rangle = |\mathbf{a}, \mathbf{a}\rangle, \qquad U|\mathbf{b}, 0\rangle = |\mathbf{b}, \mathbf{b}\rangle.$$

Now

$$U(|\mathbf{a}\rangle + |\mathbf{b}\rangle) \otimes |0\rangle = (|\mathbf{a}\rangle + |\mathbf{b}\rangle) \otimes (|\mathbf{a}\rangle + |\mathbf{b}\rangle)$$

and

$$U(|\mathbf{a}\rangle + |\mathbf{b}\rangle) \otimes |0\rangle = |\mathbf{a}\rangle \otimes |\mathbf{a}\rangle + |\mathbf{b}\rangle \otimes |\mathbf{b}\rangle.$$

This is a contradiction since in general

$$|\mathbf{a}\rangle \otimes |\mathbf{a}\rangle + |\mathbf{b}\rangle \otimes |\mathbf{b}\rangle \neq |\mathbf{a}\rangle \otimes |\mathbf{a}\rangle + \mathbf{a}\rangle \otimes |\mathbf{b}\rangle + \mathbf{b}\rangle \otimes |\mathbf{a}\rangle + \mathbf{b}\rangle \otimes |\mathbf{b}\rangle.$$

♠

This is called the *no-cloning theorem*. It means that in general it is impossible to make an exact copy of qubits and quantum registers. However it is simple to copy a quantum register with purely classical data (i.e. it is not in a superposition) with a CNOT gate for every qubit in the register.

Mozyrsky et al. [122] derived a Hamilton operator for copying the basis up and down states of a quantum two-state system – a qubit – onto n copy qubits ($n \geq 1$) initially prepared in the down state. The qubit states by quantum numbers are denoted by $q_j = 0$ (down) and $q_j = 1$ (up), for spin j. The states of the $n + 1$ spins will then be expanded in the basis

$$|q_1 q_2 \ldots q_{n-1}\rangle.$$

The copying process imposes the two conditions

$$|100\ldots0\rangle \rightarrow |111\ldots1\rangle$$

$$|000\ldots0\rangle \rightarrow |000\ldots0\rangle$$

up to possible phase factors. Therefore, a unitary transformation that corresponds to quantum evolution over the time interval Δt is not unique. Thus the Hamilton operator is not unique. One chooses a particular transformation that allows analytical calculation and, for $n = 1$, yields a controlled-NOT gates. They considered the following unitary transformation.

$$
\begin{aligned}
U &= e^{i\beta}|111\ldots1\rangle\langle100\ldots0| \\
&+ e^{i\rho}|000\ldots0\rangle\langle000\ldots0| + e^{i\alpha}|100\ldots0\rangle\langle111\ldots1| \\
&- \sum_{\{q_j\}} |q_1 q_2 q_3 \ldots q_{n-1}\rangle\langle q_1 q_2 q_3 \ldots q_{n-1}| \,.
\end{aligned}
$$

The sum in the fourth term, $\{q_j\}$, is over all the other quantum states of the system, i.e., excluding the three states

$$
|111\ldots1\rangle, \quad |100\ldots0\rangle, \quad |000\ldots0\rangle \,.
$$

The first two terms accomplish the desired copying transformation. The third term is needed for unitarity since the quantum evolution is reversible. General phase factors are allowed in these terms. Thus

$$
U|000\ldots0\rangle = e^{i\rho}|000\ldots0\rangle
$$

$$
U|100\ldots0\rangle = e^{i\beta}|111\ldots1\rangle \,.
$$

To calculate the Hamilton operator $\hat{H}$ according to

$$
U = e^{i\hat{H}\Delta t/\hbar}
$$

we diagonalize the unitary matrix U. The diagonalization is simple because we only have to work in the subspace of the three special states

$$
|111\ldots1\rangle, \quad |100\ldots0\rangle, \quad |000\ldots0\rangle \,.
$$

The part related to the state $|000\ldots0\rangle$ is diagonal. In the subspace labeled by $|111\ldots1\rangle$, $|100\ldots0\rangle$, $|000\ldots0\rangle$, in that order, the unitary matrix U is represented by the matrix

$$
U = \begin{pmatrix} 0 & e^{i\beta} & 0 \\ e^{i\alpha} & 0 & 0 \\ 0 & 0 & e^{i\rho} \end{pmatrix} \,.
$$

The eigenvalues of U are

$$
e^{i(\alpha-\beta)/2}, \qquad e^{i(\alpha+\beta)/2}, \qquad e^{i\rho} \,.
$$

Thus the eigenvalues of the Hamiton operator in the selected subspace are given by

$$E_1 = \frac{\hbar}{2\Delta t}(\alpha + \beta) + \frac{2\pi\hbar}{\Delta t}N_1$$

$$E_2 = -\frac{\hbar}{2\Delta t}(\alpha + \beta) + \frac{2\pi\hbar}{\Delta t}\left(N_2 + \frac{1}{2}\right)$$

$$E_3 = -\frac{\hbar}{\Delta t}\rho - \frac{2\pi\hbar}{\Delta t}N_3 .$$

Universal optimum cloning [33, 35] is an attempt to provide the best copy of an arbitrary quantum state given the constraints of quantum mechanics. The specific constraints specified for the copy operation are given as follows

1. The density operators of the source and destination states must be identical after the copy operation

2. All pure states should copy equally well. This can be implemented, for example, by requiring that, for a certain distance measure, the copied state is always a fixed distance d from the original pure state.

3. The distance between the state to be copied and the copy must be a minimum. The distance between the original state before and after copying must also be minimized.

Using the *Bures distance* for density operators,

$$d_B(\rho_1, \rho_2) := \sqrt{2}\left(1 - \mathrm{tr}\left(\sqrt{\rho_1^{1/2}\rho_2\rho_1^{1/2}}\right)\right)^{1/2}$$

Bužek and Hillery [35] found the following transformations satisfy the given constraints

$$U_{OQC}|0\rangle \otimes |j\rangle \otimes |q\rangle := \sqrt{\frac{2}{3}}|j\rangle \otimes |j\rangle \otimes |\alpha_j\rangle + \sqrt{\frac{1}{6}}(|0\rangle \otimes |1\rangle + |1\rangle \otimes |0\rangle) \otimes |\beta_{1-j}\rangle$$

where $|q\rangle$ is the initial state of the ancillary system used in the copying process and $|\alpha_0\rangle$ and $|\alpha_1\rangle$ are orthonormal states in the Hilbert space of the ancillary system. Using a slightly different approach Bruß et al. [33] found the same transformation.

17.7 Example Programs

In the following we give an implementation of the decomposition of a non-entangled state [165] as described earlier in the chapter.

Let us assume the state $|\mathbf{u}\rangle \in \mathbf{C}^2$ is not entangled, i.e. $u_1 u_4 = u_2 u_3$. The C++ program decompose.cpp will calculate the decomposition into $|\mathbf{x}\rangle$ and $|\mathbf{y}\rangle$ using (17.18)–(17.21) assuming $|\mathbf{u}\rangle$ is normalized. We use a two-dimensional array of data type double to represent the state $|\mathbf{u}\rangle$ and an array of two double variables to represent the real and imaginary parts of the complex numbers. Owing to the numerical calculation of $|\mathbf{x}\rangle$ and $|\mathbf{y}\rangle$ these states can contain small rounding errors.

```cpp
// decompose.cpp

#include <iostream>
#include <cmath>
using namespace std;

void factor(double x[2][2],double y[2][2],double u[4][2])
{
 double x1n,x2n,y1n,y2n,u1s,u2s,u3s,u4s;

 u1s = u[0][0]*u[0][0]+u[0][1]*u[0][1];
 u2s = u[1][0]*u[1][0]+u[1][1]*u[1][1];
 u3s = u[2][0]*u[2][0]+u[2][1]*u[2][1];
 u4s = u[3][0]*u[3][0]+u[3][1]*u[3][1];

 x1n = sqrt(u1s+u2s); x2n = sqrt(u3s+u4s);
 y1n = sqrt(u1s+u3s); y2n = sqrt(u2s+u4s);

 double au1,au2,au3,a4;

 if(u1s==0.0) au1 = 0.0; else
 au1 = acos(u[0][0]/(sqrt(u1s)));
 if(u2s==0.0) au2 = 0.0; else
 au2 = acos(u[1][0]/(sqrt(u2s)));
 if(u3s==0.0) au3 = 0.0; else
 au3 = acos(u[2][0]/(sqrt(u3s)));
 a4 = au3-au1;

 x[0][0] = x1n;        x[0][1] = 0.0;
 x[1][0] = x2n*cos(a4); x[1][1] = x2n*sin(a4);
 y[0][0] = y1n*cos(au1); y[0][1] = y1n*sin(au1);
 y[1][0] = y2n*cos(au2); y[1][1] = y2n*sin(au2);
}


void displayfactor(double u[4][2])
{
```

```
int i;
double x[2][2],y[2][2];

cout << "u = ( "; for(i=0;i<4;i++) cout << u[i][0] << "+"
    << u[i][1] << "i "; cout << ")" << endl;
factor(x,y,u);
cout << "x = ( " << x[0][0] << "+" << x[0][1] << "i " << x[1][0]
    << "+" << x[1][1] << "i )" << endl;
cout << "y = ( " << y[0][0] << "+" << y[0][1] << "i " << y[1][0]
    << "+" << y[1][1] << "i )" << endl << endl;
}


// u[4] represents 4 complex numbers
// where u[j][0] is the real part of u[j]
// and u[j][1] is the imaginary part of u[j]
int main()
{
 double u[4][2];
 // ( 0.5 0.5 0.5 0.5 )
 u[0][0] = 0.5;u[0][1] = 0;u[1][0] = 0.5;u[1][1] = 0.0;
 u[2][0] = 0.5;u[2][1] = 0;u[3][0] = 0.5;u[3][1] = 0.0;
 displayfactor(u);
 // ( 0.5 -0.5 -0.5 0.5 )
 u[0][0] = 0.5; u[0][1] = 0.0;u[1][0] = -0.5;u[1][1] = 0.0;
 u[2][0] = -0.5;u[2][1] = 0.0;u[3][0] = 0.5; u[3][1] = 0.0;
 displayfactor(u);
 // i is equivalent to (0,1) in the implementation
 // ( i/sqrt(2) 1/sqrt(2) 0 0 )
 u[0][0] = 0.0;u[0][1] = 1/sqrt(2);
 u[1][0] = 1.0/sqrt(2);u[1][1] = 0.0;
 u[2][0] = 0.0;u[2][1] = 0.0;
 u[3][0] = 0.0;u[3][1] = 0.0;
 displayfactor(u);
 // ( 0.7 0.3 2.1 0.9 ) once normalized
 double size=sqrt(0.7*0.7 + 0.3*0.3 + 2.1*2.1 + 0.9*0.9);
 u[0][0] = 0.7/size; u[0][1] = 0;
 u[1][0] = 0.3/size; u[1][1] = 0.0;
 u[2][0] = 2.1/size; u[2][1] = 0;
 u[3][0] = 0.9/size; u[3][1] = 0.0;
 displayfactor(u);
 return 0;
 }
```

We consider now three applications for quantum computing [166] and give the simulation using SymbolicC++ [169]. First we show how entangled states can be generated from unentangled states using unitary transformations. The quantum circuit is also given. Next we consider a quantum circuit for swapping two bits. The third application deals with teleporation [17, 28]. Finally, we consider the Greenberger-Horne-Zeilinger state [96]. Then we provide the SymbolicC++ [169] implementation of these applications.

In our first example we start from standard basis (unentangled states) in the Hilbert space $\mathbf{C}^4$ and transform them into the Bell states. The Bell states are defined as

$$\Phi^+ = \frac{1}{\sqrt{2}}(|00\rangle + |11\rangle) \equiv \frac{1}{\sqrt{2}}\begin{pmatrix} 1 \\ 0 \\ 0 \\ 1 \end{pmatrix}, \qquad \Phi^- = \frac{1}{\sqrt{2}}(|00\rangle - |11\rangle) \equiv \frac{1}{\sqrt{2}}\begin{pmatrix} 1 \\ 0 \\ 0 \\ -1 \end{pmatrix},$$

$$\Psi^+ = \frac{1}{\sqrt{2}}(|01\rangle + |10\rangle) \equiv \frac{1}{\sqrt{2}}\begin{pmatrix} 0 \\ 1 \\ 1 \\ 0 \end{pmatrix}, \qquad \Psi^- = \frac{1}{\sqrt{2}}(|01\rangle - |10\rangle) \equiv \frac{1}{\sqrt{2}}\begin{pmatrix} 0 \\ 1 \\ -1 \\ 0 \end{pmatrix}.$$

The Bell states also form a basis in $\mathbf{C}^4$. They are entangled. Entangled states exhibit nonlocal correlations. This means that two entangled systems which have interacted in the past and are no longer interacting still show correlations. These correlations are used for example in dense coding and quantum error-correction techniques [162, 17]. To transform the standard basis into the Bell states we apply the following two unitary transformations. The first unitary transformation is given by

$$U_H \otimes I_2$$

where I_2 is the 2×2 unit matrix. Our second unitary transformation is U_{XOR}. Applying these two unitary matrices to the states $|00\rangle$, $|01\rangle$, $|10\rangle$ and $|11\rangle$ we find

$$U_{XOR}(U_H \otimes I_2)|00\rangle = \Phi^+, \qquad U_{XOR}(U_H \otimes I_2)|01\rangle = \Psi^+,$$

$$U_{XOR}(U_H \otimes I_2)|10\rangle = \Phi^-, \qquad U_{XOR}(U_H \otimes I_2)|11\rangle = \Psi^-.$$

These operations can be represented by the quantum circuit

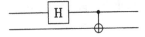

Figure 17.6: Quantum Circuit to Generate Bell States

As our second example we consider the swapping of a pair of bits. The circuit for swapping a pair of bits is given by

Figure 17.7: Quantum Circuit to Swap a Pair of Bits

This circuit is represented by the product of the three permutation matrices

$$U_{XOR} \begin{pmatrix} 1 & 0 & 0 & 0 \\ 0 & 0 & 0 & 1 \\ 0 & 0 & 1 & 0 \\ 0 & 1 & 0 & 0 \end{pmatrix} U_{XOR}.$$

Thus we find the permutation matrix U_{EXCH}. This permutation matrix cannot be represented as the Kronecker product of 2×2 matrices.

Finally we consider the Greenberger-Horne-Zeilinger (GHZ) state [96]. This state is an entangled superposition of three qubits and is given by

$$|\Psi\rangle_{GHZ} := \frac{1}{\sqrt{2}}(|000\rangle + |111\rangle) \equiv \frac{1}{\sqrt{2}}(|0\rangle \otimes |0\rangle \otimes |0\rangle + |1\rangle \otimes |1\rangle \otimes |1\rangle).$$

Thus in the Hilbert space $\mathbf{C}^8$ we have

$$|\Psi\rangle_{GHZ} = \frac{1}{\sqrt{2}}(1\,0\,0\,0\,0\,0\,0\,1)^T$$

where ^{T} stands for transpose. If we consider 000 and 111 to be the binary representation of "0" and "7", respectively, the GHZ state simply represents the coherent superposition $1/\sqrt{2}(|"0"\rangle + |"7"\rangle)$. In this state all three qubits are either 0 or 1 but none of the qubits has a well-defined value of its own. Measurement of any one qubit will immediately result in the other two qubits attaining the same value. For example

$$\left(\sqrt{2}\langle 0| \otimes I_2 \otimes I_2\right)\left(\frac{1}{\sqrt{2}}(|0\rangle \otimes |0\rangle \otimes |0\rangle + |1\rangle \otimes |1\rangle \otimes |1\rangle)\right) = |0\rangle \otimes |0\rangle.$$

The implementation in SymbolicC++ [169] is as follows. The **Matrix** class of SymbolicC++ includes the method **kron** for the Kronecker product of two matrices and the method **dsum** for the direct sum of two matrices. The overloaded operators * and + are used for matrix multiplication and addition. The identity matrix is also implemented. Thus the code for the three quantum circuits is as follows.

```
// qthree.cpp

#include <iostream>
#include "Vector.h"
#include "Matrix.h"
#include "Rational.h"
#include "Msymbol.h"
using namespace std;

typedef Sum<Rational<int> > C;

template <class T> Vector<T> Hadamard(Vector<T> v)
{
 assert(v.length() == 2);
 Matrix<T> H(2,2);
 H[0][0] = T(1)/sqrt(T(2)); H[0][1] = T(1)/sqrt(T(2));
 H[1][0] = T(1)/sqrt(T(2)); H[1][1] = T(-1)/sqrt(T(2));
 return (H*v);
}

template <class T> Vector<T> XOR(Vector<T> v)
{
 assert(v.length() == 4);
 Matrix<T> X(4,4);
 X[0][0] = T(1); X[0][1] = T(0); X[0][2] = T(0); X[0][3] = T(0);
 X[1][0] = T(0); X[1][1] = T(1); X[1][2] = T(0); X[1][3] = T(0);
 X[2][0] = T(0); X[2][1] = T(0); X[2][2] = T(0); X[2][3] = T(1);
 X[3][0] = T(0); X[3][1] = T(0); X[3][2] = T(1); X[3][3] = T(0);
 return (X*v);
```

```
}

template <class T> Vector<T> Bell(Vector<T> v)
{
 assert(v.length() == 4);

 Matrix<T> I(2,2),H(2,2),X(4,4);
 I.identity();

 H[0][0] = T(1)/sqrt(T(2)); H[0][1] = T(1)/sqrt(T(2));
 H[1][0] = T(1)/sqrt(T(2)); H[1][1] = T(-1)/sqrt(T(2));

 Matrix<T> UH=kron(H,I);

 X[0][0] = T(1); X[0][1] = T(0); X[0][2] = T(0); X[0][3] = T(0);
 X[1][0] = T(0); X[1][1] = T(1); X[1][2] = T(0); X[1][3] = T(0);
 X[2][0] = T(0); X[2][1] = T(0); X[2][2] = T(0); X[2][3] = T(1);
 X[3][0] = T(0); X[3][1] = T(0); X[3][2] = T(1); X[3][3] = T(0);

 return (X*(UH*v));
}

template <class T> Vector<T> Swap(Vector<T> v)
{
 assert(v.length()==4);
 Matrix<T> S(4,4);
 S[0][0] = T(1); S[0][1] = T(0); S[0][2] = T(0); S[0][3] = T(0);
 S[1][0] = T(0); S[1][1] = T(0); S[1][2] = T(0); S[1][3] = T(1);
 S[2][0] = T(0); S[2][1] = T(0); S[2][2] = T(1); S[2][3] = T(0);
 S[3][0] = T(0); S[3][1] = T(1); S[3][2] = T(0); S[3][3] = T(0);
 return XOR(S*XOR(v));
}

template <class T> Vector<T> Teleport(Vector<T> v)
{
 int i;
 assert(v.length() == 8);
 Vector<T> result;
 Matrix<T> NOT(2,2),H(2,2),I(2,2),X(4,4);

 NOT[0][0] = T(0); NOT[0][1] = T(1);
 NOT[1][0] = T(1); NOT[1][1] = T(0);

 H[0][0] = T(1)/sqrt(T(2)); H[0][1] = T(1)/sqrt(T(2));
 H[1][0] = T(1)/sqrt(T(2)); H[1][1] = T(-1)/sqrt(T(2));

 I.identity();
```

```
X[0][0] = T(1); X[0][1] = T(0); X[0][2] = T(0); X[0][3] = T(0);
X[1][0] = T(0); X[1][1] = T(1); X[1][2] = T(0); X[1][3] = T(0);
X[2][0] = T(0); X[2][1] = T(0); X[2][2] = T(0); X[2][3] = T(1);
X[3][0] = T(0); X[3][1] = T(0); X[3][2] = T(1); X[3][3] = T(0);

  Matrix<T> U1=kron(I,kron(H,I));
  Matrix<T> U2=kron(I,X);
  Matrix<T> U3=kron(X,I);
  Matrix<T> U4=kron(H,kron(I,I));
  Matrix<T> U5=kron(I,X);
  Matrix<T> U6=kron(I,kron(I,H));
  Matrix<T> U7=dsum(I,dsum(I,dsum(NOT,NOT)));
  Matrix<T> U8=kron(I,kron(I,H));

  result=U8*(U7*(U6*(U5*(U4*(U3*(U2*(U1*v)))))));
  for(i=0;i<8;i++)
  {
   while(result[i].put(power(sqrt(T(2)),-6),power(T(2),-3)));
   while(result[i].put(power(sqrt(T(2)),-4),power(T(2),-2)));
   while(result[i].put(power(sqrt(T(2)),-2),power(T(2),-1)));
  }
  return result;
}


// The outcome after measuring value for qubit.
// Since the probabilities may be symbolic this function
// cannot simulate a measurement where random outcomes
// have the correct distribution
template <class T>
Vector<T> Measure(Vector<T> v,unsigned int qubit,unsigned int value)
{
 assert(pow(2,qubit)<v.length());
 assert(value==0 || value==1);
 int i,len,skip = 1-value;
 Vector<T> result(v);
 T D = T(0);

 len = v.length()/int(pow(2,qubit+1));
 for(i=0;i<v.length();i++)
 {
  if(!(i%len)) skip = 1-skip;
  if(skip) result[i] = T(0);
  else D += result[i]*result[i];
 }
 result/=sqrt(D);
 return result;
}
```

```
// for output clarity
ostream &print(ostream &o,Vector<C> v)
{
 char *b2[2]={"|0>","|1>"};
 char *b4[4]={"|00>","|01>","|10>","|11>"};
 char *b8[8]={"|000>","|001>","|010>","|011>",
              "|100>","|101>","|110>","|111>"};
 char **b,i;

 if(v.length()==2) b=b2;
 if(v.length()==4) b=b4;
 if(v.length()==8) b=b8;

 for(i=0;i<v.length();i++)
  if(!v[i].is_Number() || v[i].nvalue()!=C(0))
    o << "+(" << v[i] << ")" << b[i];
 return o;
}


void main(void)
{
 Vector<C> zero(2),one(2);
 Vector<C> zz(4),zo(4),oz(4),oo(4),qreg;
 Vector<C> tp00,tp01,tp10,tp11,psiGHZ;
 Sum<Rational<int> > a("a",0),b("b",0);
 int i;

 zero[0] = C(1); zero[1] = C(0);
 one[0]  = C(0); one[1]  = C(1);
 zz = kron(Matrix<C>(zero),Matrix<C>(zero))(0);
 zo = kron(Matrix<C>(zero),Matrix<C>(one))(0);
 oz = kron(Matrix<C>(one),Matrix<C>(zero))(0);
 oo = kron(Matrix<C>(one),Matrix<C>(one))(0);

 cout << "UH|0> = "; print(cout,Hadamard(zero))<< endl;
 cout << "UH|1> = "; print(cout,Hadamard(one)) << endl;
 cout << endl;
 cout << "UXOR|00> = "; print(cout,XOR(zz)) << endl;
 cout << "UXOR|01> = "; print(cout,XOR(zo)) << endl;
 cout << "UXOR|10> = "; print(cout,XOR(oz)) << endl;
 cout << "UXOR|11> = "; print(cout,XOR(oo)) << endl;
 cout << endl;
 cout << "UBELL|00> = "; print(cout,Bell(zz)) << endl;
 cout << "UBELL|01> = "; print(cout,Bell(zo)) << endl;
 cout << "UBELL|10> = "; print(cout,Bell(oz)) << endl;
 cout << "UBELL|11> = "; print(cout,Bell(oo)) << endl;
 cout << endl;
 cout << "USWAP|00> = "; print(cout,Swap(zz)) << endl;
```

```
   cout << "USWAP|01> = "; print(cout,Swap(zo)) << endl;
   cout << "USWAP|10> = "; print(cout,Swap(oz)) << endl;
   cout << "USWAP|11> = "; print(cout,Swap(oo)) << endl;
   cout << endl;

   qreg=kron(a*zero+b*one,kron(zero,zero))(0);
   cout << "UTELEPORT("; print(cout,qreg) << ") = ";
    print(cout,qreg=Teleport(qreg)) << endl;
   cout << "Results after measurement of first 2 qubits:" << endl;
   tp00 = Measure(Measure(qreg,0,0),1,0);
   tp01 = Measure(Measure(qreg,0,0),1,1);
   tp10 = Measure(Measure(qreg,0,1),1,0);
   tp11 = Measure(Measure(qreg,0,1),1,1);
   for(i=0;i<8;i++)
   {
    while(tp00[i].put(a*a,C(1)-b*b));
    while(tp00[i].put(power(sqrt(C(1)/C(2)),-2),C(2)));
    while(tp01[i].put(a*a,C(1)-b*b));
    while(tp01[i].put(power(sqrt(C(1)/C(2)),-2),C(2)));
    while(tp10[i].put(a*a,C(1)-b*b));
    while(tp10[i].put(power(sqrt(C(1)/C(2)),-2),C(2)));
    while(tp11[i].put(a*a,C(1)-b*b));
    while(tp11[i].put(power(sqrt(C(1)/C(2)),-2),C(2)));
   }
   cout << " |00> : " ; print(cout,tp00) << endl;
   cout << " |01> : " ; print(cout,tp01) << endl;
   cout << " |10> : " ; print(cout,tp10) << endl;
   cout << " |11> : " ; print(cout,tp11) << endl;
   cout << endl;

   psiGHZ=(kron(Matrix<C>(zz),Matrix<C>(zero))/sqrt(C(2))
          +kron(Matrix<C>(oo),Matrix<C>(one))/sqrt(C(2)))(0);
   cout << "Greenberger-Horne-Zeilinger state : ";
   print(cout,psiGHZ) << endl;
   cout << "Measuring qubit 0 as 1 yields : ";
   print(cout,Measure(psiGHZ,0,1)) <<endl;
   cout << "Measuring qubit 1 as 1 yields : ";
   print(cout,Measure(psiGHZ,1,1)) <<endl;
   cout << "Measuring qubit 2 as 0 yields : ";
    print(cout,Measure(psiGHZ,2,0)) <<endl;
}
```

The program generates the following output:

```
UH|0> = +(sqrt(2)^(-1))|0>+(sqrt(2)^(-1))|1>
UH|1> = +(sqrt(2)^(-1))|0>+(-sqrt(2)^(-1))|1>

UXOR|00> = +(1)|00>
```

```
UXOR|01> = +(1)|01>
UXOR|10> = +(1)|11>
UXOR|11> = +(1)|10>


UBELL|00> = +(sqrt(2)^(-1))|00>+(sqrt(2)^(-1))|11>
UBELL|01> = +(sqrt(2)^(-1))|01>+(sqrt(2)^(-1))|10>
UBELL|10> = +(sqrt(2)^(-1))|00>+(-sqrt(2)^(-1))|11>
UBELL|11> = +(sqrt(2)^(-1))|01>+(-sqrt(2)^(-1))|10>


USWAP|00> = +(1)|00>
USWAP|01> = +(1)|10>
USWAP|10> = +(1)|01>
USWAP|11> = +(1)|11>


UTELEPORT(+(a)|000>+(b)|100>)  = +(1/2*a)|000>+(1/2*b)|001>
                                 +(1/2*a)|010>+(1/2*b)|011>
                                 +(1/2*a)|100>+(1/2*b)|101>
                                 +(1/2*a)|110>+(1/2*b)|111>
Results after measurement of first 2 qubits:
  |00> : +(a)|000>+(b)|001>
  |01> : +(a)|010>+(b)|011>
  |10> : +(a)|100>+(b)|101>
  |11> : +(a)|110>+(b)|111>


Greenberger-Horne-Zeilinger state : +(sqrt(2)^(-1))|000>
                                    +(sqrt(2)^(-1))|111>
Measuring qubit 0 as 1 yields : +(sqrt(2)^(-1)
                                *sqrt(sqrt(2)^(-2))^(-1))|111>
Measuring qubit 1 as 1 yields : +(sqrt(2)^(-1)
                                *sqrt(sqrt(2)^(-2))^(-1))|111>
Measuring qubit 2 as 0 yields : +(sqrt(2)^(-1)
                                *sqrt(sqrt(2)^(-2))^(-1))|000>
```

In 1996 Schack and Brun [141] described a powerful C++ library for solving quantum systems. The core of the library are the C++ classes State and Operator, which represent state vectors and operators in Hilbert space. However the disadvantage of this C++ library is that the constants (for example the coupling constant in a Hamilton operator) can only be treated numerically, i.e. it is of data type double. In SymbolicC++ we can treat constants either symbolically or numerically. Using the method set we can switch from a symbolic representation of a constant to a numeric representation. Using the approach of Schack and Brun it is also difficult to construct the CNOT operator on any two qubits of a state. In 1995, a year before the paper of Schack and Brun [141], Steeb [159] described a computer algebra package based on Reduce and Lisp that can handle Bose, Fermi and coupled Bose-Fermi systems. Since spin operators can be expressed with Fermi operators, the package can also deal with spin-systems. It also has the advantage that constants can be treated either numerically or symbolically.

Two other simulations are described by Ömer [124] and Pritzker [134]. Both are implemented in C++. However, these implementations can also only use numeric representations and not symbolic representations. None of them implement the Kronecker product and direct sum to aid the construction of operators such as we have used for the simulation of teleportation.

The OpenQubit simulation [134] implements the classes QState, which represents the state of the entire quantum computer, and QRegister, which refers to specific qubits from QState to be used as a quantum register. Further support for quantum algorithms are provided by four operators denoted by R_x, R_y, Ph and $CNot$ which are rotations, phase changes and the controlled NOT gate. The implementation supports the simulation of measurement. Shor's factoring algorithm has been successfully implemented using this system.

The simulation described by Ömer [124] attempts to reduce the requirements on the classical computer used for simulation by reducing the storage requirements for states, using a bitvector and by never storing zero amplitudes for states as well as using other techniques such as hashing. The class quBaseState represents the state of the quantum computer. The class quSubState references qubits from a quBaseState, and provides access to the registers of the quantum computer. The system provides a number of methods to decribe operators. The class opMatrix represents an arbitrary $2^n \times 2^n$ matrix. The class opEmbedded is used to describe operators which are applied to subspaces of the quantum system, and the class opPermutation is used for the permutation operators. The system provides operators for the identity, arbitrary single qubit transformations, the identity operation, the qubit swapping operation, controlled NOT, Toffoli gate, and phase change amongst others. Shor's algorithm has also been illustrated in this system.

Chapter 18
Measurement and Quantum States

18.1 Introduction

The interpretation of measurements in quantum mechanics is still under discussion (Healey [84], Bell [9], Redhead [136]). Besides the Copenhagen interpretation we have the many-worlds interpretations (Everett interpretations), the modal interpretations, the decoherence interpretations, the interpretations in terms of (nonlocal) hidden variables, the quantum logical interpretations.

A satisfactory interpretation of quantum mechanics would involve several things. It would provide a way of understanding the central notions of the theory which permits a clear and exact statement of its key principles. It would include a demonstration that, with this understanding, quantum mechanics is a consistent, empirically adequate, and explanatorily powerful theory. And it would give a convincing and natural resolution of the paradoxes. A satisfactory interpretation of quantum mechanics should make it clear what the world would be like if quantum mechanics were true. But this further constraint would not be neutral between different attempted interpretations. There are those, particularly in the Copenhagen tradition, who would reject this further constraint on the grounds that, in their view, quantum mechanics should not be taken to describe (microscopic) reality, but only our intersubjectively communicable experimental observations of it. It would therefore be inappropriate to criticize a proposed interpretation solely on the grounds that it does not meet this last constraint. But this constraint will certainly appeal to philosophical realists, and for them at least it should count in favour of an interpretation if it meets this constraint.

It is well known that the conceptual foundations of quantum mechanics have been plagued by a number of paradoxes, or conceptual puzzles, which have attracted a host of mutually incompatible attempted resolutions – such as that presented by Schrödinger [143], popularly known as the paradox of Schrödinger's cat, and the EPR paradox, named after the last initials of its authors, Einstein, Podolsky, and Rosen [60].

18.2   Measurement Problem

Consider a spin-$\frac{1}{2}$ particle initially described by a superposition of eigenstates of S_z, the z component of spin:

$$|\Phi\rangle = c_1|S_z =\uparrow\rangle + c_2|S_z =\downarrow\rangle, \qquad c_1, c_2 \in \mathbf{C}$$

where

$$\langle\uparrow= S_z|S_z =\uparrow\rangle = 1, \qquad \langle\downarrow= S_z|S_z =\downarrow\rangle = 1, \qquad \langle\uparrow= S_z|S_z =\downarrow\rangle = 0$$

and

$$\langle\Phi|\Phi\rangle = 1.$$

Thus

$$|c_1|^2 + |c_2|^2 = 1.$$

Let $|R =\uparrow\rangle$ and $|R =\downarrow\rangle$ denote the up and down pointer-reading eigenstates of an S_z-measuring apparatus. Thus $|R =\uparrow\rangle$ and $|R =\downarrow\rangle$ are eigenstates of the operator $\hat{R}$. According to quantum mechanics (with no wave-function collapse), if the apparatus ideally measures the particle, the combined system evolves into an entangled superposition

$$|\varphi\rangle = c_1|S_z =\uparrow\rangle \otimes |R =\uparrow\rangle + c_2|S_z =\downarrow\rangle \otimes |R =\downarrow\rangle.$$

Common sense insists that after the measurement, the pointer reading is definite. According to the orthodox value-assignment rule, however, the pointer reading is definite only if the quantum state is an eigenstate of

$$I \otimes \hat{R}$$

the pointer-reading operator, where I is the identity operator. Since $|\varphi\rangle$ is not an eigenstate of $I \otimes \hat{R}$, the pointer reading is indefinite. The interpretations of quantum mechanics mentioned above attempt to deal with this aspect of the measurement problem. However their solutions run into a technical difficulty which is called the *basis degeneracy problem*.

18.3 Copenhagen Interpretation

In the Copenhagen view, the Born rules explicitly concern the probabilities for various possible measurement results (Healey [84]). They do not concern possessed values of dynamical variables. On each system there will always be some dynamical variables which do not possess precise values. In the Copenhagen interpretation, the Born rules assign probabilities. They have the form

$$\text{prob}_\psi(A \in \Omega) = p.$$

Here p is a real number between zero and one (including those limits), A is a quantum dynamical variable, Ω is a (Borel) set of real numbers, and ψ is a mathematical representative of an instantaneous quantum state. In quantum state ψ, the probability of finding that the value of A lies in Ω is p. How is the phrase "of finding" to be understood? This probability is calculated according to the appropriate quantum algorithm. For example,

$$\text{prob}_\psi(A \in \Omega) = \langle \psi, \hat{P}^A(\Omega)\psi \rangle$$

where ψ is the system's state vector, and $\hat{P}^A(\Omega)$ is the projection operator corresponding to the property $A \in \Omega$. On the present interpretation, a quantum state may be legitimately ascribed to a single quantum system (and not just to a large ensemble of similar systems), but only in certain circumstances. A system does not always have a quantum state. These circumstances are not universal. Nevertheless, every quantum system always has a dynamical state. Consequently, there can be no general identification between a system's quantum state and its dynamical state; nor is it always true that one determines the other.

The Born rules apply directly to possessed values of quantities, and only derivatively to results of measurements of these quantities. In this view every quantum dynamical variable always has a precise real value on any quantum system to which it pertains, and the Born rules simply state the probability for that value to lie in any given interval. Thus the Born rules assign probabilities to events involving a quantum system σ of the form "The value of A on σ lies in Ω". A properly conducted measurement of the value of A on σ would find that value in Ω just in case the value actually lies in Ω.

Since the statement of the Born rules then involves explicit reference to measurement (or observation), to complete the interpretation it is necessary to say what constitutes a measurement. Proponents of the Copenhagen interpretation have typically either treated measurement (or observation) or cognates as primitive terms in quantum mechanics, or else have taken each to refer vaguely to suitable interactions involving a classical system. If measurement remains a primitive term, then it is natural to interpret it epistemologically as referring to an act of some observer which, if successful, gives him or her knowledge of some structural feature of a phenomenon. But then, quantum mechanics seems reduced to a tool for predicting what is likely to be observed in certain (not very precisely specified) circumstances, with nothing to say about the events in the world which are responsible for the results of those observations we make, and with no interesting implications for a world without observers. This instrumentalist/pragmatist conception of quantum mechanics has often gone along with the Copenhagen interpretation. On the other hand, if a measurement is a suitable interaction with a classical system, we need to know what interactions are suitable.

In what we call the weak version of the Copenhagen interpretation, the dynamical properties of an individual quantum system are fully specified by means of its quantum state. A dynamical variable A possesses a precise real value a_i on a system if and only if that system is describable by a quantum state for which the Born rules assign probability one to the value a_i of A. In that state, a measurement of A would certainly yield the value a_i. In other states, for which there is some chance that value a_i would result if A were measured, and some chance that it would not, it is denied that A has any precise value prior to an actual measurement. Within the limits of experimental accuracy, measurement of a dynamical variable always yields a precise real value as its result, and this raises the question of the significance to be attributed to this value, given that it is typically not the value the variable possessed just before the measurement, nor the value it would have had if no measurement had taken place. Thus the measured variable acquires the measured value as a result of the measurement. Then the Born rules explicitly concern the probabilities that dynamical variables acquire certain values upon measurement. By ascribing a precise real value to a variable given earlier, one concludes that after a precise measurement of a dynamical variable, a system is describable by a quantum state for which the Born rules assign probability one to the measured value of that variable.

Since, in this version, measurement effects significant changes in the dynamical properties of a system, it is important for a proponent of the interpretation to specify in just what circumstances such changes occur. One might expect that such a specification would be forthcoming in purely quantum mechanical terms, through a quantum mechanical account of measuring interactions. Such an account would show how a physical interaction between one quantum system and another, which proceeds wholly in accordance with the principles of quantum mechanics, can effect a correlation between an initial value of the measured variable on one system (the object system) and a final recording property on the other (apparatus) system. The problem of giving such an account has become known as the quantum measurement problem. A solution to the measurement problem would explain the reference to measurement in the Born rules in purely physical (quantum mechanical) terms, and would also show to what extent the projection postulate may be considered a valid principle of quantum mechanics. The key difficulty may be stated quite simply. It is that many initial states of an object system give rise to final compound object+apparatus quantum states which, in the present interpretation, imply that the apparatus fails to register any result at all. For, in such a final compound quantum state, the Born rules do not assign probability one to any recording property of the apparatus system (Healey [84]).

18.4 Hidden Variable Theories

A motivation behind the construction of such theories has been the belief that some more complete account of microscopic processes is required than that provided by quantum mechanics according to the Copenhagen interpretation (Healey [84]). The general idea has been to construct such an account by introducing additional quantities, over and above the usual quantum dynamical variables (such as de Broglie's pilot wave, Bohm's quantum potential, or fluctuations in Vigier's random ether), and additional dynamical laws governing these quantities and their coupling to the usual quantum variables. The primary object is to permit the construction of a detailed dynamical history of each individual quantum system which would underlie the statistical predictions of quantum mechanics concerning measurement results. Though it would be consistent with this aim for such dynamical histories to conform only to indeterministic laws, it has often been thought preferable to consider in the first instance deterministic hidden variable theories. A deterministic hidden variable theory would underlie the statistical predictions of quantum mechanics much as classical mechanics underlies the predictions of classical statistical mechanics. In both cases, the results of the statistical theory would be recoverable after averaging over ensembles of individual systems, provided that these ensembles are sufficiently typical: but the statistical theory would give demonstrably incorrect predictions for certain atypical ensembles.

Bell [9] showed that no deterministic hidden variable theory can reproduce the predictions of quantum mechanics for certain composite systems without violating a principle of locality. This principle is based on basic assumptions concerning the lack of physical connection between spatially distant components of such systems; and the impossibility of there being any such connection with the property that a change in the vicinity of one component should instantaneously produce a change in the behaviour of the other. Further work attempting to extend Bell's result to apply to indeterministic hidden variable theories has shown that there may be a small loophole still open for the construction of such a theory compatible with the relativistic requirement that no event affects other events outside of its future light-cone.

Existing hidden variable theories, such as that of Vigier [179], are explicitly nonlocal, and do involve superluminal propagation of causal influence on individual quantum systems, although it is held that exploiting such influences to transmit information superluminally would be extremely difficult, if not impossible. Any superluminal transmission of causal signals would be explicitly inconsistent with relativity theory. If this were so, such nonlocal hidden variable theories could be immediately rejected on this ground alone. Relativity does not explicitly forbid such transmission. Nonlocal hidden variable theories like that of Vigier can conform to the letter of relativity by introducing a preferred frame, that of the subquantum ether, with respect to which superluminal propagation is taken to occur. By doing so they avoid the generation of so-called causal paradoxes. However they violate the spirit of relativity theory by reintroducing just the sort of privileged reference frame. The principle that a fundamental theory can be given a relativistically invariant formulation seems so fundamental to contemporary physics that no acceptable interpretation of quantum mechanics should violate it.

A hidden variable theory is a separate and distinct theory from quantum mechanics. To offer such a theory is not to present an interpretation of quantum mechanics but to change the subject. One reason is that a hidden variable theory incorporates quantities additional to the quantum dynamical variables. Another is that hidden variable theories are held to underlie quantum mechanics in a way similar to that in which classical mechanics underlies the distinct theory of statistical mechanics. A final reason is that a hidden variable theory (at least typically) is held to be empirically equivalent to quantum mechanics only with respect to a restricted range of conceivable experiments, while leading to conflicting predictions concerning a range of possible further experiments which may, indeed, be extremely hard to actualize.

18.5   Everett Interpretation

Everett's interpretation has proven most influential in the development of the present interactive interpretation (Everett [64], Bell [9], Healey [84]). The Everett interpretation may be regarded as the prototype of all interactive interpretations, since it was the earliest and most influential attempt to treat measurement as a physical interaction internal to a compound quantum system, one component of which represents the observer or measuring apparatus. The Everett interpretation, like the present interactive interpretation, rejects the projection postulate. Both interpretations maintain that all interactions, including measurement interactions, may be treated as internal to a compound system, the universe, whose state evolves always in accordance with a deterministic law such as the time-dependent Schrödinger equation. Both deny that it is necessary to appeal to any extra quantum-mechanical notions such as that of a classical system, or an observer, in order to give a precise and empirically adequate quantum mechanical model of a measurement interaction. Finally, both interpretations undertake to explain how, and to what extent, quantum interactions internal to a compound system can come to mimic the effects of the projection postulate.

According to Everett, all observers correspond to quantum systems, which may be called, for convenience, apparatus systems. An observation or measurement is simply a quantum interaction of a certain type between an apparatus system α and an object system σ, which (provided this compound system is isolated) proceeds in accordance with the time-dependent Schrödinger equation governed by the Hamilton operator for the pair of systems concerned. In particular, for a good observation of a dynamical variable A whose associated self-adjoint operator $\hat{A}$ has a complete set of eigenvectors $\{ |\phi_i\rangle \}$, the interaction Hamilton operator is such that the joint quantum state immediately after the conclusion of the interaction is related to the intitial state as follows

$$|\psi^{\sigma \oplus \alpha}\rangle = |\phi_i^\sigma\rangle \otimes |\psi_{[...|}^\alpha\rangle \rightarrow |\psi'^{\sigma \oplus \alpha}\rangle = |\phi_i^\sigma\rangle \otimes |\psi_{i|...,a_i|}^\alpha\rangle$$

for each eigenvector $|\phi_i^\alpha\rangle$ of $\hat{A}$, where the $|\psi_i^\alpha\rangle$ are orthonormal vectors, $[a_i]$ stands for a recording of the eigenvalue a_i of $\hat{A}$. The dots indicate that results of earlier good observations may also be recorded in the state of α. It follows from the linearity of the Schrödinger equation that an arbitrary normalized initial object system quantum state $\sum_i c_i |\phi_i^\sigma\rangle$ with

$$\sum_i |c_i|^2 = 1$$

gives rise to the following transformation

$$\sum_i c_i |\phi_i^\sigma\rangle \otimes |\psi_{[...|}^\alpha\rangle \rightarrow \sum_i c_i (|\phi_i^\sigma\rangle \otimes |\psi_{i|...,a_i|}^\alpha\rangle).$$

Each component $|\phi_i^\sigma\rangle \otimes |\psi_{i|...,a_i|}^\alpha\rangle$ with nonzero coefficient c_i in the superposition on the right-hand side corresponds to a distinct state in which the observer has recorded the ith eigenvalue for the measured quantity on the object system, while the object system remains in the corresponding eigenstate $|\phi_i\rangle$. Moreover, all these states are equally real. Every possible result is recorded in some observer state $|\psi_{i|...,a_i|}^\alpha\rangle$, and there is no unique actual result. For a sequence of good observations by a single observer, consisting of multiple pairwise interactions between the apparatus system and each member of a set of object systems, Everett is able to show the following. If a good observation is repeated on a single object system in circumstances in which that system remains undisturbed in the intervening interval (in the sense that the total Hamilton operator commutes with the operator representing the observed quantity), then the eigenvalues recorded for the two observations are the same, in every observer state. This is exactly what would be predicted by an observer who represents each object system independently by a quantum state vector and regarded the first of each sequence of repeated measurements on it as projecting the relevant object system's quantum state onto an eigenvector corresponding to the initially recorded eigenvalue. This is the first respect in which, for each observer, a good observation appears to obey the projection postulate. Everett shows that each observer will get the right probabilities for results of arbitrary good observations on a system which has been subjected to an initial good observation, if, following this initial observation, one assigns to the system the quantum state it would have had if projection had then occurred. For the following two probabilities are demonstrably equal: the probability of result b_j in a subsequent good observation of B on σ by an observer corresponding to α who applies the projection postulate to the state of σ alone after an initial good observation of A on σ by α yielding result a_i; and the probability assuming that the state of the compound $\sigma \oplus \alpha$ evolves according to the Schrödinger equation that after the B measurement the observer state of α will record the values a_i and b_j, conditional on the observer state of α after the A measurement recording the result a_i of the initial observation. This demonstration explains how, for each observer, it is as if a good observation prepared a corresponding eigenstate of the observed system. However this still does not suffice to establish that everything is as if projection actually occurs. There are two further consequences of projection. If projection really occurred, then each of several independent observers performing repeated good observations of the same quantity on an otherwise undisturbed system would necessarily obtain the same result. Moreover, if projection really occurred, then the state of α immediately after a good observation would be one of the $|\psi_i^\alpha\rangle$. In this apparatus state the pointer position quantity has its ith eigenvalue, recording that the observed quantity had its ith eigenvalue on σ. In this apparatus state the probability is 1 that a subsequent observation of the pointer position quantity would reveal that it has its ith eigenvalue. Consequently, the result of a subsequent observation of the pointer position quantity on an undisturbed apparatus system will reveal that the pointer position quantity had at the conclusion of the initial interaction with σ the value which recorded the result of the measurement on σ.

18.6 Basis Degeneracy Problem

Many-world, decoherence, and modal interpretations of quantum mechanics suffer from a basis degeneracy problem arising from the nonuniqueness of some biorthogonal decompositions. According to the *biorthogonal decomposition theorem*, any quantum state vector describing two systems can for a certain choice of bases, be expanded in the simple form

$$\sum_i c_i |A_i\rangle \otimes |B_i\rangle$$

where the $\{|A_i\rangle\}$ and $\{|B_i\rangle\}$ vectors are orthonormal, and are therefore eigenstates of self-adjoint operators (observables) $\hat{A}$ and $\hat{B}$ associated with systems 1 and 2, respectively. This biorthogonal expansion picks out the Schmidt basis. The basis degeneracy problem arises because the biorthogonal decomposition is unique just in case all of the nonzero $|c_i|$ are different. When $|c_1| = |c_2|$, we can biorthogonally expand

$$|\varphi\rangle = c_1 |S_z = \uparrow\rangle \otimes |R = \uparrow\rangle + c_2 |S_z = \downarrow\rangle \otimes |R = \downarrow\rangle$$

in an infinite number of bases. If $c_1 = c_2$, then the biorthogonal decomposition of the apparatus with the particle-environment system is not unique, and therefore gives us no principled reasoning for singling out the pointer-reading basis. This is the basis degeneracy problem. The basis degeneracy problem arises in the context of many-world interpretations. Many-world interpretations [62] address the measurement problem by hypothesizing that when the combined system occupies state $|\varphi\rangle$, the two branches of the superposition split into separate worlds, in some sense. The pointer reading becomes definite relative to its branch. For instance, in the "up" world, the particle has spin up and the apparatus possesses the corresponding pointer reading. In this way, many-world interpreters explain why we always see definite pointer readings, instead of superpositions.

Elby and Bub [62] proved that when a quantum state can be written in the tri-orthogonal form

$$|\Psi\rangle = \sum_i c_i |A_i\rangle \otimes |B_i\rangle \otimes |C_i\rangle$$

then, even if some of the c_i are equal, no alternative bases exist such that $|\Psi\rangle$ can be rewritten

$$\sum_i d_i |A'_i\rangle \otimes |B'_i\rangle \otimes |C'_i\rangle \,.$$

Therefore the triorthogonal decomposition picks out a special basis. This preferred basis can be used to address the basis degeneracy problem. The *tridecompositional uniqueness theorem* provides many-world interpretations, decoherence interpretations, and modal interpretations with a rigorous solution to the basis degeneracy problem. Several interpretations of quantum mechanics can make use of this special basis. For instance, many-world adherents can claim that a branching of worlds occurs in the preferred basis picked out by the unique triorthogonal decomposition. Modal interpreters can postulate that the triorthogonal basis helps to pick out which observables possess definite values at a given time. Decoherence theorists can cite the uniqueness of the triorthogonal decomposition as a principled reason for asserting that pointer readings become classical upon interacting with the environment.

When the environment interacts with the combined particle-apparatus system the following state results

$$|\Psi\rangle = c_1 |S_z =\uparrow\rangle \otimes |R =\uparrow\rangle \otimes |E_+\rangle + c_2 |S_z =\downarrow\rangle \otimes |R =\downarrow\rangle \otimes |E_-\rangle$$

where $|E_\pm\rangle$ is the state of the rest of the universe after the environment interacts with the apparatus. As time passes, these environmental states quickly approach orthogonality:

$$\langle E_+|E_-\rangle \to 0.$$

In this limit, we have a triorthogonal decomposition of $|\Psi\rangle$. Even if $c_1 = c_2$, the triorthogonal decomposition is unique. In other words, no transformed bases exist such that $|\Psi\rangle$ can be expanded as

$$d_1 |S' =\uparrow\rangle \otimes |R' =\uparrow\rangle \otimes |E'_+\rangle + d_2 |S' =\downarrow\rangle \otimes |R' =\downarrow\rangle \otimes |E'_-\rangle \,.$$

Therefore, $|\Psi\rangle$ picks out a preferred basis. Many-world interpreters can postulate that this basis determines the branches into which the universe splits. For the proof we refer to the literature (Elby and Bub [62]).

18.7 Information Theoretic Viewpoint

The classical concept of information can be extended to the von Neumann entropy in quantum mechanics. From a classical viewpoint entropy can never be negative. By extending the definitions of conditional, combined and mutual entropy Adami and Cerf [1, 38, 39] have concluded that negative entropies are needed in the quantum case. The measurement problem is then explained as follows. Suppose the system is in a superposition

$$|\psi\rangle = a|\alpha\rangle + b|\beta\rangle$$

where

$$|a|^2 + |b|^2 = 1.$$

Now we introduce an ancillary system A to perform the measurement resulting in the product state

$$|\psi A\rangle = a|\alpha\rangle \otimes |\alpha\rangle + b|\beta\rangle \otimes |\beta\rangle.$$

Finally the observer must interact with the system A to observe the measured value.

$$|\psi AO\rangle = a|\alpha\rangle \otimes |\alpha\rangle \otimes |\alpha\rangle + b|\beta\rangle \otimes |\beta\rangle \otimes |\beta\rangle.$$

Since we are only interested in the measurement the original system is ignored. In the mathematical representation this involves taking the partial trace of $|\psi AO\rangle\langle\psi AO|$ with respect to the original system. This yields the mixed state

$$\mathrm{tr}_\psi|\psi AO\rangle\langle\psi AO| = |a|^2(|\alpha\rangle \otimes |\alpha\rangle)(\langle\alpha| \otimes \langle\alpha|) + |b|^2(|\beta\rangle \otimes |\beta\rangle)(\langle\beta| \otimes \langle\beta|).$$

Thus the measurement is classically correlated, but the result is random. Further measurements will retain this correlation giving the observer the illusion of the projection postulate being satisfied. The mutual information shared with the original system vanishes, thus no information is obtained about the state of the original system.

Chapter 19
Quantum State Machines

19.1 Introduction

In this chapter we introduce the quantum state machine [80, 119]. The quantum state machine is an extension of classical finite state machines used to represent the computations possible in quantum computing. Quantum state machines introduce amplitudes for transitions between states to represent the parallelism available.

19.2 Quantum Automata

Definition. A *quantum automaton* consists of

- A finite set S of states where the elements are uniquely identified with orthonormal states in a Hilbert space $\mathcal{H}$ of dimension at least $|S|$. One state $s_0 \in S$ is designated as the *start state*. We will use the one-to-one function $m : S \to \mathcal{H}$ to denote the relationship between states and elements of the Hilbert space. We will use the notation $|s\rangle = m(s)$.

- A sub-Hilbertspace $\mathcal{H}_A$ of $\mathcal{H}$, and the corresponding projection operator P_A from $\mathcal{H}$ into $\mathcal{H}_A$.

- An *alphabet* Λ of possible input symbols.

- A finite set of *transitions* for each combination of two (possibly identical) states and symbols in the alphabet. Transitions are ordered 4-tuples $(a, b, c, d_{a,b,c})$ where $a, b \in S$, $c \in \Lambda$ and $d \in \mathbf{C}$. We require that $\sum_{b,c} |d_{a,b,c}|^2 = 1$ where the sum is over all transitions from a. We will also define $d_{a,b,c}$ to be zero when no transition exists between a and b for input c. The values $d_{a,b,c}$ must also satisfy

$$\sum_{t \in S} d_{s,t,c} \overline{d_{s',t,c}} = \delta_{s,s'}$$

for every pair of states $s, s' \in S$, where $\delta_{s,s'}$ is the Kronecker delta.

The condition

$$\sum_{t \in S} d_{s,t,c} \overline{d_{s',t,c}} = \delta_{s,s'}$$

is used to enforce the unitarity of transformations. We can construct a unitary transition matrix U_c for every input symbol c from the values $d_{a,b,c}$ where a and b determine the column and row in the matrix. The above condition only refers to entries in the matrix U_c, and refers directly to the unitarity condition in terms of the entries in the matrix obtained by multiplying U_c with the complex conjugate of the transpose of U_c. Since we identify transitions with unitary operators in the Hilbert space, we can describe a computational path as the product of unitary operators. For the input symbols $a_1 a_2 \ldots a_n$ the finite quantum automaton defines the evolution of the initial state according to

$$|s_0\rangle \rightarrow |s_n\rangle := U_{a_n} U_{a_{n-1}} \ldots U_{a_1} |s_0\rangle.$$

From s_n we can define the words which are accepted. If

$$\langle s_n | P_A | s_n \rangle \neq 0$$

we say that $a_1 a_2 \ldots a_n$ is accepted, otherwise it is rejected. Thus the input symbols define a sequence of unitary operations to apply to an initial state, or a program. This can be thought of as a program of quantum operations controlled classically, which is exactly the way we have described quantum algorithms. The end of the input corresponds to a measurement, i.e. we have to determine if the machine is in a final state. We cannot define halt states, since the initial state may evolve into a superposition of halt states and states which are not. This is the *quantum halting problem* [110, 101]. The quantum finite automaton cannot crash on an input, since it simply performs the transition with amplitude (probability) 0.

We can also define the amplitude for a state s after n steps as

$$D(s, n) = \sum_{\substack{s_1, s_2, \ldots, s_n \in S \\ a_1, a_2, \ldots, a_n \in \Lambda}} \prod_{j=1}^{n} d_{s_{j-1}, s_j, a_j}.$$

It is easy to show that

$$\sum_{s \in S} |D(s, n)|^2 = 1.$$

Graphically, we can represent quantum automata in the same way as with finite automata, with the additional labelling of arcs between states with the complex amplitudes for the corresponding transition. The description of quantum automata is tied closely to unitary transformations in a Hilbert space on an initial state. It is much simpler to analyse quantum algorithms in terms of unitary operations in the Hilbert space $\mathcal{H}$. We consider quantum automata since they are language acceptors. Finite automata are also language acceptors so the classical and quantum machines can be compared to determine which types of languages they accept. This is an important question, since it would be useful to know if quantum machines can achieve more than their classical counterparts.

A *q-automaton* is a 4-tuple $(\mathcal{H}, s_0, A, U)$, where $\mathcal{H}$ is a finite dimensional Hilbert space, $s_0 \in \mathcal{H}$ is the initial state, A is a finite alphabet, and U is a mapping from A to the unitary operators acting on $\mathcal{H}$. The q-automaton is useful since it is described in terms of the Hilbert space, and the unitary operations defined on its states. For words

$$a_1 a_2 \ldots a_n \in A \times A \times \ldots \times A \ (n \text{ times})$$

we denote by $U(a_1 a_2 \ldots a_n)$ the product $U_{a_n} U_{a_{n-1}} \ldots U_{a_1}$ of the operators derived from the symbols a_1 to a_n. A *finalizing q-automaton* is a 5-tuple $(\mathcal{H}, s_0, A, U, F)$ where $(\mathcal{H}, s_0, A, U)$ is a q-automaton and F is a subspace of $\mathcal{H}$ such that either $s_0 \in F$ or $s_0 \in F^{\perp}$. We denote by P_F the projection onto the subspace F. The probability that a word w causes the q-automaton to reach a final state in F is given by

$$|\langle s_0 | U^*(w) P_F U(w) | s_0 \rangle|^2.$$

The function $U(w) s_0$ is called the *response function* of the q-automaton. A function $R : A^* \rightarrow \mathcal{H}$ is realizable by the q-automaton if $R(w) = U(w) s_0$. A word w is accepted if $U(w) s_0 \in F$. The language of the q-automaton is the set of all words accepted by the q-automaton.

For any $R : A^* \rightarrow \mathcal{H}$ we have

- R is realizable by a q-automaton.

- There exists a unitary operator U_x for every $x \in A$ such that $R(xw) = U(x) R(w)$ for $w \in A^*$.

- There exists an orthonormal basis $|\psi_j\rangle$ for $\mathcal{H}$ and an orthonormal basis $|\psi_j(x)\rangle$ for every $x \in A$ such that $\langle R(xw) | \psi_j \rangle = \langle R(x) | \psi_j(x) \rangle$.

We can define the tensor product of two finalizing q-automata $q_1 = (\mathcal{H}_1, s_1, A, U_1, F_1)$ and $q_2 = (\mathcal{H}_2, s_2, A, U_2, F_2)$ over the same input alphabet as

$$q_1 \otimes q_2 := (\mathcal{H}_1 \otimes \mathcal{H}_2, s_1 \otimes s_2, A, U_1 \otimes U_2, F_1 \otimes F_2).$$

Thus the language accepted by $q_1 \otimes q_2$ is the intersection of the languages accepted by q_1 and q_2.

We can extend the languages accepted by a finalizing q-automaton. For $0 \leq \eta < 1$ a word w is η-accepted by a q-automata $q = (\mathcal{H}, s_0, A, U, F)$ if

$$|\langle s_0|U^*(w)P_F U(w)|s_0\rangle|^2 > \eta.$$

The η-accepted language for q is the set of all words η-accepted by q.

For further results in the theory of quantum automata we refer to Gudder [80].

We proposed that Turing machines implement the classically computable functions, so we will extend Turing machines to the quantum case since this will provide a better comparison of the complexity and computability of problems in quantum computing, with that of the classical case.

19.3   Quantum Turing Machines

Definition. A *quantum Turing machine* [11, 54, 125, 147] consists of

- A finite set of states S. One state, $s_0 \in S$, is designated as the *start* state. The states in S are identified with orthonormal states in a Hilbert space $\mathcal{H}_S$ of dimension at least $|S|$. The one-to-one mapping $m_S : S \to \mathcal{H}_S$ specifies the association of states and elements of the Hilbert space. We use the notation $|s\rangle := m_S(s)$.

- An *alphabet* Λ of possible input symbols.

- An *alphabet* Γ of possible output symbols.

- The *blank* symbol Δ.

- A *tape* or memory device which consists of adjacent *cells* labelled

$$\ldots, cell[-1], \ cell[0], \ cell[1], \ldots.$$

Cells of the tape can contain a single symbol from

$$T := \Lambda \cup \Gamma \cup \{\Delta\}.$$

The input string is placed in the first cells of the tape, the rest of the cells are filled with Δ. The content of a cell is identified with orthonormal states in a Hilbert space $\mathcal{H}_T$ of dimension at least $|\Lambda| + |\Gamma| + 1$. The Hilbert space describing the tape is thus $\mathcal{H}_M := \otimes_{-\infty}^{\infty} \mathcal{H}_T$. The one-to-one mapping m_T, with $m_T : T \to \mathcal{H}_T$, associates elements in the tape cells with the elements of the Hilbert space $\mathcal{H}_T$. We use the notation $|t\rangle := m_T(t)$.

- A *tape head* that can read the contents of a tape cell, put a symbol from Γ or the Δ symbol in the tape cell and move one cell right or left. All these actions take place simultaneously. If the head is at $cell[i]$ and moves left (right) then the head will be at $cell[i-1]$ ($cell[i+1]$). The position of the tape head is identified with orthonormal states in an infinite dimensional Hilbert space $\mathcal{H}_{TH}$. The one to one mapping $m_{TH} : \mathbf{Z} \to \mathcal{H}_{TH}$ associates the tape head position (an integer specifying the cell) with elements of the Hilbert space. We use the notation $|j\rangle := m_{TH}(j)$.

- A finite set of *transitions* for states and symbols from $\Sigma \cup \Gamma \cup \{\Delta\}$. A transition is an ordered 6-tuple $(a, b, c, d, e, f_{a,b,c,d,e})$ with

$$a \in S, \quad b \in \Sigma \cup \Gamma \cup \{\Delta\}, \quad c \in S, \quad d \in \Gamma \cup \{\Delta\}, \quad e \in \{r, l\}$$

and $f_{a,b,c,d,e} \in \mathbf{C}$. Here a is the current state, b is the symbol read by the tape head, c is the next state, d is the symbol for the tape head to write in the current cell and $e = r$ ($e = l$) moves the tape head right (left). If no transition exists for (a, b, c, d, e) we define $f_{a,b,c,d,e} = 0$. We require that

1. $\displaystyle\sum_{c,d,e} f_{s,t,c,d,e}\overline{f_{s',t',c,d,e}} = \delta_{s,s'}\delta_{t,t'}$

2. $\displaystyle\sum_{a,b} f_{a,b,c,d,e}\overline{f_{a,b,c',d',e'}} = \delta_{c,c'}\delta_{d,d'}\delta_{e,e'}$

The quantum Turing machine has a tape which is infinitely long in both directions. This does not provide any additional computing power over the tape which is only infinite in one direction, but it does make the description of the quantum Turing machine simpler since we can avoid crashing the machine, which corresponds to the lack of a unitary transform to describe what happens when the machine is at $cell[0]$.

The state of a quantum Turing machine at any time is described by a normalized state in the Hilbert space

$$\mathcal{H}_{TM} := \mathcal{H}_S \otimes \mathcal{H}_{TH} \otimes \mathcal{H}_M.$$

The initial state of the machine is given by

$$|QTM_0\rangle := |s_0\rangle \otimes |0\rangle \otimes \left(\bigotimes_{-\infty}^{-1} |\Delta\rangle\right) \otimes |\psi_0\rangle \otimes \left(\bigotimes_{l}^{\infty} |\Delta\rangle\right)$$

where $|\psi_0\rangle$ is the initial contents of the tape, using l cells. The evolution of the machine is described by a unitary operator U, which in turn is specified by the transitions. Thus after n steps of execution the machine is in the state

$$U^n|QTM_0\rangle.$$

The unitary evolution U can be described in terms of the amplitudes of the transitions.

$$U = \sum_{x,a,b,c,d,e} f_{a,b,c,d,e} |c\rangle\langle a| \otimes |x + \delta_{r,e} - \delta_{l,e}\rangle\langle x| \otimes \left(\bigotimes_{-\infty}^{x-1} I_T\right) \otimes |d\rangle\langle b| \otimes \left(\bigotimes_{x+1}^{\infty} I_T\right)$$

where I_T is the identity operator for a tape cell. Considering

$$UU^* = U^*U = I = \sum_{a,b,x} |a\rangle\langle a| \otimes |x\rangle\langle x| \otimes \left(\bigotimes_{-\infty}^{x-1} I_T\right) \otimes |b\rangle\langle b| \otimes \left(\bigotimes_{x+1}^{\infty} I_T\right)$$

we obtain the constraints

$$\sum_{c,d,e} f_{s,t,c,d,e} \overline{f_{s',t',c,d,e}} = \delta_{s,s'}\delta_{t,t'}$$

and

$$\sum_{a,b} f_{a,b,c,d,e} \overline{f_{a,b,c',d',e'}} = \delta_{c,c'}\delta_{d,d'}\delta_{e,e'}.$$

We cannot determine when a quantum Turing machine halts in the same way as for quantum automata. Quantum automata relies on a finite input string which describes the running of the machine and explicitly determines when the machine halts. The tape of the quantum Turing machine cannot fulfill this role since the machine can modify any cell on the tape, the input is not "consumed". Deutsch [54] suggested reserving one cell of the tape which is always in one of two orthonormal states to indicate when the machine has halted. The cell contents can become entangled with the rest of the machine, giving a superposition of halted machines and machines which have not halted [110, 101]. If it is known that for any input of length n the quantum Turing machine will halt after $t(n)$ steps, we can use the state indicating the halt status of the machine as a control (in the same way as the controlled NOT) for the transformation U of the quantum Turing machine, and measure after $t(n)$ steps with certainty that the machine has halted. Deutsch also suggested the existence of a universal quantum Turing machine which can simulate any other quantum Turing machine. Yu Shi [147] discusses why this cannot be the case.

Chapter 20
Teleportation

20.1 Introduction

Quantum teleportation is the disembodied transport of an unknown quantum state $|\psi\rangle$ from one place to another. All protocols for accomplishing such transport require nonlocal correlations, or entanglement, between systems shared by sender and receiver. The sender is normally called *Alice* and the receiver is called *Bob*. Most attention has focused on teleporting the states of finite-dimensional systems, such as the two-dimensional polarization of a photon or the discrete level structure of an atom. First proposed in 1993 by Charles Bennett and his colleagues [17, 24, 138] quantum teleporation thus allows physicists to take a photon or any other quantum scale particle such as an atom and transfer its properties (such as the polarization) to another photon even if the two photons are on opposite sides of the galaxy. This scheme transports the particle's properties to the remote location and not the particle itself. The state of the original particle must be destroyed to create an exact reconstruction at the other end. This is a consequence of the no cloning theorem. A role in the teleportation scheme is played by an entangled ancillary pair of particles which will be initially shared by Alice and Bob.

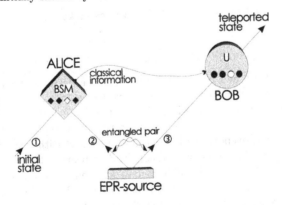

Figure 20.1: Teleportation

20.2   Teleportation Algorithm

Suppose particle 1 (in the following we assume that it is a spin-$\frac{1}{2}$ particle) which Alice wants to teleport is in the initial state

$$|\phi\rangle_1 = a|0\rangle_1 + b|1\rangle_1$$

and the entangled pair of particles 2 and 3 shared by Alice and Bob is in the state

$$|\phi\rangle_{23} = \frac{1}{\sqrt{2}}(|0\rangle_2 \otimes |1\rangle_3 - |1\rangle_2 \otimes |0\rangle_3)\,.$$

Alice gets particle 2 and Bob particle 3. This entangled state contains no information on the individual particles 2 and 3. It only indicates that the two particles will be in opposite states. Alice then performs a joint Bell-state measurement on the initial particle 1 and particle 2 projecting them also onto an entangled state. After Alice has sent the result of her measurement as classical information to Bob, he can perform a unitary transformation on the other ancillary particle resulting in it being in the state of the original particle.

Most experiments are done with photons which are spin-1 particles. The information to be teleported is the polarization state of the photon. The *Innsbruck experiment* is a simplified version of the teleportation described above. In this experiment photons are used. The photon is a particle with spin 1 and rest mass 0. If the photon moves in positive z direction, i.e. the wave vector $\mathbf{k}$ is given by $(0,0,k)^T$ we have the wave functions

$$\phi_1(z,t) = (2V)^{-1/2}(-\mathbf{e}_1 - i\mathbf{e}_2)\exp(i(kz - \omega t))$$

$$\phi_{-1}(z,t) = (2V)^{-1/2}(\mathbf{e}_1 - i\mathbf{e}_2)\exp(i(kz - \omega t))$$

where $\mathbf{e}_1 := (1,0,0)^T$ and $\mathbf{e}_2 := (0,1,0)^T$. Thus we have two transverse waves. Although the photon is a spin-1 particle the vectors $\mathbf{s}$ and $\mathbf{k}$ can only be parallel (or antiparallel). The wave ϕ_1 is in a state of positive helizity and the wave ϕ_{-1} is in a state of negative helizity. In the *Innsbruck experiment* at the sending station of the quantum teleporter, Alice encodes photon M with a specific state: 45 degree polarization. This photon travels towards a beamsplitter. Meanwhile two additional entangled photons A and B are created. Thus they have complementary polarizations. For example, if photon A is later measured to have horizontal (0 degrees) polarization, then the other photon B must collapse into the complementary state of vertical (90 degrees) polarization. Now entangled photon A arrives at the beamsplitter at the same time as the message photon M. The beamsplitter causes each photon either to continue towards detector 1 or change course and travel to detector 2. in 1/4 of all cases, in which the two photons go off into different detectors, Alice does not know which photon went to which detector. Owing to the fact that the two photons are now indistinguishable, the message photon M loses its original identity and becomes entangled with A. The polarization value for each photon is now indeterminate, but since the two photons travel towards different detectors Alice knows that the two photons must have complementary polarizations. Since message particle M must have complementary polarization to particle A, then the other entangled particle B must now attain the same polarization value as M. Therefore teleportation is successful. The receiver Bob sees that the polarization value of the particle B is 45 degrees, which is the initial value of the message photon. In the experimental version of this setup executed at the University of Innsbruck, the 45-degree polarization would always fire when detector 1 and detector 2 fired. Except in rare instances attributable to background noise, it was never the case that the 135-degree polarization detector fired in coincidence with detectors 1 and 2.

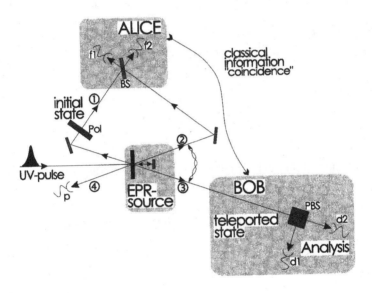

Figure 20.2: Experimental Realization of Teleportation

Teleportation can also be understood using the quantum circuit shown in the following figure.

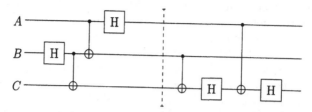

Figure 20.3: Quantum Circuit for Teleportation

In the figure A is the input $|\psi\rangle$, B the input $|0\rangle$ and C the input $|0\rangle$. Now we study what happens when we feed the product state $|\psi 00\rangle$ into the quantum circuit. From the circuit we have the following eight 8×8 unitary matrices

$$U_1 = I_2 \otimes U_H \otimes I_2, \quad U_2 = I_2 \otimes U_{XOR}, \quad U_3 = U_{XOR} \otimes I_2, \quad U_4 = U_H \otimes I_2 \otimes I_2$$

$$U_5 = I_2 \otimes U_{XOR}, \quad U_6 = I_2 \otimes I_2 \otimes U_H, \quad U_7 = I_4 \oplus U_{NOT} \oplus U_{NOT}, \quad U_8 = I_2 \otimes I_2 \otimes U_H$$

where $\oplus$ denotes the direct sum of matrices [162] and

$$U_{NOT} := \begin{pmatrix} 0 & 1 \\ 1 & 0 \end{pmatrix}.$$

Applying the first four unitary matrices $U_4 U_3 U_2 U_1$ to the input state we obtain

$$\frac{a}{2}(|000\rangle + |100\rangle + |011\rangle + |111\rangle) + \frac{b}{2}(|010\rangle - |110\rangle + |001\rangle - |101\rangle).$$

This state can be rewritten as

$$\frac{1}{\sqrt{2}}(|0\rangle + |1\rangle) \otimes (\frac{a}{\sqrt{2}}(|0\rangle \otimes |0\rangle + |1\rangle \otimes |1\rangle)) + \frac{1}{\sqrt{2}}(|0\rangle - |1\rangle) \otimes (\frac{b}{\sqrt{2}}(|0\rangle \otimes |1\rangle + |1\rangle \otimes |0\rangle))$$

Applying all eight unitary matrices $U_8 U_7 U_6 U_5 U_4 U_3 U_2 U_1$ to the input state we obtain

$$\frac{a}{2}(|000\rangle + |100\rangle + |010\rangle + |110\rangle) + \frac{b}{2}(|011\rangle + |111\rangle + |001\rangle + |101\rangle).$$

This state can be rewritten as

$$\left(\frac{1}{\sqrt{2}}(|0\rangle + |1\rangle)\right) \otimes \left(\frac{1}{\sqrt{2}}(|0\rangle + |1\rangle)\right) \otimes |\psi\rangle.$$

Thus the state $|\psi\rangle$ will be transferred to the lower output, where both other outputs will come out in the state $(|0\rangle + |1\rangle)/\sqrt{2}$. If the two upper outputs are measured in the standard basis ($|0\rangle$ versus $|1\rangle$), two random classical bits will be obtained in addition to the quantum state $|\psi\rangle$ on the lower output.

Consider the case when the qubit to be teleported is one qubit of an entangled pair. The first two qubits are entangled. Applying the teleportation algorithm to the second, third and fourth qubits yields

$$\frac{1}{\sqrt{2}}(|00\rangle + |11\rangle) \otimes |00\rangle \rightarrow \frac{1}{\sqrt{2}}(|0000\rangle + |1001\rangle).$$

The first and last qubits are now entangled, whereas the first and second are no longer entangled. Thus we have achieved *entanglement swapping*.

20.3 Example Program

The following program uses SymbolicC++ [169, 166] to implement the teleportation algorithm. It builds the matrix using the description given in the previous section. It uses the direct sum and Kronecker product [162] extensively.

```
// teleport.cpp

#include "Vector.h"
#include "Matrix.h"
#include "Rational.h"
#include "Msymbol.h"
using namespace std;

typedef Sum<Rational<int> > C;

template <class T> Vector<T> Teleport(Vector<T> v)
{
  int i;
  assert(v.length() == 8);
  Vector<T> result;

  Matrix<T> NOT(2,2);
```

```
NOT[0][0] = T(0); NOT[0][1] = T(1);
NOT[1][0] = T(1); NOT[1][1] = T(0);

Matrix<T> H(2,2);
H[0][0] = T(1)/sqrt(T(2)); H[0][1] = T(1)/sqrt(T(2));
H[1][0] = T(1)/sqrt(T(2)); H[1][1] = T(-1)/sqrt(T(2));

Matrix<T> I(2,2);
I.identity();

Matrix<T> X(4,4);
X[0][0] = T(1); X[0][1] = T(0); X[0][2] = T(0); X[0][3] = T(0);
X[1][0] = T(0); X[1][1] = T(1); X[1][2] = T(0); X[1][3] = T(0);
X[2][0] = T(0); X[2][1] = T(0); X[2][2] = T(0); X[2][3] = T(1);
X[3][0] = T(0); X[3][1] = T(0); X[3][2] = T(1); X[3][3] = T(0);

Matrix<T> U1=kron(I,kron(H,I));
Matrix<T> U2=kron(I,X);
Matrix<T> U3=kron(X,I);
Matrix<T> U4=kron(H,kron(I,I));
Matrix<T> U5=kron(I,X);
Matrix<T> U6=kron(I,kron(I,H));
Matrix<T> U7=dsum(I,dsum(I,dsum(NOT,NOT)));
Matrix<T> U8=kron(I,kron(I,H));

result=U8*(U7*(U6*(U5*(U4*(U3*(U2*(U1*v)))))));
for(i=0;i<8;i++)
{
 while(result[i].put(power(sqrt(T(2)),-6),power(T(2),-3)));
 while(result[i].put(power(sqrt(T(2)),-4),power(T(2),-2)));
 while(result[i].put(power(sqrt(T(2)),-2),power(T(2),-1)));
}
return result;
}


// The outcome after measuring value for qubit.
// Since the probabilities may be symbolic this function
// cannot simulate a measurement where random outcomes
// have the correct distribution
template <class T>
Vector<T> Measure(Vector<T> v,unsigned int qubit,unsigned int value)
{
 assert(pow(2,qubit)<v.length());
 assert(value==0 || value==1);
 int i,len,skip = 1-value;
 Vector<T> result(v);
 T D = T(0);
```

```
len = v.length()/int(pow(2,qubit+1));
for(i=0;i<v.length();i++)
{
 if(!(i%len)) skip = 1-skip;
 if(skip) result[i] = T(0);
 else D += result[i]*result[i];
}
result/=sqrt(D);
return result;
}


// for output clarity
ostream &print(ostream &o,Vector<C> v)
{
 char *b2[2]={"|0>","|1>"};
 char *b4[4]={"|00>","|01>","|10>","|11>"};
 char *b8[8]={"|000>","|001>","|010>","|011>",
              "|100>","|101>","|110>","|111>"};

 char **b,i;

 if(v.length()==2) b=b2;
 if(v.length()==4) b=b4;
 if(v.length()==8) b=b8;

 for(i=0;i<v.length();i++)
  if(!v[i].is_Number() || v[i].nvalue()!=C(0))
    o << "+(" << v[i] << ")" << b[i];
 return o;
}

 void main(void)
 {
  Vector<C> zero(2),one(2);
  Vector<C> zz(4),zo(4),oz(4),oo(4),qreg;
  Vector<C> tp00,tp01,tp10,tp11;
  Sum<Rational<int> > a("a",0),b("b",0);
  int i;

  zero[0] = C(1); zero[1] = C(0);
  one[0]  = C(0); one[1]  = C(1);

  zz = kron(Matrix<C>(zero),Matrix<C>(zero))(0);
  zo = kron(Matrix<C>(zero),Matrix<C>(one))(0);
  oz = kron(Matrix<C>(one),Matrix<C>(zero))(0);
  oo = kron(Matrix<C>(one),Matrix<C>(one))(0);

  qreg=kron(a*zero+b*one,kron(zero,zero))(0);
  cout << "UTELEPORT("; print(cout,qreg) << ") = ";
```

```
 print(cout,qreg=Teleport(qreg)) << endl;
cout << "Results after measurement of first 2 qubits:" << endl;
tp00 = Measure(Measure(qreg,0,0),1,0);
tp01 = Measure(Measure(qreg,0,0),1,1);
tp10 = Measure(Measure(qreg,0,1),1,0);
tp11 = Measure(Measure(qreg,0,1),1,1);
for(i=0;i<8;i++)
{
 while(tp00[i].put(a*a,C(1)-b*b));
 while(tp00[i].put(power(sqrt(C(1)/C(2)),-2),C(2)));
 while(tp01[i].put(a*a,C(1)-b*b));
 while(tp01[i].put(power(sqrt(C(1)/C(2)),-2),C(2)));
 while(tp10[i].put(a*a,C(1)-b*b));
 while(tp10[i].put(power(sqrt(C(1)/C(2)),-2),C(2)));
 while(tp11[i].put(a*a,C(1)-b*b));
 while(tp11[i].put(power(sqrt(C(1)/C(2)),-2),C(2)));
}
cout << " |00> : " ; print(cout,tp00) << endl;
cout << " |01> : " ; print(cout,tp01) << endl;
cout << " |10> : " ; print(cout,tp10) << endl;
cout << " |11> : " ; print(cout,tp11) << endl;
cout << endl;


}
```

The program generates the following output:

```
UTELEPORT(+(a)|000>+(b)|100>) = +(1/2*a)|000>+(1/2*b)|001>
                                +(1/2*a)|010>+(1/2*b)|011>
                                +(1/2*a)|100>+(1/2*b)|101>
                                +(1/2*a)|110>+(1/2*b)|111>
Results after measurement of first 2 qubits:
 |00> : +(a)|000>+(b)|001>
 |01> : +(a)|010>+(b)|011>
 |10> : +(a)|100>+(b)|101>
 |11> : +(a)|110>+(b)|111>
```

Chapter 21

Quantum Algorithms

21.1 Deutsch's Problem

Deutsch's problem [54] is given as follows. Suppose we have a boolean function

$$f : \{0,1\} \to \{0,1\}.$$

There are four such functions, the constant functions which map all inputs to 0 or all inputs to 1, and the varying functions which have $f(0) \neq f(1)$. In other words

$$
\begin{aligned}
f_1(0) &= 0, & f_1(1) &= 0 \\
f_2(0) &= 1, & f_2(1) &= 1 \\
f_3(0) &= 0, & f_3(1) &= 1 \\
f_4(0) &= 1, & f_4(1) &= 0.
\end{aligned}
$$

The first two functions are constant. The task is to determine for such a function if it is constant or varying using only one calculation of the function. In the classical case it is necessary to compute f twice before it is known whether it is constant or varying. For example if $f(0) = 0$, the function could be f_1 or f_3, similarly for any other single evaluation two of the functions have the same value.

In quantum computing the following solution was found [47, 178]. The function is implemented on quantum hardware with the unitary transformation U_f such that

$$U_f |x\rangle \otimes |y\rangle = |x\rangle \otimes |y \oplus f(x)\rangle$$

where $\oplus$ denotes the XOR operation. We apply the transformation to the state

$$|\psi\rangle := \frac{1}{2}(|0\rangle + |1\rangle) \otimes (|0\rangle - |1\rangle).$$

This gives

$$U_f|\psi\rangle = \frac{1}{2}(|0\rangle \otimes |0 \oplus f(0)\rangle - |0\rangle \otimes |1 \oplus f(0)\rangle)$$

$$+ \frac{1}{2}(|1\rangle \otimes |0 \oplus f(1)\rangle - |1\rangle \otimes |1 \oplus f(1)\rangle).$$

Since U_f is linear and quantum mechanics allows the superposition of states we have calculated the function f with each input twice. This feature of quantum parallelism makes it possible to solve the problem. Applying the Hadamard transform U_H to the first qubit yields

$$(U_H \otimes I_2)U_f|\psi\rangle = \frac{1}{2\sqrt{2}}(|0\rangle \otimes (|0 \oplus f(0)\rangle - |1 \oplus f(0)\rangle + |0 \oplus f(1)\rangle - |1 \oplus f(1)\rangle))$$

$$+ \frac{1}{2\sqrt{2}}(|1\rangle \otimes (|0 \oplus f(0)\rangle - |1 \oplus f(0)\rangle - |0 \oplus f(1)\rangle + |1 \oplus f(1)\rangle)).$$

If f is constant we have $f(0) = f(1)$ and we find

$$(H \otimes I_2)U_f|\psi\rangle = \frac{1}{\sqrt{2}}|0\rangle \otimes (|0 \oplus f(0)\rangle - |1 \oplus f(0)\rangle).$$

If f is varying we have

$$0 \oplus f(0) = 1 \oplus f(1),$$
$$1 \oplus f(0) = 0 \oplus f(1)$$

and we find

$$(U_H \otimes I_2)U_f|\psi\rangle = \frac{1}{\sqrt{2}}|1\rangle \otimes (|0 \oplus f(0)\rangle - |1 \oplus f(0)\rangle).$$

Thus measuring the first qubit as zero indicates the function f is constant and measuring the first qubit as one indicates the function f is varying. The function f was only calculated once using U_f and so the problem is solved.

The algorithm has been implemented using nuclear magnetic resonance techniques [43, 98].

A balanced boolean function is a boolean function which maps to an equal number of 0's and 1's. In other words $f : \{0,1\}^n \to \{0,1\}$ is balanced if

$$\left| \{\, x \in \{0,1\}^n \mid f(x) = 0 \,\} \right| = \left| \{\, x \in \{0,1\}^n \mid f(x) = 1 \,\} \right|.$$

Deutsch and Josza [47, 55] generalized the problem. Let $f : \{0,1\}^n \to \{0,1\}$ be a boolean function. Assume that f is either constant or balanced. Thus f maps only to 0, only to 1, or to an equal number of 0's and 1's for all possible inputs. The problem is to determine if f is constant or balanced using only one function evaluation. Let U_f be the unitary operator which implements the function f,

$$U_f |x\rangle \otimes |y\rangle = |x\rangle \otimes |y \oplus f(x)\rangle$$

where $x \in \{0,1\}^n$ and $y \in \{0,1\}$. We note first that

$$U_f \frac{1}{\sqrt{2}} |x\rangle \otimes (|0\rangle - |1\rangle) \;=\; \frac{1}{\sqrt{2}} |x\rangle \otimes (|0 \oplus f(x)\rangle - |1 \oplus f(x)\rangle)$$

$$= \frac{(-1)^{f(x)}}{\sqrt{2}} |x\rangle \otimes (|0\rangle - |1\rangle).$$

We also use the fact that

$$\bigotimes_n U_H |x\rangle = \frac{1}{\sqrt{2^n}} \sum_{j \in \{0,1\}^n} (-1)^{j * x} |j\rangle,$$

where $x \in \{0,1\}^n$ and $j * x$ denotes the bitwise AND of j and x followed by the XOR of the resulting bits

$$x * y = (x_0 \cdot y_0) \oplus (x_1 \cdot y_1) \oplus \ldots \oplus (x_{n-1} \cdot y_{n-1}).$$

The initial state is

$$|\psi\rangle := \frac{1}{\sqrt{2}} |0\rangle \otimes (|0\rangle - |1\rangle).$$

The first step is to apply the Walsh-Hadamard transform to the first n qubits of $|\psi\rangle$.

$$\left(\bigotimes_n U_H\right)|\psi\rangle = \frac{1}{\sqrt{2^n}} \sum_{x\in\{0,1\}^n} |x\rangle \otimes (|0\rangle - |1\rangle).$$

Applying the unitary operator U_f yields

$$U_f\left(\bigotimes_n U_H\right)|\psi\rangle = \frac{1}{\sqrt{2^n}} \sum_{x\in\{0,1\}^n} (-1)^{f(x)}|x\rangle \otimes (|0\rangle - |1\rangle).$$

Applying the Walsh-Hadamard transform on the first n qubits gives

$$\left(\bigotimes_n U_H\right)U_f\left(\bigotimes_n U_H\right)|\psi\rangle = \frac{1}{2^n} \sum_{j,x\in\{0,1\}^n} (-1)^{f(x)}(-1)^{j*x}|j\rangle \otimes (|0\rangle - |1\rangle).$$

The probability that a measurement of the first n qubits yields $|00\ldots0\rangle$ is

$$\left|\frac{1}{2^n}\sum_{x\in\{0,1\}^n}(-1)^{f(x)}\right|^2 = \begin{cases} 1 & \text{if } f \text{ is constant} \\ 0 & \text{if } f \text{ is balanced} \end{cases}.$$

Thus, after applying these transformations, measuring all $|0\rangle$ for the first n qubits indicates that f is constant and any other result indicates that f is balanced. The network representation for the algorithm is given in Figure 21.1.

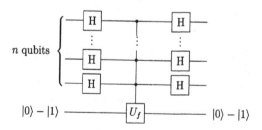

Figure 21.1: Network Representation to Solve Deutsch's Problem

21.2   Simon's Problem

Simon's problem[150] can be viewed as a generalization of Deutsch's problem. Suppose we have a function

$$f : \{0,1\}^n \to \{0,1\}^m, \quad m \geq n.$$

Furthermore, it is given that either f is one to one, or there exists a non-trivial s such that for all $x \neq y$

$$f(x) = f(y) \iff y = x \oplus s.$$

The problem is to determine if f is one to one, and if not to find s.

We start with the initial state

$$\bigotimes_{n+m} |0\rangle.$$

Next we apply the Walsh-Hadamard transform to each of the first n qubits,

$$\left(\bigotimes_n U_H \otimes \bigotimes_m I_2\right) \bigotimes_{n+m} |0\rangle = \frac{1}{\sqrt{2^n}} \sum_{j=0}^{2^n-1} |j\rangle \otimes \bigotimes_m |0\rangle.$$

Only one function evaluation is required. This is the next step in the solution.

$$U_f \frac{1}{\sqrt{2^m}} \sum_{j=0}^{2^n-1} |j\rangle \otimes \bigotimes_m |0\rangle = \frac{1}{\sqrt{2^n}} \sum_{j=0}^{2^n-1} |j\rangle \otimes |f(j)\rangle.$$

The final step is to apply the Walsh-Hadamard transform to the first n qubits again.

$$\left(\bigotimes_n U_H \otimes \bigotimes_m I_2\right) \frac{1}{\sqrt{2^n}} \sum_{j=0}^{2^n-1} |j\rangle \otimes |f(j)\rangle = \frac{1}{2^n} \sum_{j=0}^{2^n-1} \sum_{k=0}^{2^n-1} (-1)^{j*k} |k\rangle \otimes |f(j)\rangle.$$

Suppose now that for all $x \neq y$

$$f(x) = f(y) \iff y = x \oplus s.$$

The amplitude of the state $|k\rangle \otimes |f(j)\rangle$ is given by

$$\frac{1}{2^n} \left((-1)^{j*k} + (-1)^{(j \oplus s)*k} \right).$$

We have

$$
\begin{aligned}
(j \oplus s) * k &= (j_0 \oplus s_0)k_0 + (j_1 \oplus s_1)k_1 + \ldots + (j_{n-1} \oplus s_{n-1})k_{n-1} \quad \text{mod } 2 \\
&= j_0 k_0 + s_0 k_0 + j_1 k_1 + s_1 k_1 + \ldots + j_{n-1}k_{n-1} + s_{n-1}k_{n-1} \quad \text{mod } 2 \\
&= j * k + s * k \quad \text{mod } 2.
\end{aligned}
$$

Thus if $s * k = 0$,

$$(j \oplus s) * k = j * k.$$

If $s * k \neq 0$ then the amplitude of the state $|k\rangle \otimes |f(j)\rangle$ is zero. Thus measuring the $n+m$ qubits yields with certainty a number t such that $t*s = 0$. Repeating the above procedure $O(n)$ times will yield enough linearly independent t and corresponding equations of the form $t * s = 0$ so that s can be found.

If f is one to one, then each measurement will yield a random value. The resulting s determined from the equations must then be tested, for example the values $f(0)$ and $f(s)$ can checked for equality.

The expected time of the algorithm is $O(nT_f(n) + G(n))$ where $T_f(n)$ is the time required to compute f on inputs of n bits and $G(n)$ is the time required to solve n linear equations for n unknowns in $\{0,1\}$.

The algorithm does not guarantee completion in polynomial time for the worst case. A quantum algorithm which is guaranteed to complete in polynomial time (or less) is called an *exact quantum polynomial time algorithm*. Brassard and Høyer [27] discovered an exact quantum polynomial time algorithm to solve Simon's problem. Their algorithm is based on Simon's solution with the addition that after each iteration, 0 and values of t already found are removed from the superposition. Thus the time to determine the n linear equations is precisely determined.

We can eliminate the state with the first n qubits equal to t as follows. Suppose the lth bit of t is 1. We begin with the final state of an iteration of Simon's solution. We add an auxiliary qubit $|0\rangle$ and then apply the controlled NOT with qubit l as the control and the auxiliary qubit as the target which gives the state

$$
U_{CNOT}(l, n+m+1) \frac{1}{2^n} \sum_{j=0}^{2^n-1} \sum_{k=0}^{2^n-1} (-1)^{j*k} |k\rangle \otimes |f(j)\rangle \otimes |0\rangle
$$

$$
= \frac{1}{2^n} \sum_{j=0}^{2^n-1} \sum_{k=0}^{2^n-1} (-1)^{j*k} |k\rangle \otimes |f(j)\rangle \otimes |k_l\rangle
$$

$$
= \frac{1}{2^n} \sum_{j=0}^{2^n-1} \sum_{p=0}^{1} \sum_{k_l=0}^{2^n-1} (-1)^{j*(k \oplus pt)} |k \oplus pt\rangle \otimes |f(j)\rangle \otimes |p\rangle
$$

where k_l is the lth bit of k and $0t = 0$ and $1t = t$. We obtain this result by separating the sum over those k with the lth bit 0 and those with the lth bit 1. We also make use of the fact that the k form a group with $\oplus$, the bitwise XOR operation. Thus $(k \oplus pt)$ will take all the values over k.

Now we apply the operator which maps $|x\rangle$ to $|x \oplus t\rangle$, for the first n qubits, only when the auxiliary qubit is $|1\rangle$ and leaves the first n qubits unchanged otherwise. In other words we apply the operator

$$U_{\oplus t} = \left(\sum_{k=0}^{2^n-1} |k \oplus t\rangle \langle k| \right) \otimes I_m \otimes |1\rangle\langle 1| + I_{n+m} \otimes |0\rangle\langle 0|$$

which can be built using n controlled controlled NOT gates. Applying the operator $U_{\oplus t}$ yields

$$U_{\oplus t} \frac{1}{2^n} \sum_{j=0}^{2^n-1} \sum_{p=0}^{1} \sum_{k_l=0}^{2^n-1} (-1)^{j*(k\oplus pt)} |k \oplus pt\rangle \otimes |f(j)\rangle \otimes |p\rangle$$

$$= \frac{1}{2^n} \sum_{j=0}^{2^n-1} \sum_{p=0}^{1} \sum_{k_l=0}^{2^n-1} (-1)^{j*(k\oplus mt)} |k\rangle \otimes |f(j)\rangle \otimes |m\rangle$$

$$= \frac{1}{2^n} \sum_{j=0}^{2^n-1} \sum_{p=0}^{1} \sum_{k_l=0}^{2^n-1} (-1)^{j*k} |k\rangle \otimes |f(j)\rangle \otimes \left((-1)^{j*pt} |p\rangle \right)$$

Since the first $n+m$ qubits are independent of p we can discard the auxiliary qubit. Thus the final state is

$$\frac{1}{2^{n-1}} \sum_{j=0}^{2^n-1} \sum_{k_l=0}^{2^n-1} (-1)^{j*k} |k\rangle \otimes |f(j)\rangle.$$

This state is the same form as the final state of each iteration of Simon's solution, except for the fact that the probability of measuring the first n qubits as the t already found is 0 and the probability of measuring some other t' with $t' * s = 0$ is greater. This process can be repeated to eliminate all of the t values with $t * s = 0$ already found. The above modification to the algorithm does not ensure that the first n qubits will never be measured as 0. To remove this possibility, Brassard and Høyer [27] use a modification of Grover's search algorithm which, under certain conditions, succeeds with probability 1. The technique is discussed in Section 21.6.

21.3   Quantum Fourier Transform

The *discrete Fourier transform* of a set of data $x(0), x(1), \ldots, x(n-1) \in \mathbf{C}$ is given by [160, 162]

$$\hat{x}(k) = \frac{1}{n} \sum_{j=0}^{n-1} x(j) e^{-i2\pi kj/n}.$$

The inverse is given by

$$x(j) = \sum_{k=0}^{n-1} \hat{x}(k) e^{i2\pi kj/n}.$$

The discrete Fourier transform is useful in finding periodicity in a sequence. The definition can be extended for quantum states as follows when $n = 2^m$ and provided

$$|x(0)|^2 + |x(1)|^2 + \ldots + |x(n-1)|^2 = n.$$

The transform can then be written as follows

$$U_{QFT} \frac{1}{\sqrt{n}} \sum_{k=0}^{2^m-1} x(k)|k\rangle = \sum_{k=0}^{2^m-1} \hat{x}(k)|k\rangle.$$

The transform is unitary and is called the *quantum Fourier transform*. We have [47, 162]

$$U_{QFT} = \frac{1}{\sqrt{n}} \begin{pmatrix} 1 & 1 & 1 & \cdots & 1 \\ 1 & e^{-i2\pi/n} & e^{-i4\pi/n} & \cdots & e^{-i(n-1)2\pi/n} \\ 1 & e^{-i4\pi/n} & e^{-i8\pi/n} & \cdots & e^{-i(n-1)4\pi/n} \\ \vdots & \vdots & \vdots & \ddots & \vdots \\ 1 & e^{-i(n-1)2\pi/n} & e^{-i(n-1)4\pi/n} & \cdots & e^{-i(n-1)^2 2\pi/n} \end{pmatrix}.$$

The transform can be implemented using $\frac{m(m+1)}{2}$ single and double qubit gates [47, 138]. Let H_k denote the Hadamard transform acting on qubit k and $U_{PS}(k, j, \phi)$ denote the phase shift transform $U_{PS}(\phi)$ acting on qubits k and j. We can rewrite the transform as (by relabeling the sums)

$$U_{QFT} \frac{1}{\sqrt{2^m}} \sum_{k=0}^{2^m-1} x(k)|k\rangle = \frac{1}{\sqrt{2^m}} \sum_{k=0}^{2^m-1} x(k) \frac{1}{\sqrt{2^m}} \sum_{j=0}^{2^m-1} e^{-i2\pi kj/2^m} |j\rangle.$$

Thus, for $k = k_0 2^0 + k_1 2^1 + \ldots + k_{m-1} 2^{m-1}$, we have

$$
\begin{aligned}
U_{QFT}|k\rangle &= \frac{1}{\sqrt{2^m}} \sum_{j=0}^{2^m-1} \exp(-i2\pi k j/2^m)|j\rangle \\
&= \frac{1}{\sqrt{2^m}} \sum_{j_0,j_1,\ldots,j_{m-1}\in\{0,1\}} \bigotimes_{l=0}^{m-1} \exp(-i2\pi k j_l 2^{l-m})|j_l\rangle \\
&= \frac{1}{\sqrt{2^m}} \bigotimes_{l=0}^{m-1} (|0\rangle + \exp(-i2\pi k 2^{l-m})|1\rangle) \\
&= \frac{1}{\sqrt{2^m}} \bigotimes_{l=0}^{m-1} (|0\rangle + \prod_{j=0}^{m-l-1} \exp(-i2\pi k_j 2^{j+l-m})|1\rangle) \\
&= \frac{1}{\sqrt{2^m}} \bigotimes_{l=0}^{m-1} (|0\rangle + \prod_{j=l}^{m-1} \exp(-i2\pi k_j 2^{j-m})|1\rangle) \\
&= \prod_{l=0}^{m-1} \left(\prod_{j=l+1}^{m-1} U_{PS}(j,l,-2\pi 2^{j-m}) \right) H_l |k\rangle
\end{aligned}
$$

The network representation is given in Figure 21.2.

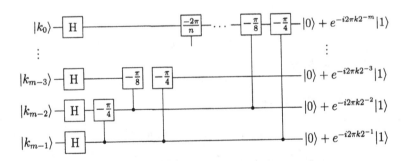

Figure 21.2: Network for the Quantum Fourier Transform

21.4   Factoring (Shor's Algorithm)

Shor [148] invented an algorithm for a quantum computer that could be used to find the prime factors of integer numbers in polynomial time. The mathematical basis of Shor's algoritm is a follows [8, 18, 118, 148, 118]. The aim is to find the prime factors of an integer N. Let $N = 21$. Then N can be written as $N = 3 \cdot 7$, where 3 and 7 are the prime factors of $N = 21$. This factorization problem can be related to finding the period r of the function ($x = 0, 1, 2, \ldots$)

$$f_{a,N}(x) := a^x \bmod N$$

where a is any randomly chosen positive integer ($a < N$) which is *coprime* with N, i.e. which has no common factors with N. If a is not coprime with N, then the factors of N are trivially found by computing the greatest common divisor of a and N.

Example. Let $N = 21$ and $a = 11$. Thus N and a have no common factors. Now for $x = 0, 1, 2, 3, 4, 5, 6$ we find

$$f_{11,21}(0) = 1, \quad f_{11,21}(1) = 11, \quad f_{11,21}(2) = 16, \quad f_{11,21}(3) = 8,$$

$$f_{11,21}(4) = 4, \quad f(11,21)(5) = 2, \quad f_{11,21}(6) = 1.$$

Thus we see that the period is $r = 6$. The period r can also be found by solving the equation

$$a^r = 1 \bmod N$$

for the smallest positive integer r. Obviously we find

$$11^6 = 1 \bmod 21$$

♣

Since

$$a^r a^x = a^x \bmod N$$

for all $x \in \mathbf{N}$, we have

$$a^r = 1 \bmod N$$

where r is even. This is equivalent to

$$(a^{\frac{r}{2}} + 1)(a^{\frac{r}{2}} - 1) = 0 \bmod N.$$

Knowing the period of $f_{a,N}$, we can factor N provided r is even and

$$a^{\frac{r}{2}} \bmod N \neq -1.$$

For the example given above these two conditions are met. When a is chosen randomly the two conditions are satisfied with probability greater than 1/2 [47].

The factors of N are then given by

$$\gcd(a^{r/2} + 1, N), \qquad \gcd(a^{r/2} - 1, N).$$

The greatest common divisor can be found using the Euclidean algorithm. The Euclidean algorithm runs in polynomial time on a classical computer. For the example given above with $N = 21$, $a = 11$ and $r = 6$ we find

$$\gcd(11^3 + 1, 21) = 3, \qquad \gcd(11^3 - 1, 21) = 7.$$

Next we describe how a quantum computer can find the period r of the number a and therefore factorize N. Shor's technique for finding the period of a periodic function consists of evaluating the function on a superposition of exponentially many arguments, computing a parallel Fourier transform on the superposition, then sampling the Fourier power spectrum to obtain the function's period.

The quantum computer is prepared with two quantum registers, X and Y, each consisting of a string of qubits initialized to the Boolean value zero

$$|\psi_0\rangle \equiv |0\rangle \otimes |0\rangle.$$

The X register is used to hold arguments of the function f whose unknown period r is sought, and the Y register is used to store values of the function. The width of the X register is chosen so that its number of possible Boolean states is comfortably greater than the square of the anticipated period r. The Y register is made sufficiently wide to store values of the function f. We first set the input register X into an equally weighted superposition of all possible states from 0 to $2^{2L} - 1$, where

$$2^{2L} - 1 \approx N^2.$$

This can be achieved by applying the Hadamard transform to each qubit of the input register X, i.e.

$$|\psi_1\rangle \equiv (U_H \otimes U_H \otimes \ldots \otimes U_H) \otimes I|0\rangle \otimes |0\rangle = \frac{1}{2^L} \sum_{x=0}^{2^{2L}-1} |x\rangle \otimes |0\rangle.$$

Applying the operator $I \otimes U_{f_{a,N}}$ to this state we obtain the state

$$|\psi_2\rangle \equiv \frac{1}{2^L} \sum_{x=0}^{2^{2L}-1} |x\rangle \otimes |f_{a,N}(x)\rangle.$$

At this stage, all the possible values of $f_{a,N}$ are encoded in the state of the second register. However, they are not all accessible at the same time. However, we are not interested in the values themselves, but only in the periodicity of the function $f_{a,N}$. Observing the second quantum register as u would yield

$$\frac{1}{\sum_{k=0}^{2^{2L}-1}(g_u(k))^2}\sum_{k=0}^{2^{2L}-1}g_u(k)|k\rangle\otimes|u\rangle$$

with

$$g_u(x):=\left\{\begin{array}{ll}1 & a^x=u\ \mathrm{mod}\ N\\0 & \mathrm{otherwise}\end{array}\right..$$

The next step is to Fourier transform the first register. This means we apply a unitary operator that maps the state onto

$$|\psi_3\rangle\equiv\frac{1}{2^{2L}}\sum_{x=0}^{2^{2L}-1}\sum_{k=0}^{2^{2L}-1}\exp(2\pi ixk/2^{2L})\,|k\rangle\otimes|f_{a,N}\rangle\,.$$

The probability for finding the state

$$|k\rangle\otimes|a^m\,\mathrm{mod}\,N\rangle\equiv|k\rangle\otimes|f_{a,N}(m)\rangle$$

is

$$P(k,a^m\,\mathrm{mod}\,N)=\left|\frac{1}{2^{2L}}\sum_{a^x=a^m}\exp(2\pi ixk/2^{2L})\right|^2$$

where the sum is over all numbers

$$0\le x\le 2^{2L}-1$$

such that

$$a^x=a^m\,\mathrm{mod}\,N\,.$$

Using $x=m+br$ with $b\in\mathbf{N}_0$ the sum becomes

$$P(k,a^m\,\mathrm{mod}\,N)=\left|\frac{1}{2^{2L}}\sum_{b=0}^{[(2^{2L}-1-m)/r]}\exp(2\pi i(br+m)k/2^{2L})\right|^2\,.$$

Factoring out the term $\exp(2\pi imk/2^{2L})$ yields

$$P(k, a^m \bmod N) = \left| \frac{1}{2^{2L}} \sum_{b=0}^{[(2^{2L}-1-m)/r]} \exp(2\pi ib\{rk\}_{2^{2L}}/2^{2L}) \right|^2$$

where $\{rk\}$ is an integer in the interval

$$-2^{2L}/2 \le \{rk\} \le 2^{2L}/2$$

which is congruent to

$$rk \bmod (2^{2L} - 1).$$

The above probability has well-defined peaks if

$$\{rk\}_{2^{2L}}$$

is small (less than r), i.e., if rk is a multiple of 2^{2L}

$$rk = d2^{2L}$$

for some $d < N$. Thus, knowing L and therefore 2^{2L} and the fact that the position of the peaks k will be close to numbers of the form $d2^{2L}$, we can find the period r using continuous fraction techniques. To explicitly construct the unitary evolution that takes the state $|\psi_1\rangle$ into the state $|\psi_2\rangle$ is a rather nontrivial task [118].

There are $r\phi(r)$ states which can be used to determine r [148], where ϕ is Euler's totient function. Each state occurs in the superposition with probability at least $1/(3r^2)$, so the probability of measuring a state which can be used to determine r is $\phi(r)/(3r)$. Using the fact that

$$\frac{\phi(r)}{r} > \frac{u}{\log_2(\log_2 r)}$$

for some u, the probability of measuring such a state is at least $u/(\log_2(\log_2 r))$. Thus repeating the algorithm $O(\log_2(\log_2 r))$ times ensures a high probability of successfully measuring a state from which r can be determined.

21.5 The Hidden Subgroup Problem

In the previous sections we discussed important algorithms which illustrate the characteristics of quantum computation. Shor [148] also discussed the problem of finding the *discrete logarithm*, i.e. suppose we have $x = a^y$ then the discrete log of x with base a is y. Kitaev was able to generalize these problems in terms of the *Abelian stabilizer problem* [102]. Let G be any group acting on a finite set X. Each element g of G acts as a map $g : X \to X$ such that for all $g_1, g_2 \in G$ we have $g_1(g_2(x)) = (g_1 \cdot g_2)(x)$ where $\cdot$ is the group operation. The stabilizer $S_G(x)$ of $x \in X$ is the subgroup of G with $s(x) = x$ for all $s \in S_G(x)$. The problem is to find the stabilizer for a given x. When G is abelian then we have the Abelian stabilizer problem.

Now the algorithms of Deutsch, Simon, Shor and Kitaev as well as others can be formulated group theoretically as a *hidden subgroup problem* [93, 100, 121]. Let f be a function from a finitely generated group G to a finite set X such that f is constant on the cosets of a subgroup K and distinct on each coset. The cosets of K are the sets

$$g \cdot K := \{ g \cdot k \,|\, k \in K \}, \quad g \in G.$$

The cosets partition G, i.e. the union of all the cosets is the set of the group G and every two cosets are equal or their intersection is empty. Thus we write $f : G \to X$ and

$$K = \{ k \in G \,|\, f(k \cdot g) = f(g), \ \forall g \in G \}.$$

The problem is, for given f and G determine the hidden subgroup K.

We describe briefly how the above mentioned problems can be expressed in terms of the hidden subgroup problem.

Deutsch's problem. We set $G = Z_2 = \{0, 1\}$ with the group operation $\cdot = \oplus$, the XOR operation. The only subgroups are $\{0\}$ and $\{0, 1\}$. If $K = \{0\}$ then f is balanced and if $K = \{0, 1\}$ then f is constant.

Simon's problem. For Simon's problem we have $G = \{0, 1\}^n$ with $\cdot = \oplus$, the bitwise XOR operation. Simon's problem requires that $f(x) = f(y)$ if and only if $x = y$ or $x = y \oplus s$. Immediately we see that $K = \{0, s\}$.

Shor's factorization algorithm. Let $f_{a,N}(x) := a^x \bmod N$ where a is given and is coprime with N and $x \in G = \mathbf{Z}_N$ the group of integers with addition modulo N. In this case $K = \{0, r, 2r, \ldots, kr\}$ where r is the period of $f_{a,N}$ and kr is the greatest multiple of r which is less than N.

Discrete logarithm. Let G be the group $Z_r \times Z_r$ where Z_r is the additive group of integers modulo r. Let $a, b \in G$ with $b = a^m$ be given. We wish to find m. Further, let $f : G \to G$ be defined by $f(x, y) = a^x b^y$. Obviously $f(x, y) = a^{x+my}$. We assume that it is known that a is of order r, i.e. $a^r = 1$. Thus $f(x_1, y_1) = f(x_2, y_2)$ if and only if

$$x_1 - x_2 = -m(y_1 - y_2) \bmod r.$$

Equivalently $f(x, y) = f(x \cdot s, y \cdot t)$ if and only if $s = -mt \bmod r$. The hidden subgroup (which is used to determine m) is

$$K = \{ (k, -km) \mid k = 0, 1, \ldots, r - 1 \}.$$

Abelian stabilizer problem. Let G be a group acting on a finite set X. Let $f : G \to X$ be defined by $f(g) = g(x)$ for a given $x \in X$. Here $K = S_G(x)$.

The quantum Fourier transform is an important component of quantum algorithms. Defining a quantum Fourier transform on an Abelian group [93, 99, 100] is necessary for a description of the algorithm to solve the Abelian subgroup problem. Suppose U_f is the unitary operation which implements f, i.e.

$$U_f(|g\rangle \otimes |0\rangle) = |g\rangle \otimes |f(g)\rangle$$

where $g \in G$. For any $k_1, k_2 \in K$ we have $f(g \cdot k_1) = f(g \cdot k_2)$. Define for $g \in G$

$$|g \cdot K\rangle := \frac{1}{\sqrt{|K|}} \sum_{k \in K} |g \cdot k\rangle.$$

Thus (f is constant on a coset of K)

$$U_f(|g \cdot K\rangle \otimes |0\rangle) = |g \cdot K\rangle \otimes |f(g)\rangle.$$

If we apply U_f to the superposition

$$|G\rangle \otimes |0\rangle = \sum_{g \in G} |g\rangle \otimes |0\rangle$$

and measure the function value, the measurement projects the first register onto one of the cosets of K (since the cosets form a partition of G). From the coset we would like to determine K. The states $|g \cdot k\rangle$ are all displaced by g with respect to the group operation. We can associate the idea of a periodic sequence $g \cdot k$ where the "period" of the sequence is the generator of the subgroup K. Thus we can try to apply a transform analogous to the quantum Fourier transform. The transform is constructed using techniques from group representation theory. For more information see Josza [99, 100].

The above mentioned problems are all defined with Abelian groups. The construction of the Fourier transform is also for Abelian groups. Non-Abelian hidden subgroup problems create more difficulties, for some results on these problems see for example Ivanyos et al. [95], and Rötteler and Beth [140].

21.6   Unstructured Search (Grover's Algorithm)

In some problems it is necessary to determine if any element in a set S satisfies a certain property. An immediate example is the problem of satisfiability. Suppose $P(x_1, x_2, \ldots, x_n)$ is a predicate on n boolean variables, P can be satisfied if some combination of assignments to the boolean variables results in P being true. If 0 represents false and 1 represents true, we can directly associate bit sequences with assignments to the n boolean variables in the predicate. The task is to determine if any bit sequence causes the predicate to be true. Classically every assignment might have to be checked until one satisfies the predicate. This process is of order $O(2^n)$. Grover [78, 79] proposed a solution that is of order $O(2^{\frac{n}{2}})$. Let

$$|\psi_0\rangle = \left(\left(\bigotimes_n U_H\right) \otimes I\right)\left(\frac{1}{\sqrt{2}}\left(\bigotimes_{n+1}|0\rangle\right) \otimes (|0\rangle - |1\rangle)\right) = \frac{1}{\sqrt{2^{n+1}}}\sum_{x=0}^{2^n}|x\rangle \otimes (|0\rangle - |1\rangle).$$

We use the notation

$$\bigotimes_n U := \overbrace{U \otimes U \otimes \ldots \otimes U}^{n-\text{times}}.$$

Now suppose the unitary operator U_P performs the operation

$$U_P(|x_1, x_2, \ldots, x_n\rangle \otimes |b\rangle) = |x_1, x_2, \ldots, x_n\rangle \otimes |b \oplus P(x_1, x_2, \ldots, x_n)\rangle.$$

Denote by X_T the set of all bit sequences of length n which satisfy P, and by X_F the set of all bit sequences of length n which do not satisfy P. Thus applying U_P to $|\psi_0\rangle$ gives

$$|\psi_1\rangle = U_P|\psi_0\rangle$$

$$= \frac{1}{\sqrt{2^{n+1}}}U_P\left(\sum_{x \in X_T}|x\rangle \otimes |0\rangle - \sum_{x \in X_T}|x\rangle \otimes |1\rangle + \sum_{x \in X_F}|x\rangle \otimes |0\rangle - \sum_{x \in X_F}|x\rangle \otimes |1\rangle\right)$$

$$= \frac{1}{\sqrt{2^{n+1}}}\left(\sum_{x \in X_T}|x\rangle \otimes |1\rangle - \sum_{x \in X_T}|x\rangle \otimes |0\rangle + \sum_{x \in X_F}|x\rangle \otimes |0\rangle - \sum_{x \in X_F}|x\rangle \otimes |1\rangle\right)$$

$$= \frac{1}{\sqrt{2^{n+1}}}\left(\sum_{x \in X_F}|x\rangle - \sum_{x \in X_T}|x\rangle\right) \otimes (|0\rangle - |1\rangle)$$

The amplitudes of the bit sequences satisfying P are negative. If the last qubit had only been $|0\rangle$, the sequences satisfying P would be marked with a $|1\rangle$ in the last qubit, but measuring would yield any sequence with equal probability, in general obtaining a sequence which satisfies P would have low probability. The state $|\psi_1\rangle$ is not much better, but can be manipulated to increase the probability of measuring a sequence which satisfies P. The next step is to increase the absolute value of the elements of X_T, i.e. those elements in the superposition with negative amplitudes. This is done with the inversion around average operation. This operation maps each amplitude a_i to $2A - a_i$ where A is the average of all the amplitudes. We note

$$2A - a_i \equiv A + (A - a_i)$$

which explains the terminology "inversion around the average". The operation is repesented by the $2^n \times 2^n$ unitary matrix U_{IA} (where IA indicates inversion about the average)

$$(U_{IA})_{ij} = \frac{2}{2^n} - \delta_{ij}, \qquad i,j = 1,2,\ldots,2^n.$$

For $n = 1$ and $n = 2$ we have the matrices

$$\begin{pmatrix} 0 & 1 \\ 1 & 0 \end{pmatrix}, \qquad \begin{pmatrix} -\frac{1}{2} & \frac{1}{2} & \frac{1}{2} & \frac{1}{2} \\ \frac{1}{2} & -\frac{1}{2} & \frac{1}{2} & \frac{1}{2} \\ \frac{1}{2} & \frac{1}{2} & -\frac{1}{2} & \frac{1}{2} \\ \frac{1}{2} & \frac{1}{2} & \frac{1}{2} & -\frac{1}{2} \end{pmatrix}.$$

The operator U_{IA} can be written as

$$U_{IA} = \begin{pmatrix} \frac{2}{2^n} & \cdots & \frac{2}{2^n} \\ \vdots & \ddots & \vdots \\ \frac{2}{2^n} & \cdots & \frac{2}{2^n} \end{pmatrix} - I_{2^n}$$

$$= \left(\bigotimes_n U_H \right) \begin{pmatrix} \frac{2}{\sqrt{2^n}} & \cdots & \frac{2}{\sqrt{2^n}} \\ 0 & \cdots & 0 \\ \vdots & \ddots & \vdots \\ 0 & \cdots & 0 \end{pmatrix} - I_{2^n}$$

$$= \left(\bigotimes_n U_H \right) \mathrm{diag}(2,0,\ldots,0) \left(\bigotimes_n U_H \right) - I_{2^n}$$

$$= \left(\bigotimes_n U_H \right) (\mathrm{diag}(2, 0, \ldots, 0) - I_{2^n}) \left(\bigotimes_n U_H \right)$$

$$= \left(\bigotimes_n U_H \right) \mathrm{diag}(1, -1, \ldots, -1) \left(\bigotimes_n U_H \right)$$

where I_{2^n} is the $2^n \times 2^n$ unit matrix. If only one state satisfies P, the inversion about average operation inverts and increases the amplitude of the state with negative amplitude while the other amplitudes decrease. The process is repeated (calculate P on the bit sequences with the amplitudes of those states satisfying P negative, and inversion about the average) $\frac{\pi}{8}\sqrt{2^n}$ times for a greater than 50% chance of obtaining the state which satisifes P [26]. The algorithm also works when more than one x satisfies P [26]. Unlike classical algorithms, applying the process further will lead to a decrease in the probability of measuring the required state. This is due to the fact that the states are normalized, and operations perform a rotation in the state space. Since iterations of the algorithm always perform the same rotation, the rotation must at some stage necessarily move away from the desired state, although it may approach the desired state again under further iteration of the algoritm. We can think of the algorithm as the rotation of a ray from the origin to the surface of the unit ball. For the case of a single qubit with real amplitudes the rotation is on the unit circle. Figure 21.3 gives the network representation of the algorithm.

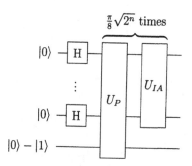

Figure 21.3: Network Representation of Grover's Algorithm

The algorithm has been generalized for the case when the amplitudes of the states in the superposition are initially in an arbitrary configuration [22].

Bennett et al [20] found lower bounds for the unstructured search problem and proved that a square root speed up (obtained by Grover's algorithm) is optimal.

Consider any quantum algorithm A for solving the unstructured search problem. First we do a test run of A on the function $f \equiv 0$. Define the query magnitude of x

to be $\sum_t |\alpha_{x,t}|^2$, where $\alpha_{x,t}$ is the amplitude with which A queries x at time t. The expectation value of the query magnitude is $T/2^n$. Thus,

$$\min_x \left(\sum_t |\alpha_{x,t}|^2 \right) \leq \frac{T}{2^n}.$$

For such x, the Cauchy-Schwarz inequality gives

$$\sum_t |\alpha_{x,t}| \leq \frac{T}{\sqrt{2^n}}.$$

Let the states of the algorithm A run on f be $|\phi_0\rangle, |\phi_1\rangle, \ldots, |\phi_T\rangle$. We run the algorithm A on the function

$$g(x) := \begin{cases} 1 & x = y \\ 0 & x \neq y \end{cases}.$$

Suppose the final state of A run on g is $|\psi_T\rangle$. Then [20] $\| |\phi_T\rangle - |\psi_T\rangle \|$ must be small and

$$|\psi_T\rangle = |\phi_T\rangle + |E_0\rangle + |E_1\rangle + \ldots + |E_{T-1}\rangle,$$

where

$$\| |E_t\rangle \| \leq |\alpha_{x,t}|,$$

and $\| |x\rangle \|$ denotes the norm $\sqrt{|\langle x|x\rangle|}$.

The algorithm has also been proposed as a fast database search. This can be achieved as follows. Suppose we search for the data relating to an item x_{id}. The predicate P_{id} will search for a match for x_{id}. For simplicity we assume that there are 2^n items stored in the database. It is simple to construct the n bit quantum register

$$|s\rangle = (\bigotimes_n U_H)(\bigotimes_n |0\rangle) = \frac{1}{\sqrt{2^n}} \sum_{j=0}^{2^n-1} |j\rangle.$$

Now we prepare the database state as a tensor product of all items in the database, the identifier of the item we are searching for, a qubit to store the search result and

$|s\rangle$. A superposition state of all the items in the database could also be used, but this would require determining for each bit sequence if the item is in the database which reduces the efficiency to no better than classical. We also assume that the database is maintained in this tensor product form since constructing the quantum database each time for a search again reduces the efficiency to no better than classical. The initial state is

$$|\psi_0\rangle = \frac{1}{\sqrt{2}}|x_{id}\rangle \otimes |s\rangle \otimes \left(\bigotimes_{j=0}^{2^n-1} (|x_j\rangle \otimes |d_j\rangle) \right) \otimes (|0\rangle - |1\rangle).$$

This state associates the data $|d_j\rangle$ with the identifier $|x_j\rangle$. We define the unitary operator U_P' by

$$U_P' \left(\frac{1}{\sqrt{2}}|x_{id}\rangle \otimes |k\rangle \otimes \left(\bigotimes_{j=0}^{2^n-1} (|x_j\rangle \otimes |d_j\rangle) \right) \otimes (|0\rangle - |1\rangle) \right) = |\psi_P\rangle$$

$$|\psi_P\rangle := \left(\frac{1}{\sqrt{2}}|x_{id}\rangle \otimes |j\rangle \otimes \left(\bigotimes_{j=0}^{2^n-1} (|x_j\rangle \otimes |d_j\rangle) \right) \otimes (|0 \oplus \delta_{x_{id},x_k}\rangle - |1 \oplus \delta_{x_{id},x_k}\rangle) \right).$$

Thus applying U_P' to $|\psi\rangle$ yields

$$U_P'|\psi\rangle = \left(\frac{1}{\sqrt{2}} \sum_{k=0}^{2^n-1} |x_{id}\rangle \otimes |k\rangle \otimes \left(\bigotimes_{j=0}^{2^n-1} |x_j\rangle \otimes |d_j\rangle \right) \right) \otimes (|0 \oplus \delta_{x_{id},x_k}\rangle - |1 \oplus \delta_{x_{id},x_k}\rangle).$$

From this point the algorithm proceeds as before. The probability of registering the second quantum register as the id representing x_{id} is greater than $\frac{1}{2}$ and identifies the element in the database to examine.

Grover's algorithm has an interesting property when $\frac{1}{4}$ of the states in the superposition are states satisfying P. We consider again the state $|\psi_1\rangle$

$$|\psi_1\rangle = U_P|\psi_0\rangle = \frac{1}{\sqrt{2^{n+1}}} \left(\sum_{x \in X_F} |x\rangle - \sum_{x \in X_T} |x\rangle \right) \otimes (|0\rangle - |1\rangle).$$

Now we have $|X_F| = \frac{3}{4}2^n$ and $|X_T| = \frac{1}{4}2^n$. Applying U_{IA} to the first n qubits (i.e. we apply $U_{IA} \otimes I$) will ensure that measurement of the first n qubits will yield a state from X_T. To see this we calculate the average A of the amplitudes of the first n qubits. We find

$$A = \frac{1}{2^n} \left(\frac{|X_F|}{\sqrt{2^n}} - \frac{|X_T|}{\sqrt{2^n}} \right)$$

$$= \frac{1}{\sqrt{2^n}} \left(\frac{3}{4} - \frac{1}{4} \right)$$

$$= \frac{1}{2\sqrt{2^n}}$$

Since each amplitude of a state in X_F is $\frac{1}{\sqrt{2^n}}$, these amplitudes become $2A - \frac{1}{\sqrt{2^n}} = 0$.

The amplitudes of states in X_T are $-\frac{1}{\sqrt{2^n}}$. These amplitudes become $2A + \frac{1}{\sqrt{2^n}} = \frac{2}{\sqrt{2^n}}$. In this case a single iteration of Grover's algorithm guarantees success.

Brassard and Høyer [27] discuss a related algorithm which works more generally, when the probability of measuring a state which satisfies P is $\frac{1}{2}$. Let $|\psi\rangle$ be the state

$$|\psi\rangle := |X_T\rangle + |X_F\rangle,$$

where $|X_T\rangle$ is the superposition of all states satisfying P and $|X_F\rangle$ is the superposition of all states not satisfying P. There is no constraint on the amplitudes of the states, except that $|\psi\rangle$ must be normalized. Obviously we must have $\langle X_T|X_F\rangle = 0$. Suppose $\langle X_T|X_T\rangle = t$, then $\langle X_F|X_F\rangle = (1-t)$. The algorithm transforms $|\psi\rangle$ to $|\psi'\rangle$ with

$$|\psi'\rangle = (2i(1-t) - 1)|X_T\rangle + i(1 - 2t)|X_F\rangle.$$

If $t = \frac{1}{2}$ then the amplitudes of all states in X_F are zero. If $t = 0$ the algorithm changes the global phase only. Let A be a quantum algorithm which evolves $|0\rangle$ (an appropriate tensor product of $|0\rangle$ qubits) into the superposition $|\psi\rangle$, represented as the unitary operator U_A. Instead of multiplying amplitudes of $|X_T\rangle$ by -1 we need to multiply them by i. We achieve this by applying

$$S_P := U_P \left(\left(\bigotimes_n I \right) \otimes U_i \right) U_P$$

to $|\psi\rangle \otimes |0\rangle$ where

$$U_i := \begin{pmatrix} 1 & 0 \\ 0 & i \end{pmatrix}$$

is the single qubit phase change gate, ignoring the global phase. Similarly we need S_0 which takes the state $|0\rangle$ (n qubits) to $i|0\rangle$. The transform we use is described by

$$G = U_A S_0 U_A^{-1} S_P.$$

We need to calculate $\langle X_T|G|X_T\rangle$ and $\langle X_F|G|X_F\rangle$. We have

$$S_P(|X_T\rangle + |X_F\rangle) = i|X_T\rangle + |X_F\rangle.$$

The following calculations lead to the desired result for $G|\psi\rangle$.

$$
\begin{aligned}
U_A|0\rangle &= |X_T\rangle + |X_F\rangle \\
U_A^{-1}|X_F\rangle &= |0\rangle - U_A^{-1}|X_T\rangle \\
U_A^{-1}S_P|\psi\rangle &= iU_A^{-1}|X_T\rangle + |0\rangle - U_A^{-1}|X_T\rangle = (i-1)U_A^{-1}|X_T\rangle + |0\rangle
\end{aligned}
$$

We need to determine $S_0 U_A^{-1}|X_T\rangle$. First we calculate $\langle 0|U_A^{-1}|X_T\rangle = \overline{\langle X_T|U_A|0\rangle} = t$. Thus we can define $|\phi\rangle := U_A^{-1}|X_T\rangle - t|0\rangle$ orthogonal to $|0\rangle$ so that

$$
\begin{aligned}
U_A^{-1}|X_T\rangle &= t|0\rangle + |\phi\rangle \\
U_A|\phi\rangle &= (1-t)|X_T\rangle - t|X_F\rangle \\
S_0 U_A^{-1} S_P|\psi\rangle &= (i-1)(it|0\rangle + |\phi\rangle) + i|0\rangle \\
U_A S_0 U_A^{-1} S_P|\psi\rangle &= it(i-1)(|X_T\rangle + |X_F\rangle) \\
&\quad + (i-1)((1-t)|X_T\rangle - t|X_F\rangle) \\
&\quad + i(|X_T\rangle + |X_F\rangle) \\
&= (i(1-t)-1)|X_T\rangle + i(1-2t)|X_F\rangle
\end{aligned}
$$

Thus we obtain the desired result. This technique is used to remove $|0\rangle$ from the superposition in the exact quantum algorithm for the solution of Simon's problem since at least one qubit has probability $\frac{1}{2}$ of being measured as $|1\rangle$ and every other qubit has either probability 0 or $\frac{1}{2}$.

21.7   Quantum Key Distribution

In cryptography systems a key is used to decipher an encrypted message. It is necessary to guarantee security when transmitting keys, otherwise the encryption system achieves nothing. In the case of public key cryptography systems, a public key transmitted securely gives a third party even less information to attempt to decode messages. Thus it would be useful to be able to transmit keys securely. In classical terms this cannot be guaranteed. A third party can obtain the key by copying the key while it is transmitted. Due to the no cloning theorem, copying of quantum states can cause disturbance in the states. Transmitting a key using quantum states can thus be used to detect a third party attempting to copy the key.

Let B_0 denote the basis $\{|0\rangle, |1\rangle\}$ and B_1 denote the basis

$$\left\{ \frac{1}{\sqrt{2}}(|0\rangle + |1\rangle), \frac{1}{\sqrt{2}}(|0\rangle - |1\rangle) \right\}.$$

Suppose Alice transmits the key and Bob is to receive the key. Alice and Bob agree to use one to one mappings $f_0 : B_0 \rightarrow \{0, 1\}$ and $f_1 : B_1 \rightarrow \{0, 1\}$ to uniquely convert between 0 and 1 and a given basis.

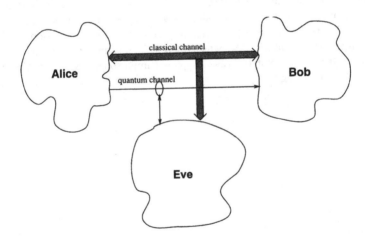

Figure 21.4: Quantum Key Distribution

For each bit in the key Alice randomly chooses a basis from B_0 and B_1 and sends the quantum state from that basis which corresponds to the bit. Bob randomly chooses a basis from B_0 and B_1 and measures the quantum state he receives relative to this basis. On average 50% of the time the basis chosen by Alice and Bob will be the same. After Bob has received all the bits, Alice and Bob communicate on an open channel to determine which quantum states were prepared and measured using the same basis. This determines which bits are used, since Alice and Bob have the same bit values in these cases. Suppose now a third party Eve attempts to obtain the key from the quantum states sent by Alice to Bob. Eve attempts to measure the states being sent from Alice to Bob by randomly choosing a basis from B_0 and B_1, which she chooses correctly 50% of the time. Eve then resends the quantum state using the basis she guesses and the value she measured. When Bob uses the same basis as Alice when Eve does not, Bob will measure the correct value 50% of the time. This means that when Alice and Bob use the same basis when Eve attempts to obtain the key Bob will obtain an incorrect value 25% of the time. Alice and Bob agree after sending all the quantum states to use a number of the states where the same basis was used to determine if someone tried to obtain the key. If enough states are used an error rate of larger than (say) 5% may be agreed to indicate the transmission was potentially influenced by a third party and is insecure. The rest of the corresponding states are used as the key if the error rate is low enough.

Other techniques have been described for quantum key distribution based on Bell's inequality [61], measurement uncertainty [12, 14, 15], and a distribution scheme where quantum states are reused [36] using entanglement. These schemes rely on the fact that the third party (Eve) cannot pretend to be Alice and Bob, i.e. Eve can only inspect the quantum states which are sent, and cannot receive a state and send another. If this were not the case Eve could impersonate Bob and Alice without either party knowing.

21.8   Dense Coding

Dense coding [16, 138, 156] uses an EPR pair and a single qubit to transmit two classical bits of information. The EPR pair is shared ahead of time between the transmitter and receiver. Only a single qubit is needed to transfer the two classical bits of information. This is possible due to entanglement. Entanglement makes it possible to transform a two qubit state to one of four orthogonal states by interacting with only one qubit. This choice can be identified with two classical bits of information. Dense coding illustrates the use of entanglement as an information resource. The scheme is also secure in the sense that communication is only possible between the the two parties which share the EPR pair.

The transmitter (Alice) and the receiver (Bob) each have one quantum subsystem which together form the quantum system of an already prepared EPR state

$$|\psi\rangle := \frac{1}{\sqrt{2}}(|00\rangle + |11\rangle) \equiv \frac{1}{\sqrt{2}}(|0\rangle \otimes |0\rangle + |1\rangle \otimes |1\rangle).$$

We let the first system denote Alice's quantum subsystem (qubit) of the EPR state and the second system denote Bob's quantum subsystem (qubit) of the EPR state. Alice can transform $|\psi\rangle$ to any one of the Bell basis states according to

$$\left(\begin{pmatrix} 1 & 0 \\ 0 & 1 \end{pmatrix} \otimes I_2 \right) |\psi\rangle = \frac{1}{\sqrt{2}}(|00\rangle + |11\rangle) \equiv \Phi^+$$

$$\left(\begin{pmatrix} 0 & 1 \\ 1 & 0 \end{pmatrix} \otimes I_2 \right) |\psi\rangle = \frac{1}{\sqrt{2}}(|10\rangle + |01\rangle) \equiv \Psi^+$$

$$\left(\begin{pmatrix} 1 & 0 \\ 0 & -1 \end{pmatrix} \otimes I_2 \right) |\psi\rangle = \frac{1}{\sqrt{2}}(|00\rangle - |11\rangle) \equiv \Phi^-$$

$$\left(\begin{pmatrix} 0 & 1 \\ -1 & 0 \end{pmatrix} \otimes I_2 \right) |\psi\rangle = \frac{1}{\sqrt{2}}(-|10\rangle + |01\rangle) \equiv \Psi^-$$

where I_2 is the 2×2 unit matrix.

Alice has two bits representing the values 0,1,2 or 3. She transforms $|\psi\rangle$ according to the following table.

Value	Initial state	Transformed state
0	$|\psi\rangle$	$\frac{1}{\sqrt{2}}(|00\rangle + |11\rangle)$
1	$|\psi\rangle$	$\frac{1}{\sqrt{2}}(|10\rangle + |01\rangle)$
2	$|\psi\rangle$	$\frac{1}{\sqrt{2}}(-|10\rangle + |01\rangle)$
3	$|\psi\rangle$	$\frac{1}{\sqrt{2}}(|00\rangle - |11\rangle)$

Table 21.1: Dense Coding: Alice's Transformations

The transformations are obviously unitary. Alice then sends her quantum qubit to Bob. Now Bob applies a controlled NOT using the first (Alice's) qubit as the control and then applies the Hadamard transform to the first qubit. Finally a controlled NOT is applied to yield the data. The following table describes the quantum state after the transformation.

Value	Initial state	Transformed state	Final state
0	$\frac{1}{\sqrt{2}}(|00\rangle + |11\rangle)$	$|00\rangle$	$|00\rangle$
1	$\frac{1}{\sqrt{2}}(|10\rangle + |01\rangle)$	$|01\rangle$	$|01\rangle$
2	$\frac{1}{\sqrt{2}}(-|10\rangle + |01\rangle)$	$|11\rangle$	$|10\rangle$
3	$\frac{1}{\sqrt{2}}(|00\rangle - |11\rangle)$	$|10\rangle$	$|11\rangle$

Table 21.2: Dense Coding: Bob's Transformations

Thus the transformed state uniquely determines the value 0,1,2 or 3.

Chapter 22

Quantum Information Theory

22.1 Introduction

The concepts of classical information theory can be extended to quantum information theory. Since in general measurement yields a result with probability, we
may suggest using these probabilities in classical information theory. However the
probabilities do not contain phase information, which cannot be neglected. Thus the
definitions are given in terms of the density operator. These probabilities depend on
the basis used for measurement. A density operator ρ over a n-dimensional Hilbert
space $\mathcal{H}$ is a positive operator with unit trace. The trace $\mathrm{tr}(A)$ is defined as

$$\mathrm{tr}(A) := \sum_{j=1}^{n} \langle \beta_j | A | \beta_j \rangle$$

where β_j for $j = 1, \ldots, n$ is any orthonormal basis in $\mathcal{H}$. Thus $\mathrm{tr}(\rho) = 1$. The
eigenvalues of a density operator are greater than zero. By the spectral theorem
every density operator can be represented as a mixture of pure states

$$\rho = \sum_{j=1}^{n} p_j |\alpha_j\rangle \langle \alpha_j|$$

where $|\alpha_j\rangle$ for $j = 1, \ldots, n$ are the orthonormal eigenvectors of ρ (which form a
basis in $\mathcal{H}$), and

$$p_j \in \mathbf{R}, \qquad p_j \geq 0, \qquad \sum_{j=1}^{n} p_j = 1.$$

The eigenvalue p_j is the probability of finding the state $|\alpha_j\rangle$.

22.2　Von Neumann Entropy

The Von Neumann entropy of a system A, represented by the density operator ρ_A, is defined as [38]

$$S(A) := -\text{tr}_A(\rho_A \log_2 \rho_A).$$

The log function is base 2, following the convention of classical information theory, and gives the qubit as the natural unit of information. Pure states have density operators which are projection operators, their eigenvalues are 0 and 1, and so the Von Neumann entropy for a pure state is always zero.

Example. We consider the Werner state.

$$\rho := \frac{5}{8}|\Phi^+\rangle\langle\Phi^+| + \frac{1}{8}(|\Phi^-\rangle\langle\Phi^-| + |\Psi^+\rangle\langle\Psi^+| + |\Psi^-\rangle\langle\Psi^-|)$$

The Von Neumann entropy is

$$-\text{tr}(\rho \log_2 \rho) = -\frac{3}{8}\log_2 \frac{1}{8} - \frac{5}{8}\log_2 \frac{5}{8} \approx 1.5487949.$$

♣

For two quantum systems, A and B, the combined entropy is given by

$$S(AB) := -\text{tr}_{AB}(\rho_{AB} \log_2 \rho_{AB}).$$

The conditional and mutual entropies ($S(A|B)$ and $S(A : B)$ respectively) on the quantum systems A and B are defined as [38, 39]

$$S(A|B) = -\text{tr}_{AB}(\rho_{AB} \log_2 \rho_{A|B})$$

and

$$S(A : B) = -\text{tr}_{AB}(\rho_{AB} \log_2 \rho_{A:B})$$

where

$$\rho_{A|B} = \lim_{n\to\infty} (\rho_{AB}^{\frac{1}{n}}(I_A \otimes \rho_B)^{-\frac{1}{n}})^n$$

$$\rho_{A:B} = \lim_{n\to\infty} ((\rho_A \otimes \rho_B)^{\frac{1}{n}}\rho_{AB}^{-\frac{1}{n}})^n$$

It can be verified that the following identities hold.

$$S(AB) \equiv S(A) + S(B|A)$$

and

$$S(A : B) \equiv S(A) + S(B) - S(AB).$$

22.3 Measures of Entanglement

22.3.1 Bell's Inequality

Bell's inequality [40, 128, 129, 145, 170] was initially used to show that a local hidden variable model of quantum mechanics is inadequate. The assumption of a local hidden variable model leads to an inequality that must be satisfied, but experimentally is shown to be violated. As an example, consider a correlation experiment in which a particle of total spin zero decays into two particles each with spin $\frac{1}{2}$. A rotatable polarizer and detector are set up for each particle, sufficiently far from the source, so that the correlation between the spin orientations can be investigated. The first polarizer, with angular setting α, only lets the first particle through if its spin in the direction $\mathbf{n}_a$ takes the value $\frac{\hbar}{2}$. The second polarizer, with angular setting β, lets the second particle through only if its spin in the direction $\mathbf{n}_a$ takes the value $\frac{\hbar}{2}$. The detectors respond if the spin is positive, otherwise it is negative. A measure of the correlation is $N(\alpha; \beta)$, defined by the relative number of experiments resulting in the first particle registering as positive with the first polarizer at angle α, and the second as positive with the second polarizer at angle β. We have

$$N(\alpha; \beta) = N_+(\alpha, \gamma, \beta) + N_-(\alpha, \gamma, \beta)$$

where $N_s(\alpha, \gamma, \beta)$ is the relative number of experiments where the first particle has positive spin with the polarizer at angle α and negative with β, and spin s with the polarizer at angle γ. In a local hidden variable theory, these quantities are available. Since

$$N(\alpha, \gamma) = N_+(\alpha, \beta, \gamma) + N_-(\alpha, \beta, \gamma),$$

$$N(\gamma, \beta) = N_+(\gamma, \alpha, \beta) + N_-(\gamma, \alpha, \beta),$$

we have $N(\alpha, \gamma) \geq N_-(\alpha, \beta, \gamma)$ and $N(\gamma, \beta) \geq N_+(\gamma, \alpha, \beta)$. Thus

$$N(\alpha, \beta) \leq N(\alpha, \gamma) + N(\gamma, \beta).$$

For coplanar detectors, $\alpha = 0$, $\beta = \frac{\pi}{2}$ and $\gamma = \frac{\pi}{4}$ the inequality is not satisfied.

A variant of Bell's inequality derived by Clauser, Horne, Shimony and Holt is given by [127]

$$|\langle AB \rangle + \langle AB' \rangle + \langle A'B \rangle - \langle A'B' \rangle| \leq 2$$

The operators A and A' (respectively B and B') are normalized, noncommuting and can be measured by an observer. The expectation values can be calculated if the quantum state is known, and it is also experimentally observable, by repeating the measurements sufficiently many times with identically prepared initial pairs of quantum systems. The validity of this inequality for all combinations of independent measurements on both systems is necessary, although not sufficient, for the existence of a local hidden variable model.

The nonlocal nature of entangled states can be illustrated without a statistical formulation [25, 186]. Consider the GHZ (Greenberger-Horne-Zeilinger) state

$$|\psi\rangle := \frac{1}{\sqrt{2}}(|001\rangle + |110\rangle).$$

We consider the basis $\alpha = \{|L\rangle, |R\rangle\}$ and the basis $\beta = \{|H\rangle, |V\rangle\}$ described by

$$\alpha: \qquad |0\rangle = \frac{1}{\sqrt{2}}(|L\rangle + |R\rangle), \quad |1\rangle = \frac{1}{\sqrt{2}}(|L\rangle - |R\rangle),$$

$$\beta: \qquad |0\rangle = \frac{1}{\sqrt{2}}(|H\rangle + |V\rangle), \quad |1\rangle = \frac{i}{\sqrt{2}}(|H\rangle - |V\rangle).$$

Expressing only one qubit of $|\psi\rangle$ in the basis α and the rest in the basis β, and expressing each qubit in the basis α yields

$$
\begin{aligned}
|\psi\rangle &= \tfrac{i}{2}(|LHH\rangle + |RVH\rangle - |LVV\rangle - |RHV\rangle) & \alpha\beta\beta \\
&= \tfrac{i}{2}(|HLH\rangle + |VRH\rangle - |VLV\rangle - |HRV\rangle) & \beta\alpha\beta \\
&= \tfrac{1}{2}(|HVL\rangle + |VHL\rangle - |HHR\rangle - |VVR\rangle) & \beta\beta\alpha \\
&= \tfrac{1}{2}(|LLL\rangle + |RRL\rangle - |LRR\rangle - |RLR\rangle) & \alpha\alpha\alpha.
\end{aligned}
$$

Measuring in the basis β for each qubit must yield a duplicate result. Measuring two qubits in the β basis as given in the previous equations allows us to deduce the result of measuring the other qubit in the α basis. Let β_j denote the result after measuring qubit j in the β basis. As an example, from the first equality, if $\beta_2 = \beta_3$ then the first qubit is $|L\rangle$ and $|R\rangle$ otherwise. Thus we construct the following table.

Outcomes in β basis	Outcomes in α basis
$\beta_1 = \beta_2 = \beta_3$	LLR
$\beta_1 = \beta_2 \neq \beta_3$	RRR
$\beta_1 = \beta_3 \neq \beta_2$	RLL
$\beta_2 = \beta_3 \neq \beta_1$	LRL

None of the results obtained are consistent with the final equation for $|\psi\rangle$ in the α basis for all qubits.

22.3.2 Entanglement of Formation

Here we introduce a measure of entanglement for mixed states. It is called the *entanglement of formation* [19]. Another measure is distillable entanglement [19]. Let A and B be two quantum systems. We define for the coupled system $A \otimes B$

$$\rho_A := \mathrm{tr}_B(\rho_{AB})$$

where tr_B denotes the *partial trace* over B, i.e we use $I \otimes |\beta_j\rangle$ as the basis for the trace where $|\beta_j\rangle$ is an orthonormal basis in B. The measure of entanglement $E(AB)$ is then defined as

$$E(AB) := S(\rho_A).$$

This describes the entanglement for pure states.

Example. For the state $|\psi\rangle := \frac{1}{\sqrt{3}}(|00\rangle + |01\rangle + |10\rangle)$ we find

$$\rho_A = \mathrm{tr}_B|\psi\rangle\langle\psi| = \tfrac{1}{3}(2|0\rangle\langle 0| + |0\rangle\langle 1| + |1\rangle\langle 0| + |1\rangle\langle 1|).$$

Thus

$$E(\psi) = -(\tfrac{1}{2} + \tfrac{\sqrt{5}}{6})\log_2(\tfrac{1}{2} + \tfrac{\sqrt{5}}{6}) - (\tfrac{1}{2} - \tfrac{\sqrt{5}}{6})\log_2(\tfrac{1}{2} - \tfrac{\sqrt{5}}{6}) \approx 0.55$$

where we used the eigenvalues $\tfrac{1}{2} \pm \tfrac{\sqrt{5}}{6}$ of ρ_A.   ♣

For mixed states we take the minimum entanglement as defined above over all possible ensembles of pure states realizing ρ_{AB}. Thus if

$$\rho_{AB} = \sum_j p_j |\psi_j\rangle\langle\psi_j|,$$

we have

$$E(AB) := \min \sum_j p_j S(\rho_{A,j})$$

where

$$\rho_{A,j} = \mathrm{tr}_B(|\psi_j\rangle\langle\psi_j|).$$

Example. For the Werner state

$$\rho_W := \frac{5}{8}|\Phi^+\rangle\langle\Phi^+| + \frac{1}{8}(|\Phi^-\rangle\langle\Phi^-| + |\Psi^+\rangle\langle\Psi^+| + |\Psi^-\rangle\langle\Psi^-|)$$

it has been shown that $E(W) \approx 0.117$ [19]. If we use the definition given for pure states we obtain 1.

22.3.3   Conditions on Entanglement Measures

The states which are not entangled can be written as a convex combination of tensor products of pure states. We define the set of states which are not entangled or *separable* for a particular tensor product space $\mathcal{H}_1 \otimes \mathcal{H}_2$ as

$$S = \{\rho \mid \rho = \textstyle\sum_j p_j \rho_{1,j} \otimes \rho_{2,j}, \quad \textstyle\sum_j p_j = 1,\ p_j \geq 0,\ p_{1,j} \in \mathcal{H}_1,\ p_{2,j} \in \mathcal{H}_2 \}.$$

Thus we can define the entanglement of a state as the minimum distance to any state in the set S [176, 177]. The *Hilbert-Schmidt norm* is defined as

$$\|A\|_{HS} := \sqrt{\operatorname{tr}(A^* A)}.$$

Thus a measure of entanglement may be defined as [184]

$$E_{HS}(\rho) := \min_{\sigma \in S} \|\rho - \sigma\|_{HS}^2$$

where the minimum is taken over all states $\sigma \in S$ which are not entangled.

Three processes are used to increase correlations between two quantum subsystems, i.e. to distill locally a subensemble of highly entangled states from an original ensemble of less entangled states.

1. *Local general measurements.* The measurements are performed separately by two parties. They are described by the operators A_j and B_j with the condition

$$\sum_j A_j^* A_j = I \quad \text{and} \quad \sum_j B_j^* B_j = I.$$

   The joint action is described by

$$\sum_{j,k} A_j \otimes B_k.$$

2. *Classical communication.* The actions of the two parties can be correlated. This can be described by a complete measurement on the whole space. If ρ_{AB} describes the initial state shared by the two parties, local general measurement and classical communication gives

$$\sum_j (A_j \otimes B_j) \rho_{AB} (A_j^* \otimes B_j^*).$$

3. *Post-selection.* The general measurement is not complete. The density matrix describes a subensemble of the original ensemble with appropriate normalization.

A manipulation involving these techniques constitutes a *purification procedure.* There are some properties we would like a measure E of entanglement to have [176, 177],

1. $E(\rho) = 0$ if and only if $\rho \in \mathcal{S}$.

2. For any unitary transforms U_1 in $\mathcal{H}_1$ and U_2 in $\mathcal{H}_2$,

$$E((U_1 \otimes U_2)\rho(U_1^* \otimes U_2^*)) = E(\rho).$$

3. The expected entanglement cannot increase due to local general measurement, classical communication and post-selection described by $\sum_j V_j^* V_j = I$, i.e.

$$\sum_j \text{tr}(\sigma_j) E(\sigma_j / \text{tr}(\sigma_j)) \leq E(\sigma)$$

where

$$\sigma_j = V_j \sigma V_j^*.$$

So if we define the measure of entanglement according to the distance, as above, we require that the "distance" measure ensures that the entanglement measure obeys these three properties. Thus the distance does not have to be a metric.

Vedral et al. [176, 177] also give sufficient conditions for the distance measure to define a measure of entanglement. Ozawa has shown [126] that the Hilbert-Schmidt norm does not satisfy these conditions. Thus the Hilbert-Schmidt norm may not be useful for the definition of an entanglement measure.

A related question is to determine which states belong to $\mathcal{S}$. The *Peres-Horodecki criterion* [90, 91, 127] gives a necessary and, under certain conditions, sufficient criterion for separability.

22.4   Quantum Coding

From classical information theory we have the noiseless and noisy coding theorems. Schumacher [144] developed the quantum equivalents which we present here. As in the classical case we wish to transmit information over some sort of channel. An important difference is due to the no-cloning theorem. Quantum states cannot be copied across the channel since this would violate the no-cloning theorem. Instead, for a quantum transmission channel, transposition can be used. For example teleportation can be used to transmit information. The sender transposes the source system M with another system X which can be conveyed to the transmitter. The system X serves as the channel. The receiver then transposes X with the system M'. If M and M' are identical, the inverse operation can be used to transfer the information. In other words, suppose U transposes the state from M to X then U^{-1} transposes the state from X to M'.

$$U(|\psi_M\rangle \otimes |0_X\rangle) = |0_M\rangle \otimes |\psi_X\rangle$$

$$U^{-1}(|0_{M'}\rangle \otimes |\psi_X\rangle) = |\psi_{M'}\rangle \otimes |0_X\rangle$$

Suppose the state to be transmitted is $|\alpha_M\rangle$. In transmission some information may be lost (for example due to noise, or a channel with lower dimension than the source system), so the end state must be represented by a density operator $w(\alpha)_{M'}$ (this is due to the fact that we need to trace over the portion of the system which is lost, and the original state may have been entangled). The probability that $w(\alpha)_{M'}$ represents $|\alpha_{M'}\rangle$ is given by

$$\mathrm{tr}(|\alpha_M\rangle\langle\alpha_M|w(\alpha)_{M'}).$$

Thus a measure of the fidelity would be

$$F(M, M') = \sum_a p(a)\mathrm{tr}(|a_M\rangle\langle a_M|w(a)_{M'})$$

where $p(a)$ is the probability that a is sent and

$$\sum_a p(a) = 1.$$

Two results are required before we can prove the quantum noiseless coding theorem.

Lemma 1. Let the ensemble of signals in M be described by the density operator ρ with

$$\rho = \sum_a p(a)|a_M\rangle\langle a_M|.$$

If the quantum channel C has dimension d and any projection Γ onto a d dimensional subspace of M has the property

$$\mathrm{tr}(\rho\Gamma) < \eta$$

for some fixed η, then $F(M, M') < \eta$.

Since the channel has only dimension d, the final decoded state $w(a)_{M'}$ is only supported on a d dimensional subspace of M'. Let Γ denote the projection onto this subspace. In other words $w(a)_{M'}$ results from a unitary transformation of the separable state $w(a)_C \otimes 0_{M'-C}$ where $0_{M'-C}$ is the initial state introduced once the state is transmitted. Let $w(a)_{M',\Gamma}$ denote $w(a)_{M'}$ in the subspace. The d eigenstates of $w(a)_{M',\Gamma}$ (denoted by $|\phi(a)_1\rangle, \dots |\phi(a)_d\rangle$) form an orthonormal basis in this subspace. Let $\lambda(a)_1, \dots, \lambda(a)_d$ denote the eigenvalues corresponding to these eigenstates. Then $w(a)_{M',\Gamma}$ can be expressed as

$$w(a)_{M',\Gamma} = \sum_{k=0}^{d} \lambda(a)_k |\phi(a)_k\rangle\langle\phi(a)_k|.$$

Denote by $|\psi(a)_k\rangle$ the state $|\phi(a)_k\rangle \otimes |0_{\Gamma\perp}\rangle$ which is the state $|\phi(a)_k\rangle$ extended in M'. The projection operator Γ is given by

$$\Gamma = \sum_{k=0}^{d} |\psi(a)_k\rangle\langle\psi(a)_k|.$$

Now

$$\mathrm{tr}(|a_M\rangle\langle a_M|w(a)_{M'}) = \mathrm{tr}(|a_M\rangle\langle a_M| \sum_{k=0}^{d} \lambda(a)_k |\psi(a)_k\rangle\langle\psi(a)_k|)$$

$$\leq |a_M\rangle\langle a_M| \sum_{k=0}^{d} \mathrm{tr}(|\psi(a)_k\rangle\langle\psi(a)_k|)$$

$$= \mathrm{tr}(|a_M\rangle\langle a_M|\Gamma).$$

Thus for the fidelity we have

$$F(M, M') = \sum_a p(a)\mathrm{tr}(|a_M\rangle\langle a_M|w(a)_{M'}) \leq \mathrm{tr}(\rho\Gamma) < \eta.$$

Lemma 2 Let the ensemble of signals in M be described by the density operator ρ. If the quantum channel C has dimension d and there exists a projection Γ onto a d dimensional subspace of M which has the property

$$\mathrm{tr}(\rho\Gamma) > 1 - \eta$$

for some fixed $0 \leq \eta \leq 1$, then there exists a transposition scheme with fidelity $F(M, M') > 1 - 2\eta$.

The proof is by construction. Let Γ_G be such a projection, projecting to the subspace G. Let $\Gamma_{G\perp}$ denote the projection onto the subspace of M orthogonal to G. This allows us to rewrite $|a_M\rangle$ in a more usable form.

$$\begin{aligned}|a_M\rangle &= \langle a_M|\Gamma_G|a_M\rangle\Gamma_G|a_M\rangle + \langle a_M|\Gamma_{G\perp}|a_M\rangle\Gamma_{G\perp}|a_M\rangle \\ &= \gamma_G^a|a_G\rangle + \gamma_{G\perp}^a|a_{G\perp}\rangle\end{aligned}$$

where

$$\gamma_k^a = (\langle a_M|\Gamma_k|a_M\rangle)^{\frac{1}{2}}$$

and

$$|a_k\rangle = \frac{\Gamma_k|a_M\rangle}{\gamma_k^a}.$$

Obviously

$$|\gamma_G^a|^2 + |\gamma_{G\perp}^a|^2 = 1.$$

Let the dimension of M be N, $|1\rangle, \ldots, |d\rangle$ denote an orthonormal basis in G, and $|d+1\rangle, \ldots, |N\rangle$ denote an orthonormal basis in $G^\perp$. Also let $|1_C\rangle, \ldots, |d_C\rangle$ be an orthonormal basis in C and $|(d+1)_E\rangle, \ldots, |N_E\rangle$ be an orthonormal basis in E (the system representing the information lost during transmission). The states in G will be used for transmission over the channel C. The initial state is prepared as $|a_M\rangle \otimes |0_C\rangle \otimes |0_E\rangle$ where $|0_E\rangle$ is the initial state for the system which represents the loss of information, and we require $\langle 0_E|k_E\rangle = 0$ for $k = d+1, \ldots, N$. The following unitary transformation is used to prepare the state for transmission.

$$U_{MC}|k\rangle \otimes |0_C\rangle \otimes |0_E\rangle = \begin{cases} |0\rangle \otimes |k_C\rangle \otimes |0_E\rangle & k = 1, \ldots, d \\ |0\rangle \otimes |0_C\rangle \otimes |k_E\rangle & k = d+1, \ldots, N \end{cases}$$

Applying the transformation to $|a_M\rangle$ gives

$$|c\rangle := U_{MC}|a_M\rangle \otimes |0_C\rangle \otimes |0_E\rangle$$

$$= (\gamma_G^a U_{MC} \sum_{k=1}^{d} \langle k|a_G\rangle|k\rangle + \gamma_{G\perp}^a U_{MC} \sum_{k=d+1}^{N} \langle k|a_{G\perp}\rangle|k\rangle) \otimes |0_C\rangle \otimes |0_E\rangle$$

$$= \gamma_G^a \sum_{k=1}^{d} \langle k|a_G\rangle|0\rangle \otimes |k_C\rangle \otimes |0_E\rangle + \gamma_{G\perp}^a \sum_{k=d+1}^{N} \langle k|a_{G\perp}\rangle|0\rangle \otimes |0_C\rangle \otimes |k_E\rangle$$

$$= \gamma_G^a|0\rangle \otimes |a_C\rangle \otimes |0_E\rangle + \gamma_{G\perp}^a|0\rangle \otimes |0_C\rangle \otimes |a_E\rangle$$

where we used

$$|a_C\rangle := \sum_{k=1}^{d} \langle k|a_G\rangle|k_C\rangle, \qquad |a_E\rangle := \sum_{k=d+1}^{N} \langle k|a_{G\perp}\rangle|k_E\rangle.$$

Only system C is used in transmission so systems M and E must be traced over yielding

$$\mathrm{tr}_{M,E}(|c\rangle\langle c|) = |\gamma_G^a|^2|a_C\rangle\langle a_C| + |\gamma_{G\perp}^a|^2|0_C\rangle\langle 0_C|.$$

After the transmission the system is augmented with an initial state for the receiving system, in terms of two systems equivalent to M and $G^\perp$. Thus we work with the operator

$$v(a)_{M'} = |\gamma_G^a|^2|0\rangle\langle 0| \otimes |a_C\rangle\langle a_C| \otimes |0_E\rangle\langle 0_E| + |\gamma_{G\perp}^a|^2|0\rangle\langle 0| \otimes |0_C\rangle\langle 0_C| \otimes |0_E\rangle\langle 0_E|.$$

For the decoding step we use U_{MC}^{-1} which gives

$$w(a)_{M'} = U_{MC}^{-1}v(a)_{M'}U_{MC}$$

$$= |\gamma_G^a|^2|a_G\rangle\langle a_G| \otimes |0_C\rangle\langle a_C| \otimes |0_E\rangle\langle 0_E| + |\gamma_{G\perp}^a|^2|0\rangle\langle 0| \otimes |0_C\rangle\langle 0_C| \otimes |0_E\rangle\langle 0_E|.$$

The fidelity is given by

$$F(M, M') = \sum_a p(a)\mathrm{tr}(|a_M\rangle\langle a_M|w(a)_{M'})$$

$$= \sum_a p(a)\mathrm{tr}(|\gamma_G^a|^2|a_M\rangle\langle a_M|a_G\rangle\langle a_G| + |\gamma_{G\perp}^a|^2|a_M\rangle\langle a_M|0\rangle\langle 0|)$$

$$= \sum_a p(a)\mathrm{tr}(\overline{\gamma_G^a}|\gamma_G^a|^2|a_M\rangle\langle a_G| + |\gamma_{G\perp}^a|^2\langle a_M|0\rangle|a_M\rangle\langle 0|)$$

$$= \sum_a p(a)(|\gamma_G^a|^4 + |\gamma_{G\perp}^a|^2(\langle a_M|0\rangle)^2)$$

$$\geq \sum_a p(a)|\gamma_G^a|^4$$

$$= \sum_a p(a)(1 - |\gamma_{G\perp}^a|^2)^2$$

$$\geq 1 - 2\sum_a p(a)|\gamma_{G\perp}^a|^2$$

It is given that

$$\text{tr}(\rho\Gamma_G) = \sum_a p(a)|a_M\rangle\langle a_M|\Gamma_G$$

$$= \sum_a p(a)|\gamma_G^a|^2$$

$$= 1 - \sum_a p(a)|\gamma_{G\perp}^a|^2$$

$$> 1 - \eta$$

Thus

$$\eta > \sum_a p(a)|\gamma_{G\perp}^a|^2, \qquad \sum_a p(a)|\gamma_G^a|^2 > 1 - \eta.$$

This gives the fidelity

$$F(M, M') > 1 - 2\eta.$$

Suppose we have a density matrix ρ representing the ensemble of states which may be transmitted. We can write ρ as

$$\rho = \sum_a \lambda_a|a\rangle\langle a|$$

where λ_a and $|a\rangle$ are the eigenvalues and orthonormal eigenstates of ρ. The Von Neumann entropy is then

$$S(\rho) = -\sum_a \lambda_a \log_2 \lambda_a.$$

For the density matrix $\otimes_N \rho$ of N identical and independent systems the eigenvalues and orthonormal eigenstates are given by the products of N eigenvalues and eigenvectors of ρ.

If we interpret the eigenvalue λ_a as the probability that the eigenstate $|a\rangle$ is transmitted then the Von Neumann entropy is the classical Shannon entropy of these probabilities, and so following page 206, the number of sequences of N eigenstates which are likely to be transmitted is bounded above by $2^{N(S(\rho)+\delta)}$ and below by $(1-\epsilon)2^{N(S(\rho)-\delta)}$.

Quantum Noiseless Coding Theorem

1. If the quantum channel C has dimension at least $2^{S(\rho)+\delta}$ then there exists $N_0(\delta, \epsilon)$ such that for all $N > N_0$ sequences of eigenstates of ρ of length N can be transmitted via C with fidelity greater than $1 - \epsilon$.

There exists $N_0(\delta, \frac{\epsilon}{2})$ such that for all $N > N_0$

$$\text{tr}(\bigotimes_N \rho)\Gamma > 1 - \frac{\epsilon}{2}$$

where Γ is a projection onto a subspace with dimension $\dim(C)$. Γ projects to a subspace containing no more than $2^N S(\rho) + \delta$ likely eigenstates of $\bigotimes_N \rho$, where the orthonormal basis is given by a subset of the orthonormal eigenstates of $\bigotimes_N \rho$. The sum of the corresponding eigenvalues is greater than $1 - \frac{\epsilon}{2}$. Thus the fidelity

$$F(M, M') > 1 - \epsilon.$$

2. If the quantum channel C has dimension at most $2^{S(\rho)-\delta}$ then there exists $N_0(\delta, \epsilon)$ such that for all $N > N_0$ sequences of eigenstates of ρ of length N cannot be transmitted with fidelity greater than ϵ.

There exists $N_0(\delta, \epsilon)$ such that for all $N > N_0$

$$\text{tr}(\bigotimes_N \rho)\Gamma < \epsilon.$$

where Γ is any projection onto a subspace with dimension $\dim(C)$. Following the reasoning on page 206 the sum of the eigenvalues in any $\dim(C)$ subspace are bounded by ϵ, which gives the above result. Thus

$$F(M, M') < \epsilon.$$

♠

22.5 Holevo Bound

The quantum noiseless coding theorem discussed in the previous section dealt only with pure states, further work has been done in considering mixed states for coding [86, 87, 88, 132].

Let A denote an input alphabet, C the set of quantum states in the Hilbert space of the quantum communication channel and

$$c : A \to C$$

the mapping from the input alphabet to quantum states. The quantum states are represented by density operators. Further let $p(a)$ for $a \in A$ denote the probability that a will be required to be sent over the quantum channel. The channel capacity is then given by

$$\max_p \left(S(\sum_{a \in A} p(a)c(a)) - \sum_{a \in A} p(a)S(c(a)) \right)$$

where the maximum is over all probability distributions and $\sum_{a \in A} p(a) = 1$. The quantity

$$H(A, C, c) := S(\sum_{a \in A} p(a)c(a)) - \sum_{a \in A} p(a)S(c(a))$$

is called the *Holevo information*. Decoding is described by the positive operators X_b with

$$\sum_{b \in B} X_b = I,$$

where b is the output from the alphabet B. We denote by $p(b|a)$ the probability that b is the output (where X_b is identified with b) if the input was a. The *Shannon information* is given by

$$I_S(p, A, X) := \sum_{b \in B} \sum_{a \in A} p(a)p(b|a) \log_2 \left(\frac{p(a|b)}{\sum_{a' \in A} p(a')p(b|a')} \right).$$

With these quantities we can describe the *quantum entropy bound*,

$$\max_X I_S(p, A, X) \leq H(A, C, c)$$

where equality is achieved if and only if the operators $p(a)c(a)$ commute.

Chapter 23

Quantum Error Detection and Correction

23.1 Introduction

The algorithms discussed in Chapter 21 rely on having isolated systems to store information. In practical applications this is not possible, and the environment interacts with the systems causing *decoherence*. Suppose the data is contained in the state $|\mathbf{x}\rangle$, and the environment is described by $|\mathbf{E}\rangle$. The initial state of the entire system is described by the tensor product of the states $|\mathbf{x}\rangle \otimes |\mathbf{E}\rangle$, which evolves according to some unitary operation U. The state $|\mathbf{x}\rangle$ evolves according to the unitary operation U_x which describes the algorithm. In classical error correction codes, all that needs to be corrected are bit flips (see Chapter 10). In the quantum case errors such as bit flips, phase changes and rotations complicate the error correction techniques. Since arbitrary errors in an encoding of information cannot be corrected, only certain types of errors are assumed to occur (this was also the case in classical error correction, where an upper bound in the number of bits that could be flipped in a transmission was assumed). The types of errors depend on the implementation. For example, suppose the types of errors (which we assume are distinguishable due to an encoding) are described by the unitary basis $E_1, \ldots, E_n$ so that all errors are a linear combination [138]

$$E = a_1 E_1 + \ldots + a_n E_n, \qquad E^\dagger E = I.$$

We use the state $|\mathbf{x}\rangle \otimes |0\rangle$, where $|\mathbf{x}\rangle$ is an encoded quantum state with the necessary property that it can be used to determine if any error of $E_1, \ldots, E_n$ has occurred, and the second quantum register will hold the number of the type of error which occurred. Further let S denote the operator for the *error syndrome* [138].

$$S(E_j \otimes I)|\mathbf{x}\rangle \otimes |0\rangle := |\mathbf{x}\rangle \otimes |j\rangle.$$

Now the encoded state with errors is given by

$$(E \otimes I)|\mathbf{x}\rangle \otimes |0\rangle = \sum_{j=1}^{n} E_j |\mathbf{x}\rangle \otimes |0\rangle.$$

Applying the operator for the error syndrome gives

$$S(E \otimes I)|\mathbf{x}\rangle \otimes |0\rangle = \sum_{j=1}^{n} E_j |\mathbf{x}\rangle \otimes |j\rangle.$$

Measuring the second register identifies the error. Suppose the measurement corresponds to $|k\rangle$, then the error is easily repaired since

$$(E_k^{-1} \otimes I)((E_k \otimes I)|\mathbf{x}\rangle \otimes |k\rangle) = |\mathbf{x}\rangle \otimes |k\rangle.$$

This illustrates that the additional difficulties in quantum error correction can, to some degree, be overcome by the properties of quantum mechanics itself. In classical error correction codes, duplication is used to overcome errors. This simple approach cannot be directly applied in quantum error correcting codes since this would involve a violation of the no cloning theorem. In the following section a code is introduced which involves duplication of certain properties of a state, and does not violate the no-cloning theorem. These duplications are specific to certain types of errors. The code words for $|0\rangle$ and $|1\rangle$ must be orthogonal to make sure they are distinguishable.

Further error correcting techniques introduced are fault tolerant error correction codes [58, 149, 157] which allow for some errors occuring in the error correction process, and fault tolerant quantum gates [149].

23.2   The Nine-qubit Code

The nine-qubit code invented by Shor [75] can be used to correct a bit flip error and a phase error. The code does this by duplicating the states from the orthonormal basis and the phase of the state, and then using correction by majority, assuming that only one such error has occurred. Only two duplications are needed so that a "majority out of three correction" scheme can be used. The coding applies to one qubit. Overcoming a bit flip error is achieved by the mapping

$$|0\rangle \rightarrow |000\rangle$$

$$|1\rangle \rightarrow |111\rangle$$

Thus the qubit $|\psi_0\rangle = \alpha|0\rangle + \beta|1\rangle$ is mapped to

$$|\psi_1\rangle = \alpha|000\rangle + \beta|111\rangle.$$

Thus a single bit flip error can be corrected by a majority value correction scheme. First the additional syndrome register must be added. We apply the operator

$$S_0 = (I \otimes U_S)U_{XOR(1,5)}U_{XOR(1,6)}U_{XOR(2,4)}U_{XOR(2,6)}U_{XOR(3,4)}U_{XOR(3,5)}$$

where $U_{XOR(i,j)}$ denotes the CNOT operator working with the ith qubit as the control and the jth qubit as the target, and

$$U_S = |000\rangle\langle000| + |111\rangle\langle111| + |100\rangle\langle011| + |010\rangle\langle101| +$$
$$|001\rangle\langle110| + |110\rangle\langle001| + |101\rangle\langle010| + |011\rangle\langle100|.$$

So to correct the error we simply apply $U_{XOR(3,1)}U_{XOR(4,2)}U_{XOR(5,3)}$. It is simple to extend this to correct phase errors in one qubit. The mapping is extended to

$$|0\rangle \rightarrow (|000\rangle + |111\rangle) \otimes (|000\rangle + |111\rangle) \otimes (|000\rangle + |111\rangle)$$
$$|1\rangle \rightarrow (|000\rangle - |111\rangle) \otimes (|000\rangle - |111\rangle) \otimes (|000\rangle - |111\rangle).$$

The bit flip errors are corrected for each of the three 3-qubit registers in the same way as above. The phase error (i.e. at most one sign change) is dealt with in exactly the same way using the subspace described by $\{|000\rangle + |111\rangle, |000\rangle - |111\rangle\}$ instead of $\{|0\rangle, |1\rangle\}$. It is important to note that the total phase is ignored. Using this procedure we can correct both bit flips and sign changes if they occur. The operators I, σ_x, σ_y and σ_z described by

$$I = \begin{pmatrix} 1 & 0 \\ 0 & 1 \end{pmatrix}, \qquad \sigma_x = \begin{pmatrix} 0 & 1 \\ 1 & 0 \end{pmatrix}, \qquad \sigma_y = \begin{pmatrix} 0 & -i \\ i & 0 \end{pmatrix}, \qquad \sigma_z = \begin{pmatrix} 1 & 0 \\ 0 & -1 \end{pmatrix}$$

form a basis for all 2×2 matrices since

$$\begin{pmatrix} a & b \\ c & d \end{pmatrix} \equiv \frac{1}{2}(a+d)I + \frac{1}{2}(a-d)\sigma_z + \frac{1}{2}(b+c)\sigma_x + \frac{i}{2}(b-c)\sigma_y.$$

Furthermore the unit matrix I effects no error on a qubit, σ_x effects a bit flip, σ_z effects a sign change and σ_y effects a bit flip and sign change. All these errors can be corrected by the nine-qubit code. Thus any linear combination of these errors can be corrected. Consider the arbitrary phase change

$$R_\theta = \begin{pmatrix} 1 & 0 \\ 0 & e^{i\theta} \end{pmatrix} \equiv \frac{1}{2}(1 + e^{i\theta})I + \frac{1}{2}(1 - e^{i\theta})\sigma_z$$

which can also be corrected by this scheme. Thus the scheme can correct any one-qubit error.

23.3   The Seven-qubit Code

A slightly more efficient code is a seven qubit code invented by Steane [133, 154]. The code uses classical Hamming codes to do the error correction for both phase errors and bit flip errors. The classical seven-bit Hamming code can correct any single-bit error. First we consider the mapping

$$|0\rangle \rightarrow |0_c\rangle = \frac{1}{\sqrt{8}} \sum_{d \in C_{H_7}, \ d_H(d,0) \ \text{even}} |d\rangle$$

$$|1\rangle \rightarrow |1_c\rangle = \frac{1}{\sqrt{8}} \sum_{d \in C_{H_7}, \ d_H(d,0) \ \text{odd}} |d\rangle$$

This coding is obviously orthogonal and can correct any single bit flip error. But we require that the coding can correct phase errors as well. This can be done by noting that a phase change is a bit flip when we apply the Hadamard gate

$$U_H|0\rangle = \frac{1}{\sqrt{2}}(|0\rangle + |1\rangle)$$

$$U_H|1\rangle = \frac{1}{\sqrt{2}}(|0\rangle - |1\rangle)$$

and use the basis $\{|0'\rangle = U_H|0\rangle, |1'\rangle = U_H|1\rangle\}$. Applying the Hadamard transform to all seven qubits in the code gives

$$|0\rangle \rightarrow |0_c\rangle \rightarrow |0_C\rangle = \frac{1}{4}(|0_c\rangle + |1_c\rangle)$$

$$|1\rangle \rightarrow |1_c\rangle \rightarrow |1_C\rangle = \frac{1}{4}(|0_c\rangle - |1_c\rangle)$$

Thus the code can still correct any single qubit error.

This code still requires seven qubits to encode a single qubit. Thus a 128 qubit system requires 896 qubits to operate reliably. With such a large number of qubits it is possible that interactions with the environment involving not only single qubits can become a larger problem. Thus it is desirable to encode a qubit with as few qubits as possible, in the next section we show that a qubit can be encoded reliably with less that 7 qubits.

23.4 Efficiency and the Five-qubit Code

A measure of efficiency of a code is how many qubits the code needs to correct an arbitrary single qubit error. From the above we know that any code correcting arbitrary errors needs to be able to correct the three errors described by σ_x, σ_y and σ_z. These errors map codeword subspaces into subspaces. To distinguish these errors the subspaces must be orthogonal. Suppose the code consists of n qubits. Three errors can occur for each qubit or no errors occur. Furthermore the two states must be orthogonal in each of these subspaces. Thus [108]

$$2(3n + 1) \leq 2^n.$$

The requirement of orthonormality of the encoding is used to aid a derivation of the code. Thus we obtain the constraints

$$\sum_{k=0}^{31} |\mu_k|^2 = \sum_{k=0}^{31} |\nu_k|^2 = 1$$

$$\sum_{k=0}^{31}\sum_{l=0}^{31} \overline{\mu_k}\nu_l \langle k|E|l\rangle = \sum_{k=0}^{31}\sum_{l=0}^{31} \overline{\nu_k}\mu_l \langle k|E|l\rangle = 0, \qquad E \in I, \sigma_x, \sigma_y, \sigma_z.$$

where μ_k and ν_k are the amplitudes of the encodings for $|0\rangle$ and $|1\rangle$, respectively. For the code they obtain [108]

$$|0\rangle \rightarrow \frac{1}{\sqrt{8}}(|b_1\rangle|00\rangle - |b_3\rangle|11\rangle + |b_7\rangle|10\rangle + |b_5\rangle|01\rangle)$$

$$|1\rangle \rightarrow \frac{1}{\sqrt{8}}(-|b_2\rangle|00\rangle - |b_4\rangle|11\rangle + |b_8\rangle|10\rangle - |b_6\rangle|01\rangle)$$

where

$$
\begin{aligned}
|b1\rangle &= |000\rangle + |111\rangle \\
|b2\rangle &= |000\rangle - |111\rangle \\
|b3\rangle &= |100\rangle + |011\rangle \\
|b4\rangle &= |100\rangle - |011\rangle \\
|b5\rangle &= |010\rangle + |101\rangle \\
|b6\rangle &= |010\rangle - |101\rangle \\
|b7\rangle &= |110\rangle + |001\rangle \\
|b8\rangle &= |110\rangle - |001\rangle.
\end{aligned}
$$

The code was discovered by assuming that the absolute value of the non-zero amplitudes were equal and real. Thus a solution would be described exclusively by the signs of the amplitudes. A computer search was used to find the code. A surprising feature of the scheme is that the error correcting technique is the exact reverse of the encoding technique [108], i.e we apply the same transformations but in the reverse order. The following figure illustrates the encoding process

$$|qabcd\rangle \rightarrow |q'a'b'c'd'\rangle.$$

For the decoding we follow the process from right to left giving

$$|q'a'b'c'd'\rangle \rightarrow |qabcd\rangle.$$

In the figure the π is a controlled phase change (multiplication with -1), the other

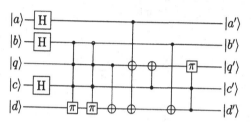

Figure 23.1: Encoding for the 5-qubit Error Correction Code

gates have the usual meanings. A filled connection (circle) indicates the operation is only applied when the corresponding qubit is $|1\rangle$ and an empty connection (circle) indicates the operation is only applied when the corresponding qubit is $|0\rangle$. The error syndrome and the result of the error on the state $\alpha|0\rangle + \beta|1\rangle$ (where $|\alpha|^2 + |\beta|^2 = 1$) is listed in the Table 23.4. Another 5 qubit code [19, 58] also found by a computer search is given by

$$|0\rangle \rightarrow |\tilde{0}\rangle := \tfrac{1}{4}(+|00000\rangle$$
$$+|11000\rangle + |10001\rangle + |00011\rangle + |00110\rangle + |01100\rangle$$
$$-|10100\rangle - |01001\rangle - |10010\rangle - |00101\rangle - |01010\rangle$$
$$-|11110\rangle - |11101\rangle - |11011\rangle - |10111\rangle - |01111\rangle)$$

$$|1\rangle \rightarrow |\tilde{1}\rangle := \tfrac{1}{4}(+|11111\rangle$$
$$+|00111\rangle + |01110\rangle + |11100\rangle + |11001\rangle + |10011\rangle$$
$$-|01011\rangle - |10110\rangle - |01101\rangle - |11010\rangle - |10101\rangle$$
$$-|00001\rangle - |00010\rangle - |00100\rangle - |01000\rangle - |10000\rangle).$$

We note that

$$|\tilde{1}\rangle = \bigotimes_{j=1}^{5} U_{XOR}|\tilde{1}\rangle,$$

and the signs are chosen to satisfy the orthonormality constraints.

Syndrome $|a'b'c'd'\rangle$	Resulting state
$|0000\rangle$	$+\alpha|0\rangle + \beta|1\rangle$
$|1101\rangle$	$-\alpha|1\rangle + \beta|0\rangle$
$|1111\rangle$	$-\alpha|0\rangle + \beta|1\rangle$
$|0001\rangle$	$+\alpha|0\rangle - \beta|1\rangle$
$|1010\rangle$	$+\alpha|0\rangle - \beta|1\rangle$
$|1100\rangle$	$+\alpha|0\rangle - \beta|1\rangle$
$|0101\rangle$	$+\alpha|0\rangle - \beta|1\rangle$
$|0011\rangle$	$-\alpha|0\rangle - \beta|1\rangle$
$|1000\rangle$	$-\alpha|0\rangle - \beta|1\rangle$
$|0100\rangle$	$-\alpha|0\rangle - \beta|1\rangle$
$|0010\rangle$	$-\alpha|0\rangle - \beta|1\rangle$
$|0110\rangle$	$-\alpha|1\rangle - \beta|0\rangle$
$|0111\rangle$	$-\alpha|1\rangle - \beta|0\rangle$
$|1011\rangle$	$-\alpha|1\rangle - \beta|0\rangle$
$|1110\rangle$	$-\alpha|1\rangle - \beta|0\rangle$
$|1001\rangle$	$-\alpha|1\rangle - \beta|0\rangle$

Table 23.1: Error Syndrome for the 5 Qubit Error Correction Code

23.5 Stabilizer Codes

The Pauli spin matrices (including the unit matrix) with additional phases of ± 1 and $\pm i$

$$\mathcal{P} := \{ I, \sigma_x, \sigma_y, \sigma_z, -I, -\sigma_x, -\sigma_y, -\sigma_z, iI, i\sigma_x, i\sigma_y, i\sigma_z, -iI, -i\sigma_x, -i\sigma_y, -i\sigma_z \}$$

form a group with respect to matrix multiplication

$\times$	I	σ_x	σ_y	σ_z
I	I	σ_x	σ_y	σ_z
σ_x	σ_x	I	$i\sigma_z$	$-i\sigma_y$
σ_y	σ_y	$-i\sigma_z$	I	$i\sigma_x$
σ_z	σ_z	$i\sigma_y$	$-i\sigma_x$	I

Any two elements in the group either commute or anticommute, i.e for $A, B \in \mathcal{P}$

- if $[A, B] = AB - BA \neq 0$ it follows that $[A, B]_+ = AB + BA = 0$,

- if $[A, B]_+ \neq 0$ it follows that $[A, B] = 0$.

A consequence is that the set

$$\mathcal{P}_n := \{ \otimes_{j=1}^n A_j \,|\, A_j \in \mathcal{P} \quad j = 1, 2, \ldots, n \}$$

also forms a group, where $n \in \mathbf{Z}^+$.

Let S be an Abelian subgroup of $\mathcal{P}_n$. We define the quantum code C_S as the set of states

$$C_S := \{ \, |\psi\rangle \mid U|\psi\rangle = |\psi\rangle \quad \forall U \in S \, \}.$$

The set S is called the *stabilizer* of the code C_S. The set S must be Abelian since for $M, N \in S$ and $|\psi\rangle \in C_S$

$$MN|\psi\rangle = M|\psi\rangle = |\psi\rangle$$

$$NM|\psi\rangle = N|\psi\rangle = |\psi\rangle.$$

Thus

$$[M, N]|\psi\rangle = 0.$$

For any $M \in S$, the inverse $M^{-1} = M^*$ is also in S. If S contains $-\bigotimes_{j=1}^{n} I$ then the code is trivial, i.e the code contains only the zero element of the underlying Hilbert space. Obviously a subset of $\mathcal{P}_n$ forms a basis for all operations on n qubits (i.e. the elements of $\mathcal{P}_n$ unique up to phase). Suppose $E \in \mathcal{P}_n$ such that $[E, M]_+ = 0$ for some $M \in S$. For $|\phi\rangle, |\psi\rangle \in C_S$ we have

$$\langle\phi|E|\psi\rangle = \langle\phi|EM|\psi\rangle = -\langle\phi|ME|\psi\rangle = -\langle\phi|M^*E|\psi\rangle = -\langle\phi|E|\psi\rangle.$$

Thus $\langle\phi|E|\psi\rangle = 0$. So if for every $E, F \in \mathcal{P}_n$ there exists $M \in S$ such that $[M, F^*E]_+ = 0$, where E and F introduce errors in at most t qubits, the code C_S can correct all t qubit errors. If for any errors E and F we have $F^*E \in S$, then the errors E and F can be corrected, but not distinguished. A code with this property is called a *degenerate quantum code*, otherwise it is called a *nondegenerate quantum code*. The construction of these codes is described by Gottesman [74] and Calderbank et al. [37]. If a code encodes k qubits into a Hilbert space of dimension 2^n and corrects up to t errors we have the *quantum Hamming bound*

$$\left(\sum_{j=0}^{t} 3^j \binom{n}{j} \right) 2^k \leq 2^n.$$

Let $|\psi\rangle$ be a codeword from C_S. We suppose that E, an error, has operated on $|\psi\rangle$. Let $M \in S$, E and M either commute or anticommute. Suppose E and M commute, the state $E|\psi\rangle$ is an eigenstate of M

$$ME|\psi\rangle = EM|\psi\rangle = E|\psi\rangle$$

with eigenvalue 1, if E and M anticommute the state $E|\psi\rangle$ is an eigenstate of M

$$ME|\psi\rangle = -EM|\psi\rangle = -E|\psi\rangle$$

with eigenvalue -1. Thus the eigenvalues corresponding to the eigenstate $E|\psi\rangle$ of the operators in S give the error syndrome. Measuring the state with respect to M does not destroy the information, and is used to determine the error syndrome.

Chapter 24

Quantum Hardware

24.1 Introduction

Computation is ultimately a physical process. In practice, the range of physically realizable devices determines what is computable and the resources, such as computer time, required to solve a given problem. Computing machines can exploit a variety of physical processes and structures to provide distinct trade-offs in resource requirements. An example is the development of parallel computers with their trade-off of overall computation time against the number of processors employed.

Hardware used to implement quantum algorithms have certain requirements.

- Storage. Qubits must be stored for long enough for a required algorithm to complete and a result to be obtained. The discovery of quantum error correcting codes decreases the hardware requirements at the cost of using more qubits.

- Isolation. Quantum registers must be sufficiently isolated from the environment to minimize decoherence errors. If error correcting codes are used the requirement reduces slightly to ensuring that only correctable errors can occur.

- Measurement. Quantum registers must be measured to obtain results. This process must be efficient and reliable.

- Gates. Algorithms involve the manipulation and controlled manipulation of qubits, which are described by some set of gates. These gates must be efficiently implementable in the hardware system.

- Reliability. Algorithms must run reliably. Fault-tolerant gates and error correcting codes can be used to satisfy the requirement provided that the hardware only introduces errors which can be corrected.

A number of different approaches have been suggested.

24.2   Trapped Ions

One method proposed to implement a quantum computing device is using an ion trap [132, 155, 158]. Each qubit is carried by a single ion held in a linear Paul trap. The quantum state of each ion is a linear combination of the ground state 0⟩ and a long-lived metastable exited state $|1\rangle$. A linear combination of the two states

$$a|0\rangle + be^{i\omega t}|1\rangle, \qquad |a|^2 + |b|^2 = 1$$

remains coherent for a time comparable to the lifetime of the excited state, with oscillating relative phase. To measure a qubit, a laser tuned to a transition from the ground state to a short lived excited state is used to illuminate the ion. An ion in the state $|0\rangle$ repeatedly absorbs and reemits the laser light. An ion in the state $|1\rangle$ will remain dark. Due to Coulomb repulsion, the ions are sufficiently separated to be addressed by pulsed lasers. A laser tuned to the frequency ω of the transition focused on the appropriate ion induces Rabi oscillations between $|0\rangle$ and $|1\rangle$. Using the appropriate laser pulse timing and phase, any one qubit unitary transformation can be applied to the ion.

The Coulomb repulsion between ions is used to achieve the interaction between ions. The mutual Coulomb repulsion between the ions results in a spectrum of coupled normal modes of vibration for the trapped ions. When the laser is correctly tuned, then the absorption or emission of a laser photon by a single ion causes a normal mode involving many ions to recoil coherently. The vibrational mode of lowest frequency ν is the center-of-mass mode. The ions can be laser cooled to temperatures below that required for the center-of-mass mode, to levels such that each vibrational mode is likely to occupy its quantum-mechanical ground state. A laser tuned to the frequency $\omega - \nu$ and applied to an ion for the time required to rotate $|1\rangle$ to $|0\rangle$ and the the center-of-mass oscillation to transition from its ground state to the first excited state, causes the information of the qubit to be transferred to the collective state of motion of all the ions. Similarly, the information should be transferred to another ion while returning the center-of-mass oscillator to its ground state. Thus two ions can interact, and two qubit operations can be performed.

Measurement of arbitrary observables of trapped ions is described by Gardiner et al. [70]. A method of performing quantum computations without the need for cooling of the trapped ions has also been proposed [131].

Monroe et al. [120] demonstrated the operation of a two-qubit controlled NOT quantum logic gate. The two qubits are stored in the internal and external degrees of freedom of a single trapped ion, which is first laser cooled to the zero-point energy. In their implementation, the target qubit $|S\rangle$ is spanned by two $^2S_{1/2}$ hyperfine ground states of a single $^9Be^+$ ion, abbreviated by the equivalent spin-1/2 states $|\downarrow\rangle$ and $|\uparrow\rangle$, which are separated in frequency by $\omega_0/2\pi \approx 1.25$ GHz. The control qubit $|n\rangle$ is spanned by the first two quantized harmonic oscillator states of the trapped ion ($|0\rangle$ and $|1\rangle$), separated in frequency by the vibrational frequency $\omega_x/2\pi \approx 11$ MHz of the harmonically trapped ion. Manipulation between the four basis eigenstates spanning the two qubit register is achieved by applying a pair of off-resonant laser beams to the ion, which drives stimulated Raman transitions between basis states. When the difference frequencey δ is set near ω_0 transitions are coherently driven between internal states $|S\rangle$ while preserving $|n\rangle$. For $\delta \approx \omega_0 - \omega_x$ (respectively $\delta \approx \omega_0 + \omega_x$) transitions are coherently driven between $|1\rangle \otimes |\downarrow\rangle$ and $|0\rangle \otimes |\uparrow\rangle$ (respectively $|0\rangle \otimes |\downarrow\rangle$ and $|1\rangle \otimes |\uparrow\rangle$).

24.3 Cavity Quantum Electrodynamics

Instead of using the vibrational modes, as mentioned in the previous section, neutral atoms can be trapped in a small high finesse optical cavity which interact because of the coupling due to the normal modes of the electromagnetic field in the cavity [32, 132, 158].

As described in the previous section, it is again possible to induce a transition in one atom conditioned on the state of another atom. Alternatively, a qubit may be stored in the polarization of a photon. The atoms are used to cause interactions between photons. It has already been demonstrated that the circular polarization of one photon can influence the phase of another photon. The first photon is stored in the cavity where left polarization ($|L\rangle$) does not couple to the atom and right polarization ($|R\rangle$) couples strongly. The second photon traverses the cavity where again only the right polarization couples strongly with the atom. The second photon acquires a phase shift $e^{i\Delta}$ only if both polarizations are right circular, i.e the following transform is implemented

$$
\begin{aligned}
|L\rangle_1 \otimes |L\rangle_2 &\rightarrow |L\rangle_1 \otimes |L\rangle_2 \\
|L\rangle_1 \otimes |R\rangle_2 &\rightarrow |L\rangle_1 \otimes |R\rangle_2 \\
|R\rangle_1 \otimes |L\rangle_2 &\rightarrow |R\rangle_1 \otimes |L\rangle_2 \\
|R\rangle_1 \otimes |R\rangle_2 &\rightarrow e^{i\Delta}|R\rangle_1 \otimes |R\rangle_2
\end{aligned}
$$

Briegel et al. [32] propose two other methods to implement a phase shift gate. The first method involves moving the potentials of the traps towards each other in a state-dependent way while leaving the shape of the potential unchanged. This results in two kinds of phase shifts. A single particle kinetic phase shift and an interaction phase shift due to coherent interactions between two atoms.

The second method involves changing the shape of the potentials with time, depending on the internal states of the particles. The atoms are initially trapped in two displaced wells. The barrier between the wells is removed (quickly) for atoms in a state $|b\rangle$ while atoms in state $|a\rangle$ experience no change. The atoms are allowed to oscillate for some time and then the barrier is raised (again quickly) such that the atoms are trapped again in their initial positions. The atoms acquire a kinematic phase due to the oscillations within their respective wells and an interaction phase due to the collision.

Both methods implement the phase change gate similar to the transform given above, except for some additional overall phase introduced in the transform.

24.4   Quantum Dots

A *quantum dot* is a small metal or semiconductor box that holds a discrete number of electrons. This number can be changed by adjusting electric fields in the neighbourhood of the dot, for example by applying a voltage to a nearby metal gate. We refer to quantum dots that have either zero or one electron in them. In a solid most electrons are tightly bound to the atom. A few others are not bound to any atom and are only confined in the quantum dots when present. We denote by $\bigcirc$ an empty quantum dot and by $\odot$ a quantum dot with a single electron.

Advanced semiconductor growth techniques such as molecular beam epitaxy allow the fabrication of semiconductor sandwich structures with interfaces of very high precision. The various layers in the sandwich can be made to have different properties by selecting an appropriate material during growth. Differences in the bandgaps of different materials can be utilized in this way to create an effective electronic potential energy which electrons experience. The layers can be grown sufficiently thin so that quantum mechanical confinement becomes important. A quasi two-dimensional electron gas forms in this thin layer. Patterning techniques such as optical and electron-beam lithography can be used to reduce this to a quasi one-dimensional (quantum wire) or quasi zero-dimensional (quantum dot) system. A different scheme to reduce the quasi dimensionality is using electrostatic confinement. A metallic pattern on top of the sandwich with a negative bias confines the electron gas, two parallel strips create a quantum wire and two staggered patterns of parallel strips can be used to create quantum dots. Other techniques include using scanning tunneling microscopes and chemical self assembly.

Lent and Porod [130] suggest creating a cell of 5 quantum dots in the shape of an "X" containing 2 electrons. The two electrons will be positioned in two opposite corner quantum dots due to Coulomb repulsion. There are two such configurations which can be identified with 0 and 1.

Figure 24.1: Two Possible Configurations for Quantum Dot Cells

The ground state of the system will be an equal superposition of the two basic configurations with electrons at opposite corners. The quantum dots are labelled as 0 for the top left corner, 1 for the top right corner, 2 for the bottom left corner, 3 for the bottom right corner and 0 for the center. Thus the cell polarity can be defined as

$$p = \frac{(n_1 + n_3) - (n_2 + n_4)}{2}$$

where $n_i \in \{0, 1\}$ denotes the number of electrons in quantum dot i and

$$\sum_{i=0}^{4} n_i = 2.$$

We can identify the polarity of -1 with binary 0 and a polarity of 1 with binary 1, where the polarities are of the configurations given above. Polarities between -1 and 1 are interpreted as superposition states, for example the ground state has a net polarity of 0. The electrons can move between quantum dots if they are close enough due to quantum tunneling. If two cells must interact, they must be sufficiently distant to prevent electrons tunneling between them. Their interaction is due to Coulomb interaction between the electrons. Suppose two adjacent cells are configured as follows

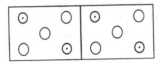

Forcing the first cell into the configuration with polarity 1, causes the second cell to reconfigure to a minimum energy configuration.

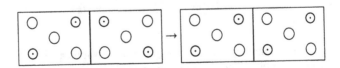

This allows signals to propagate along a series of quantum dots as quantum wires. Classical operations are also possible using these cells, for example the majority

function can be implemented by

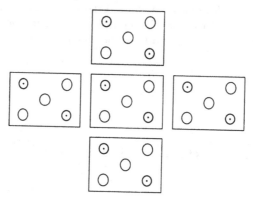

Forcing any three outer cells into some configurations causes the last outer cell to be forced to the majority configuration to achieve the lowest energy state. A majority gate can be used to construct any other classical gate. The OR gate can be constructed by fixing the polarity of one of the inputs to 1, and the AND gate by fixing the polarity of one of the inputs to -1. A full adder has been constructed using these principles.

Measurement can be achieved by passing a current through a very narrow constriction near a corner of a cell. An electron in the corner would prevent current from flowing.

Another approach using quantum dots aimed specifically at quantum computing has been proposed [34]. This approach does not use cells of quantum dots, but rather the electron spin of an electron contained in a quantum dot. The manipulation of more than one qubit is required for computation. This is achieved by coupling quantum dots. Due to the Coulomb interaction and the Pauli exclusion principle the ground state for two qubits is an entangled spin state. The system is described by the Heisenberg Hamiltonian

$$H_s(t) = J(t)\mathbf{S}_1 \cdot \mathbf{S}_2$$

where $J(t)$ is the exchange coupling between $\mathbf{S}_1$ and $\mathbf{S}_2$. If the exchange coupling is pulsed such that

$$\frac{1}{\hbar}\int J(t)dt = \frac{J_0\tau_s}{\hbar} = \pi \quad (\text{mod } 2\pi)$$

then the associated unitary evolution describes the swapping U_{SW} of the quantum states of the two qubits. The XOR operation is obtained as

$$U_{XOR} = \exp(\frac{i\pi}{2}S_1^z)\exp(-\frac{i\pi}{2}S_2^z)U_{SW}^{\frac{1}{2}}\exp(i\pi S_1^z)U_{SW}^{\frac{1}{2}}$$

a combination of $U_{SW}^{\frac{1}{2}}$ and single qubit rotations. The XOR operation combined with single qubit rotations is a universal set of quantum gates. Thus any quantum algorithm can be implemented using these operations. With only two quantum dots a gate operation can be performed using uniform magnetic fields. For more qubits local magnetic fields are necessary. The requirement is reduced by noting that using the swap operation, a qubit state can be transferred to a qubit where an operation can take place and then back to the original position without influencing the other qubit states.

24.5 Nuclear Magnetic Resonance Spectroscopy

Most implementations of quantum computing devices utilize submicroscopic assemblies of quantum spins which can be difficult to prepare, isolate, manipulate and observe.

In *nuclear magentic resonance* [50, 51, 103, 132] implementations, the qubit is identified with a nuclear spin in a molecule. A spin can be aligned ($|0\rangle$) or antialigned ($|1\rangle$) with an applied magnetic field giving the basis of computation. The spins take a long time to relax or decohere. The technique emulates a quantum computation using a large number of spins. Spin-active nuclei in each molecule of a liquid sample are largely isolated from the spins in all other molecules, each molecule is effectively an independent quantum computer. The computation is possible due to the existence of *pseudo-pure states*, whose transformation properties are identical to those of true pure states. Results of computations are then determined by, for example, thermodynamic averaging. The method is chosen to average out unwanted fluctuating properties so that only underlying coherent properties are measured. Alternatively methods such as optical pumping and dynamic nuclear polarization can be used to cool the system to a ground state. This leads to an *ensemble quantum computer*.

Using a pulsed rotating magnetic field with frequency ω determined by the energy splitting between the spin-up and spin-down states, Rabi oscillations of the spin are induced. The appropriate timing of the pulses can perform any unitary transform on a single spin. All spins are exposed to the rotating magnetic field but only those on resonance respond. The spins have dipole-dipole interactions which can be exploited to perform two-qubit operations. The XOR (controlled NOT) operation has been implemented using Pound-Overhauser double resonance and also using a spin-coherence double resonance pulse sequence.

Average Hamiltonian theory can be used to implement quantum gates. The evolution of the state at a time T is solved in terms of the time independent average Hamiltonian $\overline{H}(T)$. The total Hamiltonian $H_{tot}(t) = H_{int} + H_{ext}(t)$ is separated into a time invariant internal Hamiltonian H_{int} and a time dependent Hamiltonian

$H_{ext}(t)$. After a period of evolution the overall dynamics is described by

$$U(T) = T \exp\left(-i \int_0^T H_{tot}(\tau)d\tau\right) = e^{-i\overline{H}T}$$

where T is the Dyson time ordering operator. For sufficiently small T the Magnus expansion can be used to determine $\overline{H}$. The coupling between qubits is always active, thus it is useful to have an operation to suppress the undesirable couplings. This can be achieved by an experimental method for "tracing out" or averaging out unwanted degrees of freedom. The CNOT operation can be expressed as

$$U_{XOR} = \exp(\frac{i\pi}{4}e^{\frac{i\pi}{4}I\otimes\sigma_x}) \exp(\frac{i\pi}{4}\sigma_z \otimes I) \exp(\frac{i\pi}{4}\sigma_z \otimes \sigma_x)$$

where $\exp(\frac{i\pi}{4}\sigma_z \otimes \sigma_x)$ is implemented as

$$\exp(\frac{i\pi}{4}\sigma_z \otimes \sigma_x) \equiv \exp(-\frac{i\pi}{4}\sigma_y \otimes I) \exp(\frac{i\pi}{4}\sigma_z \otimes \sigma_z) \exp(\frac{i\pi}{4}I \otimes \sigma_y)$$

since the only two-body Hamiltonian available in liquid state nuclear magnetic resonance spectroscopy is the scalar coupling $\sigma_z \otimes \sigma_z$.

Using nuclear magnetic resonance techniques Deutsch's algorithm [43, 98], Grover's algorithm [44, 173] and a generalization of Shor's algorithm [174] have been implemented. Maximally entangled states using this technique have also been achieved.

Chapter 25
Internet Resources

In the following we give a collection of web sites which provide information about quantum computing. The web sites provide tutorials, information on experimental implementations and electronic versions of papers.

`http://issc.rau.ac.za`
The web site for the International School for Scientific Computing. The school offers courses in scientific computing including a course on classical and quantum computing.

`http://xxx.lanl.gov`
The web site of the Los Alamos National Laboratory pre-print archive. The site provides access to pre-prints in the fields of physics, mathematics, nonlinear sciences and computer science. A search engine is also provided.

`http://www.qubit.org`
The Centre for Quantum Computation, part of the University of Oxford, conducts theoretical and experimental research into all aspects of quantum information processing, and into the implications of the quantum theory of computation for physics itself.

`http://www.theory.caltech.edu/~preskill/ph229`
Quantum Information and Computation course notes. Overview of classical complexity theory, quantum complexity, efficient quantum algorithms, quantum error-correcting codes, fault-tolerant quantum computation, physical implementations of quantum computation.

`http://www.openqubit.org`
A quantum computation simulation project on Intel based architectures. The project goal is to develop a system for describing and testing quantum computing algorithms.

http://squint.stanford.edu/

A collaboration between researchers at Stanford University and U.C. Berkeley, involving the experimental and theoretical study of quantum-mechanical systems, and how they can be utilized to process and store information.

http://qso.lanl.gov/qc/

An overview of the work done at Los Alamos on quantum computation and cryptography is provided. A number of papers are also provided in electronic form.

http://theory.caltech.edu/~quic/

Quantum Information and Computation (QUIC). A collaboration of groups at MIT, Caltech and USC investigating experimental, theoretical, and modelling quantum computation.

http://www.research.ibm.com/quantuminfo/

Quantum Information and Information Physics at IBM Research Yorktown. The group's work main work is in quantum information and computation theory, but they also study other aspects of the relation between physics and information processing.

http://www.iro.umontreal.ca/labs/theorique/index.html.en

Laboratory for Theoretical and Quantum Computing of the Computer Science Department of the University of Montreal. Includes a bibliography of quantum cryptography.

http://www.fysel.ntnu.no/Optics/qcr/

Quantum cryptography in Norway. As the first large task of the project, they are building the demonstrator of a point-to-point quantum key distribution channel. Some have already been built and tested by other research groups. The basic principles are well known, but what remains a challenge is approaching practical applications. They are working in this direction.

http://www.nd.edu/~qcahome/

Quantum-dot Cellular Automata. A web site exploring the possibilities of using quantum-dots to form quantum wires and to construct gates. The web site provides tutorials, simulations and electronic versions of some papers on the subject.

Bibliography

[1] Adami C. and N.J. Cerf, "What Information Theory Can Tell Us About Quantum Reality",
http://xxx.lanl.gov, quant-ph/9806047.

[2] Ammeraal L., *STL for C++ Programmers*, John Wiley, Chichester, 1997.

[3] Ash R. B., *Information Theory*, Dover Publications, New York, 1990.

[4] Bac Fam Quang and Perov V. L., "New evolutionary genetic algorithms for NP-complete combinatorial problems", *Biological Cybernetics* **69**, 229–234, 1993.

[5] Balakrishnan A. V., *Applied Functional Analysis*, Second Edition Springer-Verlag, New York, 1981.

[6] Barenco A., "A Universal Two-Bit Gate for Quantum Computation",
http://xxx.lanl.gov, quant-ph/9505016

[7] Barenco A. *et al.*, "Elementary gates for quantum computation",
http://xxx.lanl.gov, quant-ph/9503016

[8] Barenco A., "Quantum Physics and Computers", *Contemporary Physics* **37**, 375–389, 1996.

[9] Bell J. S., *Speakable and unspeakable in quantum mechanics*, Cambridge University Press, Cambridge, 1989.

[10] Ben-Ari M., *Mathematical Logic for Computer Science*, Prentice Hall, New York, 1993.

[11] Benioff P., "Models of Quantum Turing Machines",
http://xxx.lanl.gov, quant-ph/9708054

[12] Bennett C. H. and G. Brassard, "Quantum cryptography: Public-key distribution and coin tossing", *Proceedings of IEEE International Conference on Computers, Systems and Signal Processing*, Bangalore, India, 175–179 (1984).

[13] Bennett C. H., In *Emerging Syntheses in Science*, ed. D. Pines, Addison-Wesley, Reading, MA, 1988.

[14] Bennett C. H., G. Brassard, and N. D. Mermin, "Quantum cryptography without Bell's theorem", *Phys. Rev. Lett.* **68**, 557–559 (1992).

[15] Bennett C. H., "Quantum cryptography using any two nonorthogonal states", *Phys. Rev. Lett.* **68**, 3121–3124 (1992).

[16] Bennett C. H. and S.J. Wiesner, "Communication via one- and two-particle operations on Einstein-Podolsky-Rosen states", *Phys. Rev. Lett.* **69**, 2881–2884 (1992).

[17] Bennett H. C, G. Brassard, Crépeau C., R. Jozsa, A. Peres and W. K. Wootters, "Teleporting an Unknown Quantum State via Dual Classical and Einstein-Podolsky-Rosen Channels", *Phys. Rev. Lett.* **70** 1895–1899 (1993).

[18] Bennett H. C., "Quantum Information and Computation", *Physics Today*, October, 24–30, 1995.

[19] Bennett H. C., D. P. DiVincenzo, J. A. Smolin and W. K. Wootters, "Mixed State Entanglement and Quantum Error Correction", http://xxx.lanl.gov/quant-ph/9604024

[20] Bennett H. C., E. Bernstein, G. Brassard and U. Vazirani, "Strengths and Weaknesses of Quantum Computing", *SIAM Journal on Computing* **26**, 1510–1523, 1997.

[21] Berthiaume A., "Quantum Computation", http://andre.cs.depaul.edu/Andre/publicat.htm

[22] Biron D., O. Biham, E. Biham, M. Grassl and D. A. Lidar, "Generalized Grover Search Algorithm for Arbitrary Initial Amplitude Distribution", http://xxx.lanl.gov, quant-ph/9801066

[23] Böhm A., *Quantum Mechanics*, Springer-Verlag, 1936.

[24] Bouwmeester D. et al., "Experimental Quantum Teleportation", *Nature* **390**, 575, 1997.

[25] Bouwmeester D., J.-W. Pan, M. Daniell, H.Weinfurter and A. Zeilinger, "Observation of three-photon Greenberger-Horne-Zeilinger entanglement", http://xxx.lanl.gov, quant-ph/9810035

[26] Boyer M., Brassard G., Hoyer P. and A. Tapp, "Tight bounds on quantum searching", http://xxx.lanl.gov, quant-ph/9605034

[27] Brassard G. and P. Høyer, "An Exact Quantum Polynomial-Time Algorithm for Simon's Problem", http://xxx.lanl.gov, quant-ph/9704027

[28] Brassard G., Braunstein S. L. and R. Cleve, *Physica D* **120** 43–47 (1998).

[29] Braunstein S. L., "Error Correction for Continuous Quantum Variables", *Physical Review Letters* **80**, 4084–4087, 1998.

[30] Braunstein S. L. and H. J. Kimble, "Teleportation of Continuous Quantum Variables", *Physical Review Letters* **80**, 869–872, 1998.

[31] Braunstein S. L. and H. J. Kimble, "Dense coding for continuous variables", *Physical Review A* **61**, 042302-1–042302-4, 2000.

[32] Briegel H.-J., T. Calarco, D. Jaksch, J. I. Cirac and P. Zoller, "Quantum computing with neutral atoms", http://xxx.lanl.gov, quant-ph/9904010

[33] Bruß D., D. P. DiVincenzo, A. Ekert, C. A. Fuchs, C. Macchiavello and J. A. Smolin, "Optimal universal and state-dependent quantum cloning", http://xxx.lanl.gov, quant-ph/9705038

[34] Burkard G., D. Loss and D. P. DiVincenzo, "Coupled quantum dots as quantum gates", http://xxx.lanl.gov, quant-ph/9808026

[35] Bužek V., and M. Hillery, "Universal Optimal Cloning of Qubits and Quantum Registers", http://xxx.lanl.gov, quant-ph/9801009

[36] Cabello A., "Quantum Key Distribution Based on Entanglement Swapping", http://xxx.lanl.gov, quant-ph/9911025

[37] Calderbank A. R., E. M. Rains, P. W. Shor, and N. J. A. Sloane, "Quantum Error Correction and Orthogonal Geometry", http://xxx.lanl.gov, quant-ph/9605005

[38] Cerf N.J. and C. Adami, "Negative entropy and information in quantum mechanics", http://xxx.lanl.gov, quant-ph/9512022

[39] Cerf N.J. and C. Adami, "Quantum mechanics of measurement", http://xxx.lanl.gov, quant-ph/9605002

[40] Cerf N.J. and C. Adami, "Entropic Bell Inequalities", http://xxx.lanl.gov, quant-ph/9608047

[41] Chaitin G. J., *Information, Randomness and Incompleteness*, World Scientific, Singapore, 1987.

[42] Chartrand C. and L. Lesniak, *Graphs and Digraphs*, Third Edition, Chapman and Hall, London, 1996.

[43] Chuang I. L., L. M. K. Vandersypen, Xinlan Zhou, D. W. Leung and S. Lloyd, "Experimental realization of a quantum algorithm", http://xxx.lanl.gov, quant-ph/9801037

[44] Chuang I. L., N. Gershenfeld and M. Kubinec, "Experimental Implementation of Fast Quantum Searching", http://squint.stanford.edu/qc/nmrqc-grover/index.html

[45] Cichocki A. and Unbehauen R., *Neural Networks for Optimization and Signal Processing*, John Wiley, Chichester, 1993.

[46] Cirac J. I. and P. Zoller, "Quantum Computations with Cold Trapped Ions", *Physical Review Letters* **74**, 4091–4094, 1995.

[47] Cleve R., A. Ekert, C. Macchiavello and M. Mosca, "Quantum Algorithms Revisited", *Proc. Roy. Soc. Lond. A* **454**, 339–354, 1998.

[48] Cohen D. E., *Computability and Logic*, John Wiley, New York, 1987.

[49] Cohen D.I.A., *Introduction to Computer Theory*, Revised Edition, Wiley, New York, 1991.

[50] Cory D. G., M. D. Price and T. F. Havel, "Nuclear Magnetic Resonance Spectroscopy: An Experimentally Accessible Paradigm for Quantum Computing", http://xxx.lanl.gov, quant-ph/9709001

[51] Cory D. G. *et al.*, "NMR Based Quantum Information Processing: Achievements and Prospects", http://xxx.lanl.gov, quant-ph/0004104

[52] Cybenko G., Approximation by superpositions of a sigmoidal function, *Mathematics of Control, Signals and Systems* **2**, 303–314, 1989

[53] H. T. Davis, *Introduction to Nonlinear Differential and Integral Equations* Dover Publications, New York, 1962.

[54] Deutsch D., "Quantum theory, the Church-Turing principle and the universal quantum computer", *Proc. Royal Soc. London A* **400**, 97–117, 1985.

[55] Deutsch D. and R. Jozsa, *Proc. Royal Soc. London A* **439**, 553, 1992.

[56] Deutsch D., A. Barenco and A. Ekert, "Universality in Quantum Computation", http://xxx.lanl.gov, quant-ph/9505018

[57] DiVincenzo D. P., "Two-bit gates are universal for quantum computation", http://xxx.lanl.gov, condmat/9407022

[58] DiVincenzo D. P. and P. W. Shor, "Fault-Tolerant Error correction with efficient Quantum Codes", http://xxx.lanl.gov, quant-ph/9605031

[59] Dirac P. A. M., *The Principles of Quantum Mechanics*, Clarendon Press, Oxford, 1958.

[60] Einstein A., B. Podolski and N. Rosen, "Can quantum mechanical description of reality be considered complete ?", *Physical Review* **47**, 777–780, 1935.

[61] Ekert A., "Quantum cryptography based on Bell's theorem", *Phys. Rev. Lett.* **67**, 661–663 (1991).

[62] Elby A. and J. Bub, "Triorthogonal uniqueness theorem and its relevance to the interpretation of quantum mechanics", *Physical Review A* **49**, 4213–4216, 1994.

[63] Epstein R. L. and Carnielli W. A., *Computability*, Wadsworth & Brooks/Cole, Pacific Grove, California (1989).

[64] Everett III H., "Relative state formulation of quantum mechanics" *Review of Modern Physics* **29** 454–462, 1957.

[65] Fausett L., *Fundamentals of Neural Networks: Architecture, Algorithms and Applications*, Prentice Hall, Englewood Cliffs, N. J., 1994

[66] Ferreira C., "Gene Expression Programming: a New Adaptive Algorithm for Solving Problems",
http://xxx.lanl.gov, cs.AI/0102027

[67] Feynman R. P., A. J. G. Hey (Editor) and R. W. Allen (Editor), *Feynman Lectures on Computation*, Perseus books, 1996.

[68] Feynman R. P., R. B. Leighton and M. Sands, *The Feynman Lectures on Physics Volume III*, Addison-Wesley, Reading, MA, 1966.

[69] Funahashi K.-I., "On the approximate realization of continuous mappings by neural networks", *Neural Networks*, **2**, 183–192, 1989

[70] Gardiner S. A., J. I. Cirac and P. Zoller, "Measurement of Arbitrary Observables of a Trapped Ion",
http://xxx.lanl.gov, quant-ph/9606026

[71] Glimm J. and A. Jaffe, *Quantum Physics*, Springer-Verlag, New York, 1981.

[72] Goldberg D. E., *Genetic Algorithms in Search, Optimization and Machine Learning*, Addison-Wesley, Reading, MA, 1989.

[73] Goldberg D. E. and R. Lingle, "Alleles, Loci, and the TSP", in Greffenstette, J. J. (Editor), *Proceedings of the First International Conference on Genetic Algorithms*, Lawrence Erlbaum Associates, Hillsdale, NJ, 1985.

[74] Gottesman D., "A Class of Quantum Error-Correcting Codes Saturating the Quantum Hamming Bound",
http://xxx.lanl.gov, quant-ph/9604038

[75] Gottesman D., "An Introduction to Quantum Error Correction",
http://xxx.lanl.gov, quant-ph/0004072

[76] Grassberger P., *Int. Journ. Theor. Phys.* **25**, 907, 1986.

[77] Grassmann W. K. and J.-P. Tremblay, *Logic and Discrete Mathematics: A Computer Science Perspective*, Prentice Hall, New Jersey, 1996.

[78] Grover L. K., "Quantum computers can search arbitrarily large databases by a single query",
http://xxx.lanl.gov, quant-ph/9706005

[79] Grover L. K., "Quantum Mechanics helps in searching for a needle in a haystack",
http://xxx.lanl.gov, quant-ph/9706033

[80] Gudder S., "Quantum Automata: An Overview", *Int. Journ. Theor. Phys.* **38**, 2261, 1999.

[81] Hardy Y., W.-H. Steeb and R. Stoop, "Jacobi Elliptic Functions, Nonlinear Evolution Equations and Recursion", *International Journal of Modern Physics C* **11**, 27–31, 2000.

[82] Hassoun M. H., *Fundamentals of Artificial Neural Networks*, The MIT Press, Cambridge Massachusetts, 1995.

[83] Haykin S., *Neural Networks*, Macmillan College Publishing Company, New York, 1994.

[84] Healey R., *The philosophy of quantum mechanics* Cambridge University Press, Cambridge, 1990.

[85] Hebb D. O., *The Organization of Behaviour*, John Wiley, New York, 1949.

[86] Holevo A. S., "The Capacity of Quantum Channel with General Signal States",
http://xxx.lanl.gov, quant-ph/9611023

[87] Holevo A. S., "Coding Theorems for Quantum Communication Channels",
http://xxx.lanl.gov, quant-ph/9708046

[88] Holevo A. S., "Coding Theorems for Quantum Channels",
http://xxx.lanl.gov, quant-ph/9809023

[89] Holland J. H. *Adaptation in Natural and Artificial Systems*, University of Michigan Press, Ann Arbor, 1975.

[90] Horodecki M., P. Horodecki and R. Horodecki, "Separability of mixed states: necessary and sufficient conditions",
http://xxx.lanl.gov, quant-ph/9605038

[91] Horodecki P., M. Lewenstein, G. Vidal and I. Cirac, "Operational criterion and constructive checks for the separability of low rank density matrices.",
http://xxx.lanl.gov, quant-ph/0002089

[92] Hornik K., M. Stinchcombe and H. White, "Multilayer feedforward networks are universal approximators", *Neural Networks* **2**, 359–366, 1989

[93] Høyer P., "Conjugated Operators in Quantum Algorithms",
ftp://ftp.imada.sdu.dk/pub/papers/pp-1997/34.ps.gz

[94] Huberman B. A. and T. Hogg, *Physica D* **22**, 376, 1986.

[95] Ivanyos G., F. Magniez and M. Santha, "Efficient quantum algorithms for some instances of the non-Abelian hidden subgroup problem", http://xxx.lanl.gov, quant-ph/0102014

[96] Jianwei Pan and A. Zeilinger, *Phys. Rev. A* **57** 2208–2212 (1998).

[97] Jones N. D., *Computability Theory: An Introduction*, Academic Press, New York, 1973.

[98] Jones J. A. and M. Mosca., "Implementation of a Quantum Algorithm to Solve Deutsch's Problem on a Nuclear Magnetic Resonance Quantum Computer", http://xxx.lanl.gov, quant-ph/9801027

[99] Jozsa R., "Quantum Algorithms and the Fourier Transform", http://xxx.lanl.gov, quant-ph/9707033

[100] Jozsa R., "Quantum factoring, discrete logarithms and the hidden subgroup problem", http://xxx.lanl.gov, quant-ph/0012084

[101] Kieu T. D. and M. Danos, "The halting problem for universal quantum computers", http://xxx.lanl.gov, quant-ph/9811001

[102] Kitaev A. Y., "Quantum measurements and the Abelian Stabilizer Problem", http://xxx.lanl.gov, quant-ph/9511026

[103] Knill E., I. Chuang and R. Laflamme, "Effective Pure States for Bulk Quantum Computation", http://xxx.lanl.gov, quant-ph/9706053

[104] Knuth D. E., *The Art of Computer Programming*, Volume 1, Fundamental Algorithms, Addison-Wesley, Reading Massachusetts 1981.

[105] Knuth D. E., *The Art of Computer Programming*, Volume 2, Seminumerical Algorithms, Addison-Wesley, Reading Massachusetts 1981.

[106] Kolmogorov A. N., "Three approaches to the quantitative definition of information", *Probl. Inform. Transmission* **1**, 1–7, 1965.

[107] Koza J. R., *Genetic Programming*, The MIT Press, Cambridge Massachusetts, 1993.

[108] Laflamme L., C. Miquel, J. P. Paz, and W. H. Zurek, "Perfect Quantum Error Correction Code", http://xxx.lanl.gov, quant-ph/9602019

[109] Lempel A. and J. Ziv, "On the Complexity of Finite Sequences", *IEEE Transactions on Information Theory* **22**, 75–81, 1976.

[110] Linden N. and S. Popescu, "The Halting Problem for Quantum Computers", http://xxx.lanl.gov, quant-ph/9806054

[111] Lloyd S. and H. Pagels, "Complexity as thermodynamic depth", *Ann. Phys.* **188**, 186–213, 1988.

[112] Lloyd S, and Braunstein S. L., "Quantum Computation over Continuous Variables", *Physical Review Letters* **82**, 1784–1787, 1999.

[113] Lopez-Ruiz R., Mancini H. L. and X. Calbet, "A statistical measure of complexity", *Phys. Lett. A* **209**, 321–326, 1995.

[114] Lovász L., *Computation Complexity*, http://zoo.cs.yale.edu/classes/cs460/Spring98/complex.ps

[115] Mallozzi J. S. and N. J. De Lillo, *Computability with Pascal*, Prentice Hall, New Jersey, 1984.

[116] Michalewicz Z., *Genetic Algorithms + Data Structure = Evolution Programs*, Third Edition, Springer-Verlag, Berlin, 1996.

[117] Minsky M. L., *Computation: Finite and Infinite Machines*, Prentice Hall, New York, 1967.

[118] Miquel C., J. P. Paz and R. Perazzo, "Factoring in a dissipative quantum computer", *Physical Review A* **54**, 2605–2613, 1996.

[119] Moore C. and J. P. Crutchfield, "Quantum Automata", http://xxx.lanl.gov, quant-ph/9707031

[120] Monroe C., D. M. Meekhof, B. E. King, W. M. Itano and D. J. Wineland, "Demonstration of a Fundamental Quantum Logic Gate", *Physical Review Letters* **75**, 4714–4717, 1995.

[121] Mosca M. and A. Ekert, "The Hidden Subgroup Problem and Eigenvalue Estimation on a Quantum Computer", http://xxx.lanl.gov, quant-ph/9903071

[122] Mozyrsky D., V. Privman and M. Hillary, "A Hamiltonian for quantum copying", *Physics Letters A* **226**, 253–256, 1997.

[123] Nielsen M. A. and I. L. Chuang, "Programmable Quantum Gate Arrays", *Physical Review Letters* **79**, 321–324, 1997.

[124] Ömer B., http://tph.tuwien.ac.at/~oemer

[125] Ozawa M., "Quantum Turing machines: Local transition, preparation, measurement and halting", http://xxx.lanl.gov, quant-ph/9809038

[126] Ozawa M., "Entanglement measures and the Hilbert-Schmidt distance", http://xxx.lanl.gov, quant-ph/0002036

[127] Peres A., "Quantum Entanglement: Criteria and Collective Tests",
http://xxx.lanl.gov, quant-ph/9707026

[128] Peres A., "All the Bell Inequalities",
http://xxx.lanl.gov, quant-ph/9807017

[129] Popescu S., "Bell's inequalities and density matrices. Revealing 'hidden' non-locality",
http://xxx.lanl.gov, quant-ph/9502005

[130] Porod W., "Quantum-dot devices and quantum-dot cellular automata",
International Journal of Bifurcation and Chaos **7**, 2199–2218, 1997.

[131] Poyatos J. F., J. I. Cirac and P. Zoller, "Quantum gates with 'hot' trapped ions",
http://xxx.lanl.gov, quant-ph/9712012

[132] Preskill J, *Quantum Information and Computation*,
http://www.theory.caltach.edu/~preskill/ph229

[133] Preskill J., "Fault-tolerant quantum computation",
http://xxx.lanl.gov, quant-ph/9712048

[134] Pritzker Y., http://www.openqubit.org

[135] Prugovečki E., *Quantum Mechanics in Hilbert Space*, Second Edition, Academic Press, New York, 1981.

[136] Redhead M., *Incompleteness, Nonlocality, and Realism* Clarendon Press, Oxford, 1990.

[137] Richtmyer R. D., *Principles of Advanced Mathematical Physics*, Volume I, Springer-Verlag, New York, 1978.

[138] Rieffel E. and W. Polak, "An Introduction to Quantum Computing for Non-Physicists",
http://xxx.lanl.gov, quant-ph/9809016.

[139] Rojas R., *Neural Networks*, Springer-Verlag, Berlin 1996

[140] Rötteler M. and T. Beth, "Polynomial-Time Solution to the Hidden Subgroup Problem for a Class of non-abelian Groups",
http://xxx.lanl.gov, quant-ph/9812070

[141] Schack R. and T. A. Brun, "A C++ library using quantum trajectories to solve quantum master equations",
http://xxx.lanl.gov, quant-ph/9608004

[142] Schommers W. (Editor), *Quantum Theory and Picture of Reality*, Springer-Verlag, Berlin, 1989.

[143] Schrödinger E., "Discussion of Probability Relations Between Separated Systems", *Proceedings of the Cambridge Philosophical Society* **31**, 555–563, 1935.

[144] Schumacher B. "Quantum coding", *Physical Review A* **51**,2738–2747, 1995.

[145] Schwabl F., *"Quantum Mechanics"*, Second revised edition, Springer-Verlag, Berlin, 1995.

[146] Sewell G. L., *Quantum Theory of Collective Phenomena*, Clarendon Press, Oxford, 1986.

[147] Yu Shi, "On quantum generalization of the Church-Turing universality of computation",
http://xxx.lanl.gov, quant-ph/9805083

[148] Shor P. W., "Proceedings of the 35th Annual Symposium on the Foundations of Computer Science", edited by S. Goldwasser (Los Alamitos, CA: IEEE Compuer Society Press), p. 124, 1994.

[149] Shor P. W., "Fault-Tolerant Quantum Computation",
http://xxx.lanl.gov, quant-ph/9605011

[150] Simon D., "On the power of quantum computation", *SIAM J. on Computing* **26**, 1474–1483, 1997.

[151] Skahill K., *VHDL for Programmable Logic*, Addison-Wesley, Reading Massachusetts, 1996.

[152] Stakgold I., *Boundary Value Problems of Mathematical Physics*, Volume I, MacMillan, New York 1967.

[153] Stallings W., *Computer Organization and Architecture: designing for performance*, Fourth Edition, Prentice Hall, 1996.

[154] Steane A., "Multiple-Particle Interference and Quantum Error Correction",
http://xxx.lanl.gov, quant-ph/9601029

[155] Steane A., "The Ion Trap Quantum Information Processor",
http://xxx.lanl.gov, quant-ph/9608011

[156] Steane A., "Quantum computing",
http://xxx.lanl.gov, quant-ph/9708022

[157] Steane A., "Efficient fault-tolerant quantum computing",
http://xxx.lanl.gov, quant-ph/9809054

[158] Steane A. and D. M. Lucas, "Quantum computing with trapped ions, atoms and light",
http://xxx.lanl.gov, quant-ph/0004053

[159] Steeb W.-H., "Bose-Fermi Systems and Computer Algebra", *Found. Phys. Lett.* **8** 73–82, 1995.

[160] Steeb W.-H., *Problems and Solutions in Theoretical and Mathematical Physics*, Volume I, World Scientific, Singapore, 1996.

[161] Steeb W.-H. and F. Solms, "Complexity, chaos and one-dimensional maps", *South African Journal of Science* **92**, 353–354, 1996.

[162] Steeb W.-H., *Matrix Calculus and Kronecker Product with Applications and C++ Programs*, World Scientific, Singapore, 1997.

[163] Steeb W.-H., *Hilbert Spaces, Wavelets, Generalized Functions and Modern Quantum Mechanics*, Kluwer Academic Publishers, Dordrecht, 1998.

[164] Steeb W.-H., *The Nonlinear Workbook*, World Scientific, Singapore, 1999.

[165] Steeb W.-H. and Y. Hardy, "Entangled Quantum States and a C++ Implementation", *International Journal of Modern Physics C* **11**, 69–77, 2000.

[166] Steeb W.-H. and Y. Hardy, "Quantum Computing and SymbolicC++ Simulations", *International Journal of Modern Physics C* **11**, 323–334, 2000.

[167] Steeb W.-H. and Y. Hardy, "Entangled Quantum States", *International Journal of Modern Physics C* **39**, 2765, 2000.

[168] Suzuki J., "A Markov Chain Analysis on Simple Genetic Algorithms" *IEEE Transactions on Systems, Man and Cybernetics* **25**, 655–659, 1995.

[169] Tan K.S., W.-H. Steeb, and Y. Hardy, *SymbolicC++ (2nd extended and revised edition)*, Springer Verlag, London, 2000.

[170] Terhal B. M., "Bell Inequalities and The Separability Criterion", http://xxx.lanl.gov, quant-ph/9911057

[171] van der Lubbe J. C. A., *Basic Methods of Cryptography*, Cambridge University Press, Cambridge, 1998.

[172] Valafar H. *Distributed Global Optimization and Its Applications*, Purdue University, PhD. Thesis, 1995.

[173] Vandersypen L. M. K., M. Steffen, M. H. Sherwood, C. S. Yannoni, G. Breyta and I. L. Chuang, "Implementation of a three-quantum-bit search algorithm", http://xxx.lanl.gov, quant-ph/9910075

[174] Vandersypen L. M. K., M. Steffen, G. Breyta, C. S. Yannoni, R. Cleve and I. L. Chuang, "Experimental realization of order-finding with a quantum computer", http://xxx.lanl.gov, quant-ph/0007017

[175] van Loock P. and Braunstein S. L., "Unconditional teleportation of continuous-variable entanglement", *Physical Review A* **61**, 010302-1–010302-4, 1999.

[176] Vedral V., M. B. Plenio, M. A. Rippin and P. L. Knight, "Quantifying Entanglement", http://xxx.lanl.gov, quant-ph/9702027

[177] Vedral V. and M. B. Plenio "Entanglement Measures and Purification Proce-
     dures",
     http://xxx.lanl.gov, quant-ph/9707035

[178] Vedral V. and Plenio M. B., "Basics of Quantum Computation",
     http://xxx.lanl.gov, quant-ph/9802065.

[179] Vigier J. P., "Non-locality Causality and Aether in Quantum Mechanics",
     Astronomische Nachrichten **303**, 55–80, 1982.

[180] Vose M. D., "Modelling of Genetic Algorithms", *Foundations of Genetic Al-
     gorithms 2*, 63–73, 1992.

[181] Weidmann J., *Linear Operators in Hilbert Space*, Springer-Verlag, New York,
     1980.

[182] Werner R. F., *Physical Review A* **40**, 4277, 1989.

[183] Wilf H. S., *Algorithms and Complexity*,
     http://www.cis.upenn.edu/~wilf, 1994.

[184] Witte C. and M. Trucks, "A new entanglement measure induced by the
     Hilbert-Schmidt norm", *Phys. Lett. A.* **257**, 14–20, 1999.

[185] Yosida K., *Functional Analysis*, Springer-Verlag, Berlin, 1978.

[186] Zukowski M., "Violations of Local Realism in the Innsbruck GHZ experiment",
     http://xxx.lanl.gov, quant-ph/9811013

Index